Managing Diversity and Equality in Construction

Also available from Taylor & Francis

Communication in Construction: Theory and practice
Andrew Dainty, David Moore and
Michael Murray

Hb: 0–415–32722–9
Pb: 0–415–32723–7

Taylor & Francis

Construction Collaboration Technologies:
The extranet evolution
Paul Wilkinson

Hb: 0–415–35858–2
Pb: 0–415–35859–0

Taylor & Francis

Construction Project Management
Peter Fewings

Hb: 0–415–35905–8
Pb: 0–415–35906–6

Taylor & Francis

Human Resource Management in Construction
Projects
Martin Loosemore, Andrew Dainty and
Helen Lingard

Hb: 0–415–26163–5
Pb: 0–415–26164–3

Spon Press

Occupational Health and Safety in Construction
Project Management
Helen C. Lingard and Steve M. Rowlinson

Hb: 0–419–26210–5

Spon Press

Information and ordering details

For price availability and ordering visit our website **www.tandf.co.uk/ builtenvironment**

Alternatively our books are available from all good bookshops.

Managing Diversity and Equality in Construction

Initiatives and practice

**Edited by
Andrew W. Gale and
Marilyn J. Davidson**

Taylor & Francis
Taylor & Francis Group

LONDON AND NEW YORK

First published 2006
by Taylor & Francis
2 Park Square, Milton Park, Abingdon, Oxon OX14 4RN

Simultaneously published in the USA and Canada
by Taylor & Francis
270 Madison Ave, New York, NY 10016

*Taylor & Francis is an imprint of the Taylor & Francis Group,
an informa business*

© 2006 Taylor & Francis

Typeset in Sabon by
Newgen Imaging Systems (P) Ltd, Chennai, India
Printed and bound in Great Britain by
TJ International Ltd, Padstow, Cornwall

British Library Cataloguing in Publication Data
A catalogue record for this book is available from the British Library

Library of Congress Cataloging in Publication Data
A catalog record for this book has been requested

ISBN10: 0–415–28869–X (hbk)
ISBN10: 0–203–08861–1 (ebk)

ISBN13: 978–0–415–28869–9 (hbk)
ISBN13: 978–0–203–08861–6 (ebk)

Contents

Illustrations

Figures

Tables

Contributors

Barbara M. Bagilhole, Loughborough University, UK
Chrissie Chadney, Willmott Dixon, Letchworth Garden City, UK
Linda Clarke, University of Westminster, UK
Andrew R.J. Dainty, Loughborough University, UK
Caroline Davey, University of Salford, UK
Marilyn J. Davidson, The University of Manchester, UK
Valerie Francis, University of Melbourne, Australia
Andrew W. Gale, The University of Manchester, UK
Clara Greed, University of the West of England, Bristol, UK
Annie Hopley, National Housing Federation (North), UK
Helen Lingard, Queensland University of Technology, Australia
Elisabeth Michielsens, University of Westminster, UK
Rita Newton, University of Salford, UK
Marcus Ormerod, University of Salford, UK
Sandi Rhys Jones, OBE, RhysJones Consultants, London, UK
Anne Ruff, Middlesex University, UK
Malcolm Sargeant, Middlesex University, UK
Dianne Sodhi, People First Housing Association, Manchester, UK
Somerville Peter, Lincoln University, UK
Andy Steele, University of Salford, UK
Christine Wall, London Metropolitan University, UK
Suzanne Wilkinson, University of Auckland, New Zealand

Chapter 1

An introduction to diversity

Andrew W. Gale and Marilyn J. Davidson

Introduction

Managing diversity initiatives seek to fully develop the potential of each employee and turn the different sets of skills that each employee brings into a business advantage. Through the fostering of difference, team creativity, innovation and problem-solving can be enhanced (Davidson and Fielden, 2003). The focus is therefore much more on the individual rather than the group or team. Having a diverse workforce not only enables organizations to understand and meet customer demands better, but also helps attract investors and clients as well as reduce the costs associated with discrimination.

The literature shows that there is research evidence for the proposition of 'value in diversity' (e.g. Bank, 1999; Polzer *et al.*, 2002). Nevertheless, this is criticized by some authors (e.g. Jehn, 1997) who argue that diversity has a negative effect on group process and performance. However, it is powerfully argued by O'Reilly *et al.* (1997) that diversity may be managed to avoid negative consequences (e.g. Priem *et al.*, 1995; Wittenbaum and Strasser, 1996). According to Polzer *et al.* (2002) well managed heterogeneous groups are far more creative than homogeneous ones. Early interpersonal congruence of groups has been found to have a significant long-lasting influence on creative task performance beyond simply nullifying the detrimental effects of diversity on social integration, group identification and emotional conflict (Jehn *et al.*, 1999; Polzer *et al.*, 2002). According to Davidson and Fielden:

> Managing diversity initiatives seek to fully develop the potential of each employee and turn the different sets of skills that each employee brings into a business advantage. Through the fostering of difference, team creativity, innovation and problem-solving can often be enhanced. The focus is, therefore, much more on the individual rather than the group. Having a diverse workforce not only enables organizations to understand and meet customer demand better, but also helps attract investors and clients, as well as reduce the costs associated with discrimination.
>
> (2003, p. xxii)

Historically, when organizations refer to diversity policy they are often referring to forms of practices and procedures that were previously labelled 'equal opportunities' (EO), and tended to be particularly aimed at gender and race. Furthermore, the failure of EO legislation and policies in aiming to reduce discrimination in organizations, was its emphasis on specific groups (e.g. women) having to assimilate once in the organization, rather than a change in organizational culture to one of inclusive diversity (Davidson and Burke, 2000). In addition, the singling out of certain groups for special consideration (e.g. positive action programmes, affirmative action legislation) sometimes led to degrees of resistance or backlash (Davidson and Fielden, 2003; Burke, 2005). The differences between EO policies (EOPs) and diversity policy, was aptly described by Kirton:

> The general thrust of EOPs has been for organizations to develop procedures and practices to eliminate discriminatory behaviour by line managers and other gatekeepers and in so doing reduce the disadvantage experienced by individual members of the groups covered by the policy. One of the criticisms of this approach has been that it is negative, in the sense that failure to comply is associated with penalties (imposed by legislation) and positive actions (for example, the disciplinary of anyone found contravening policy). In other words, the positive benefits of compliance and commitment to the ideals or goals of the policy are not effectively sold to organizational members. In contrast, diversity policy seeks not only to recognize workforce diversity, but to value it rather than see it as a problem requiring a remedy.
>
> (2003, p. 4)

The concept of diversity in the context of the construction industry workforce is emerging as an important aspect of human resources management on site and in contractors', architects', consultants' and clients' organizations at all levels (Sweet, 2005) This is largely due to the rapidly changing aspirations of those employed in the construction industry, legislation and trends in society. Action research (funded by the UK government) undertaken by the authors (see Appendices) was designed to meet the increasing interest and concern of industrial organizations and society with respect to the under-representation of women, black and minority ethnic groups in the construction sector. In the United Kingdom, the construction industry is a significant employer. It employs 1.4 million people and is responsible for 10 per cent of UK GNP but women only constitute 9.2 per cent (DTI 2003) of that, with only 1 per cent (CITB 2003) of these in manual trades. Women make up only 4 per cent of the membership of the construction professions (Davey et al., 1998) and black and minority ethnic workers are under-represented in the construction industry and so are people with disabilities. The construction workforce has only 1.9 per cent black and Asian workers compared with 6.4 per cent for the working population as a whole – more than 70 per cent fewer black and Asian workers than the UK industry mean. (Somerville *et al.*, 2002). Hence, despite more than three decades of EO legislation, one of the largest industrial sectors in the United Kingdom has managed to perpetrate the status quo as a predominate white, male culture.

The construction industry is very visible. The general population observes construction at work on a daily basis by simply seeing its production sites from the street. Construction has a very strong image and there are many stereotypes associated with the industry. All of these factors make construction a very public feature of every day life. As a major employer in an increasingly competitive global economy, employment in the construction industry is likely to become more and more scrutinized in the future. It is therefore very important that those involved in the construction industry are aware of the concept of diversity and the issues surrounding diversity.

Aims of this book

This book aims to cover topics associated with the implementation of equality and diversity in relation to gender, black and minority ethnic groups and disability. Diversity encompasses a wider range of factors such as sexual orientation, age, appearance, personality, social class and so forth. However, to date, research into issues of diversity and equality in the construction sector have tended to be focused specifically on gender discrimination and more recently race discrimination and disability.

The book draws on research findings, good practice guidelines (Appendices I and II) and case studies from several different countries and regions, and is written for practitioners in construction as well as students of construction. It should also be of value to those engaged in recruiting, training, supervising and managing operative labour, technical and professional staff.

The book has been structured in such a way as to facilitate ease of access to the material it contains. It is in four parts:

1 Diversity and the Law
2 Gender and Equality
3 Race, Disability and Equality
4 Managing and Implementing Diversity.

Each chapter has been written by specialists in their fields and at the end of each chapter there are discussion questions. Also, the reader will find examples and small case studies illustrating issues and points raised in the chapters.

Overview

This book brings together current debates, findings and case-study practices around the issues associated with diversity management in the construction industry. In Chapters 2 and 3 in Part I, 'Diversity and the law', Anne Ruff and Malcolm Sargeant highlight how government policy and legislation continue to evolve with influences far wider than the UK context. In particular, the European Union has an increasingly significant impact on UK legislation. For example, legislation on racial discrimination came into being in 1965 and was strengthened in 2003 due to a EU Directive in 2000. However, whilst British legislation requires public bodies to monitor EO policies and implementation, the private sector is still able to avoid this at the present time.

EU legislation and Directives also seek to prevent discrimination on the grounds of belief, religion and sexual orientation. Sex equality has been the subject of UK legislation since 1975 and much more recent legislation designed to reduce discrimination on grounds of disability has been enacted. This legislation requires both public and private sector organizations to be proactive in their behaviours to reduce disability discrimination. Undoubtedly, the impact of UK and European legislation and directives will become increasingly significant to construction employers with respect to recruitment practices and management practices in general.

According to Anne Ruff (Chapter 2), organizations in the construction industry that fail to follow discrimination legislation are more likely to risk being taken to employment tribunals. Furthermore, a failure to adhere to the changes introduced by the Race Relations (amendment) Act 2000, may result in a decreased likelihood of procuring contracts from public bodies. Interestingly, she isolates the specific issues which may well dominate the construction industry in the near future as including the 'canteen' and 'rugby club macho culture', bullying and harassment at work, recruitment practices and the obligation not to discriminate on grounds of disability.

In Chapter 3, Malcolm Sergeant concludes that the diverse workforce found in the construction sector requires a variety of contractual relationships away from the traditional model of full-time employer/employee. Both the EU and the UK Government have accepted that flexible working arrangements should be encouraged. Moreover, he believes this could be helped by extending the employment protection to cover both full-time and flexible workers, as well as the need for construction industry employers to realize the importance of equality and diversity.

In Part II of the book, entitled 'Gender and equality', the emphasis shifts to specifically examining gender equality issues in the construction industry with a particular focus on women. Clara Greed, in Chapter 4 provides a detailed discussion on the continued social exclusion of women in the construction industry, despite more than 15 years of initiatives aimed at attracting more women into the industry. She presents recent statistics on the percentage of women in construction, and illustrates the lack of women at senior levels and the continued discriminatory and macho culture of the construction industry as a whole. The concept and politics of social exclusion are emphasized as an important issue with many interrelated aspects and influences. While acknowledging that the industry is undoubtedly trying to improve itself, the problems and ineffectiveness of a large number of initiatives associated with trying to tackle social exclusion are acknowledged and discussed.

In Chapter 5, drawing on their own research findings, Andrew Dainty and Barbara Bagilhole examine the factors that influence the career dynamics of British female and male construction professionals. Following the continual theme of social exclusion, they present evidence confirming the discrimination and prejudice inherent in work practices and from male managers in particular, who act as 'gatekeepers' hampering women's career progression. Indeed they assert that the construction industry, championed by senior leaders, has a moral duty to modify its employment practices and culture if it wants to continue marketing itself as an appropriate employer for women.

The following three chapters focus on women in specific occupations within construction. Suzanne Wilkinson (Chapter 6) describes the underrepresentation and experiences of women in civil engineering from the time they enter higher education through to the challenges they encounter once they enter the industry. In Western countries, women still constitute only 10–20 per cent of all students studying civil engineering courses and often encounter negative attitudes of male students and staff, feelings of isolation and eroded confidence. Furthermore, not only do the female graduates have to make more job applications compared to their male counterparts, but around 15 per cent (Issacs, 2001) of them choose other employment (Wilkinson, 1996). Wilkinson also refers to important research by such authors as Bennett *et al.* (1999) who have reported that while there have been few significant differences between female and male construction students on employment choice, differences raised more by women include commitment to EO, raising profiles of successful women and flexible working/childcare/career breaks. Moreover, the lack of flexibility related to contracting is also highlighted as a significant 'turn-off' for female civil engineers.

Pay inequalities, 'dinosaur' attitudes of men against women in the construction industry and the predominant white, male, 'macho' cultures are further endorsed by Clara Greed in Chapter 7 in her review of the barriers facing women in Surveying and Planning. While there has been a growth in the numbers of women entering planning and surveying, they are still in a minority. Furthermore, with 70 per cent of planners found in the public sector (RICS, 2003), she points out how this results in a disproportionate 'male' influence on the design and nature of the built environment within every local authority in the country (Greed, 1999). As a consequence, the lack of a diverse culture in the occupation of planning results in evidence suggesting that women suffer disadvantage within a built environment that is developed by men, primarily for other men (e.g. Darke *et al.*, 2000; Little, 2002). Drawing on recent developments in both the European Union and United Kingdom, this chapter presents strategies for generating change to improve the situation for women in the form of a 'toolkit' aimed at mainstreaming gender issues into planning and practice (Reeves and Greed, 2003).

In Chapter 8 Linda Clarke, Elisabeth Michielsens and Christine Wall examine the reasons for the massive degree of 'sex-typed' gender segregation of women in manual trades. In the UK construction industry women constitute less than 1 per cent of the manual workforce – a proportion that has decreased further in the past decade (CITB, 2003). Based on the results of their empirical studies on experiences of tradeswomen in UK local authority repair and maintenance departments, they present cases of good practice as well as personal case studies of individual interviewees. In common with the majority of contributors in this book, they conclude that to change the position of women in manual trades requires policy regulation – without it, the male-segregated nature of construction will continue to be reinforced. Women's entry into manual trades was largely due to the goodwill of individual employers, good training schemes, good employment conditions and effective EO policies in terms of recruitment and employment, as well as support mechanisms. Indeed, they proposed that in such strongly sex-typed occupations as manual trades, proactive recruitment and special support to ensure retention rules were vital to provide the additional leverage to integrate women.

The final chapter in Part II, further develops the argument for the provision of a work environment that would be more supportive of employees in terms of improving the work–life balance. In line with numerous authors throughout this book, Helen Lingard and Valerie Francis suggest that the inflexible employment arrangements adopted by construction organizations are discriminatory and serve to disadvantage or exclude certain groups of employees. Presenting evidence from predominantly Australian research studies, they identify organizational work–life balance policies which could be implemented by construction companies and highlight how employees needs are likely to vary depending on their life style. This is undoubtedly based on the diversity model whereby use of such policies must be perceived as available to all employees, irrespective of sex, age or ethnic origin. Nevertheless, they conclude that, as with the majority of recommendations proposed throughout this book, work–life balance policies will only promote diversity if they are accompanied by cultural change within the construction organizations.

Part III, entitled 'Race, disability and equality', shifts the emphasis from focusing on women and gender discrimination in the construction industry to diversity issues related to race and disability. Andy Steel and Dianne Sodhi, in Chapter 10, provide a detailed review of empirical academic research on race and ethnic minorities in the construction industry. In their introduction they emphasize the projected skills shortage (Mackenzie *et al.*, 2000) in the industry and the fact that the Construction Industry Training Board (CITB) maintained that 'the industry must look to increase recruitment of ethnic minorities and women if it is to survive and grow (Building, 2001). Historically, however, the industry has had an even poorer record of attracting (Black and Minority Ethnic) BMEs than female recruits. Compared with the UK industry average, the construction industry employs 70 per cent fewer BME people (CITB and Royal Holloway, University of London, 1999). Furthermore, there are marked disparities in the representation of different BME groups in the construction industry with a much smaller representation from members of the Pakistani and Bangladeshi communities compared to African, Caribbean, Indian and those of mixed and other backgrounds. It is estimated that only 2 per cent of construction professionals are from ethnic minorities (Building E = Quality 1996).

In order to address the question as to why so few black and minority ethnic people fail to enter the construction industry, these authors review the evidence relating to the barriers. These include discussions of the image of the industry, the prevalence of overt racism, the lack of career development opportunities and employment uncertainty, the lack of information about the industry, and the organizational culture of the industry. Once again, the recommendations propose a move away from the former EO model of positive action policies and initiatives towards the urgent need for organizational culture changes so that equality of opportunity becomes a core value that embraces diversity. The authors also make the important point that under the Race Relations (Amendment Act) 2000, clients and particularly public bodies, have the financial muscle to ensure that construction companies comply to the law.

The following Chapter 11 examines the largely un-researched area in terms of issues facing the construction industry, in relation to disabled people. Marcus Ormerod and Rita Newton mention that almost one in five people of working age reading this book will be disabled. In order to explain the nature of disability and attitudes towards it, the authors initially outline models of disability. The authors then present a two fold process aimed at breaking down the barriers of disability through the necessity for more effectively designed buildings and employment practices (including the implementation of anti-discriminatory policies) that support social inclusion. However, they emphasize both these processes need to be based within the context of the construction industry itself and the UK legislative framework (1995 Disability Discrimination Act). In the final part of their chapter they present useful case study examples illustrating some of the challenges facing construction industry employers in relation to their legal obligations towards disabled people, as well as a list of schemes and support services. While there are few statistics available on the employment of disabled people within the UK construction industry (LMT, 2002) the evidence suggests that the construction industry is not proactively encouraging the recruitment of disabled people. In the words of Ormerod and Newton 'If the construction industry is to create truly inclusive environments then it needs to be working with, rather than for, a diverse society'.

Part IV of the book concentrates specifically on strategies and initiatives related to the management and implementation of diversity in the construction industry. In Chapter 12, Sandi Rhys Jones examines research, initiatives and good practice case studies relating to equality of opportunity in the social housing sector. Here, she highlights how the social housing commitment to meeting the diverse needs of local communities has resulted in a more successful record of EO than the construction industry, which serves it. Nevertheless, she quotes the House Builders Federation (2001), which found that even though the house building sector employed far more women that the construction industry as a whole (28 per cent compared to 8.6 per cent in the construction industry), only 0.1 per cent were registered disabled and 0.5 per cent were from black and minority ethnic groups. Undoubtedly, while the construction industry can learn useful lessons from the diversity programmes successfully initiated by the social housing sector, Rhys Jones concludes that even in social housing, there is a need for quicker progress in terms of commitment and putting policy into practice.

The theme of presenting good practice case study material is continued in Chapter 13 and Chrissie Chadney describes in detail the formulation and successful implementation of a diversity programme in a UK contracting organization. She details the development of the programme over a five year period, which included a culture change programme and a major emphasis on recruiting, retaining, promoting, developing and rewarding the very best employees. This resulted in the percentage of both female and black and minority ethnic employees being double that of the average percentage of other contractors. The author concludes that the success of this programme included a 'low key' approach to policy and practice change and a commitment for the change to be based on business needs and on the valuing of individual contributions of the workforce that is, a model of diversity.

The final Chapter 14, reviews the numerous UK government initiatives for change within the construction industry in relation to equality of opportunity and diversity. Sandi Rhys Jones discusses the relative effectiveness of these initiatives including the Respect for People Toolkits, the Construction Best Practice Program and the Re-thinking Construction Demonstration Projects. The author proposes that existing programmes need sustaining and modifying where necessary and that there is a need for a well resourced central information source for diversity issues in the construction industry. Moreover, Rhys Jones also suggests this should be part of the existing Set 4 Women Program or as a specific unit, and besides offering support, the Government should provide financial incentives to companies and organizations which actively initiate diversity programmes.

Conclusions

What is clear from the material presented throughout this book, is that the construction industry in the United Kingdom perpetuates a predominant white, male able-bodied culture, which actively discriminates against change towards attracting a more socially inclusive, diverse workforce. Employers in the construction industry include: contractors, sub-contractors, designers, architects, consultants and, of course, clients. As an integral component of the labour market, employers are significant influencers of vertical and horizontal discrimination in the market (Hakim, 1979). Vertical segregation is demonstrated by the disproportionate distribution of sex, race, disability, etc., at different levels in an industry sector as a whole, professional discipline or trade group and horizontal segregation is concerned with the disproportionate distribution across industry sectors. The construction industry is not homogeneous. There are big variations between professions as illustrated by the following percentages of women in membership of professional bodies: RTPI 26.3 per cent, RICS 10.4 per cent, RIBA 10.1 per cent, ISE 4.7 per cent, ICE 4.7 per cent and CIOB 3.2 per cent (Watts, 2003). Also, the structure of the construction industry is fragmented and in the case of contractors has a very long tail of small firms. Arguably the construction industry is a sector of the economy constituting several industries based on market, size of project and/or type of work. What is clear, however, is that women, black and minority ethnic groups and people with disabilities form extremely small proportions of whatever part of the construction industry are observed. The overall working population contains 6.4 per cent black and minority ethnic people compared with 1.9 per cent in the construction sector (CITB and Royal Holloway University of London, 1999).

How employers recruit, reward and manage the workforce has direct and indirect influences on the dynamics of the labour market (Wilson, 2003). As a major employer the construction industry will probably become more and more scrutinized by the media and politicians as the availability of employment opportunities is more and more influenced by global trends in employment. Therefore, employers have an active role to play in addressing apparent discrimination in the construction industry.

The model of full-time employment is predominant in the construction industry in comparison to many other sectors of the economy where part-time, flexible, job-share, etc., models are becoming more familiar and mainstream (EOC, 2005). Employers have a clear opportunity to initiate change in models of employment, and the concept of work–life balance relates caring responsibilities to work pressures in the context of a more egalitarian society. Traditional roles which typically were gendered, in which women tend to undertake childcare and caring for older adults undermines the ability of women to take up work in jobs requiring rigid working hours. Flexible and 'family friendly' patterns of working arrangements facilitate higher participation levels in the workforce and broaden the population from which employers can recruit, as well as increase loyalty and social inclusion in the workforce (Powel and Groves, 2003).

The concept and politics of social exclusion is an important issue with many interrelated aspects and influences. The construction industry is both the producer of an improved environment, in terms of housing, infrastructure and facilities, as well as being a major employer. Construction could be a central player in any initiative, of which there are many, to address the problem of social exclusion.

Clearly, there are problems with the large number of initiatives associated with trying to tackle social exclusion in the construction industry and there is a growing danger of initiative overload. Whilst small appears to be beautiful in the sense that small, local grass roots groups are more effective on one level, the scale of change developed from small initiatives is often insignificant and difficult to transfer and replicate.

It is also evident that the role of the social housing sector is often underestimated. It constitutes Registered Landlords (RLs) and Housing Associations (HAs). These organizations receive large volumes of capital funding from Government via the Housing Corporation. As clients they procure housing using contractors, hence they directly and indirectly employ the construction workforce. It is not difficult to see how the RLs and HAs could be used by Government to influence practices in the construction industry, particularly when HAs, for example, have within their terms of reference statements concerning equality and diversity.

In addition, education and training also influences and maintains a gate-keeping role in relation to social exclusion, particularly at the tertiary, further and higher education stages. Women, black and minority ethnic groups and people with disabilities are still a small minority on vocational courses at universities (Gale, 1995; Wilkinson, 1996; Devgun and King, 1999; Issacs, 2001). Learning on the job and the craft traditions in the construction industry are different from the concept of formal qualifications in skills. Interestingly women, for example, are almost totally dependent on proving their ability through qualification in order to enter the workforce.

The image of the construction industry is also cited by several authors in this book as a problem. Rhys Jones for example, concludes that its reputation for being dirty, dangerous, sexist and racist is a major barrier to attracting a more diverse workforce (Chapter 14). This is particularly relevant when taking into account the fact that research indicates that a child is predisposed towards a particular field of employment at a very early age (Wilson, 2003).

In addition to gender discrimination, there is also plenty of evidence in this book to suggest the existence of racial discrimination in the construction industry (Loosemore and Chau, 2002). Under-representation of black and minority ethnic people is roughly the same in construction manual trades and professions (c.1.6–2 per cent), some five to six times less than the proportions on construction related courses in further and higher education – evidence of something happening to prevent or discourage students from entering the industry workforce (UCAS, 1998).

The construction industry has also had a largely negative image in relation to people with disabilities. Clearly however, the Disability Discrimination Act (1995) has led to renewed interest in people with disabilities in relation to employment, in the design and construction of buildings. The Disability Rights Commission (2003) recently revealed that young disabled people in particular, often have to deal with prejudice and discrimination. Obviously there is a need to educate educators, trainers, employers and clients on the different models of disability, as this has a bearing on our understanding and attitudes towards disability.

People from all parts of the construction industry and those with an interest in it, frequently talk about the need to improve its image. In an earlier research by Gale (1994) one professional career advisor remarked that she felt that she had an ethical duty to advise honestly and that if the reality of any profession or occupation was undesirable due to sex or racial discrimination, then she was obliged to reflect this honestly in her advice to the parents and young people she worked with. The point is that if the reality is undesirable then by some means creating a 'better' image is unethical and ultimately of no use anyway. The key thing here is that changing the culture and reality of construction should be the focus of concern – the improved image will undoubtedly follow.

Organizational or industrial cultures are in large part a product of their significant history. A realistic and theoretically sound understanding of culture, its determinants and dynamics should be considered before pronouncements concerning the improvement of the construction culture. Further, as mentioned earlier, there are many fragments constituting the construction sector and these have their own cultural profiles. Often when discussing culture people consider only the tangible and superficial artifacts. As Schein (1987) argues, there are two underlying invisible levels underpinning culture and that these are concerned with values, beliefs and fundamental relationships between people, society, right and wrong. The culture of the various components of the construction industry, are therefore not necessarily subject to rapid change, but a gradual introduction of mainstreaming can bring about effective change. The concept of mainstreaming concerns the extent to which practices and procedures designed to reduce discrimination form an integral part of good practice in diversity management processes (Holton, 2005). The alternative to main-streaming is the marginalization of initiatives and practices through organizations diverting attention to specific parts of an organization and therefore avoiding the impact of such initiatives. Interestingly, researchers such as Mattis (2002) propose that organizationally diverse practices, programmes and policies are designed as 'best practice' when they are judged by experts to be above and beyond the mainstream compared to other organizations. Organizations in the construction industry should continually benchmark their diversity outcomes against other organizations, as well as create the business case for diversity.

We believe that the construction industry will become increasingly visible to the public and politicians. This will be driven by an increased pressure for access to the labour market in a global economy and social trends leading hitherto under-represented groups to demand inclusion through construction employment. Also, it is highly likely that legislation from both UK and European government will put managing diversity into the mainstream for managers in construction.

Hopefully, this book will help the reader gain a much clearer insight into the practicalities and issues of effective and successful diversity management in the construction industry. In the words of Davidson and Fielden:

> the future success of the management of diversity in organizations lies not solely in legislation or corporate policy. We strongly propose that the effectiveness of diversity programs is reliant also on the attitudes, perceptions and behaviour of individuals at all levels (from the top down) of the workforce, combined with an appropriate organizational and social cultural climate.
>
> (2003, p. xxv)

References

Bank, J. (1999) 'Dividends of diversity', *Management Focus*, Issue No. 12 (Summer), Cranfield School of Management.

Bennett, J.F., Davidson, M.J. and Gale, A.W. (1999) 'Women in construction: a comparative investigation into the expectations and experiences of female and male construction undergraduates and employees', *Women in Management Review*, 14(7): 273–291.

Building (2001) 'CITB launches drive to hire minorities', *Building* (Editorial), 7 September.

Building E = Quality (1996) A Discussion Document – Minority Ethnic Construction professionals and Urban Regeneration.

CITB (1991) Technological Change and Construction Skills in the 1990s, Research Document, Construction Industry Training Board, Bircham Newton

CITB (Construction Industry Training Board) (2003) CITB Skills Foresight Report.

CITB and Royal Holloway University of London (1999) 'The under-representation of black and Asian people in construction, Construction Industry Training Board', Bircham Newton.

Darke, J., Ledwith, S. and Woods, R. (2000) *Women and the City: Visibility and Voice in Urban Space*, Palgrave: Oxford, pp. 114–130.

Davidson, M.J. and Burke, R.J. (eds) (2000) *Women in Management: Current Research Issues*, Volume 11. Sage, London.

Davidson, M.J. and Fielden, S.L. (eds) (2003) *Individual Diversity and Psychology in Organizations*. John Wiley, Chichester.

Devgun, B. and King, P. (1999) *Where Have All The Girls Gone?*, New Zealand Engineering, September, Wellington, New Zealand.

DTI (2003) *Construction Statistics Annual 2003 Edition*, The Stationary Office, London.

EOC (2005) *The Fact About Women Is*, EOC, Manchester, UK.

Gale, A.W. (1994) 'Women in construction: an investigation into some aspects of image and knowledge as determinants of the under representation of women in construction management in the British construction industry'. Unpublished PhD Thesis, University of Bath.

Gale, A.W. (1995) 'Women in construction', in Langford, D., Fellows, R., Hancock, M. and Gale, A.W. (eds) *Human Resource Management In Construction*, Longman Scientific and Technical: Essex.

Greed, C. (1999) *The Changing Composition of the Construction Professions* Faculty of the Built Environment, Occasional Paper No. 5 (Bristol, University of the West of England)

Hakim, C. (1979) 'Occupational segregation: a comparative study of the degree and pattern of the differentiation between men's and women's work in Britain, the United States and other countries', Department of Employment Research Paper No. 9. HMSO, London.

Holton, V. (2005) 'Diversity Reporting: How European Business is Reporting on Diversity and Equal Opportunities', *Women in Management Review*, 2(1): 72–77.

Isaacs, B. (2001) 'Mystery of the missing women engineers: a solution', *ASCE Journal Of Professional Issues In Engineering Education And Practice*, 127(2): 85–90.

Jehn, K.A., Nothcraft, G.B. and Neale, M.A. (1999) 'Why some differences make a difference: A field study of Conflict and Performance in workgroups', *Administrative Science Quarterly*, 44(4): 741–763.

Kirton, G. (2000) 'Developing strategic approaches to diversity policy', in Davidson, M.J. and Fielding, S.L. (eds) *Individual Diversity and Psychology in Organizations*, Chichester: John Wiley, 3–18.

Labour Market Trends (2002) 'Labour market experiences of people with disabilities'. Office for National Statistics, London.

Little, J. (2002) *Gender and Rural Policy: Identity, Sexuality and Power in the Countryside*. Pearsons, Harlow.

Loosemore, M. and Chau, D.W. (2002) 'Racial discrimination towards Asian operatives in the Australian construction industry', *Construction Management and Economics*, 20(1): 91–102.

Mackenzie, S., Kilpatrick, A.R. and Akintoye, A. (2000) 'UK Construction Skills Shortage response strategies and an analysis of industry perceptions', *Construction Management and Economics*, 18: 853–862.

Mattis, M.C. (2002) 'Best practices for retaining and advancing women professionals and managers', in Burke, R.J. and Nelson, L. (eds) *Advancing Women's Careers*, Oxford: Blackwell.

O'Reilly, C.A., Williams, K.Y. and Barsade, S. (1997) 'Demography and group performance: does diversity help?', Research Paper No. 1426, Graduate School of Business. Stanford University, 40 pages.

Polzer, J.T., Milton, L.P. and Swann, W.B. (2002) 'Capitalizing on diversity: interpersonal congruence in small working-groups', WP 02-003.

Powell, G.N. and Graves, L.M. (2003) *Women and Men in Management*. (3rd Edition), Sage, London.

Priem, R., Harrison, D. and Muir, N. (1995) 'Structured conflict and consensus outcomes in group decision making', *Journal of Management*, 21: 691–710.

Reeves, D. and Greed, C. (2003) *Gender Equality and Plan Making: The Gender Mainstreaming Toolkit*, with contributors, Linda Davies, Caroline Brown and Stephanie Duhr (London, RTPI). Final web version edited by C. Sheridan and D. Reeves for RTPI and available at www.rtpi.org.uk

RICS (2003) *Raising the Ratio*, Investigation of composition of the surveying profession (Led by Louise Ellman and Sarah Sayce) London: University of Kingston on Thames and see www.rics.org.uk

Schein, E.H. (1987) *Organisational Culture and Leadership*, Jossey–Boss, San Francisco, CA.

Somerville, P., Steele, A. and Sodhi, D. (2002) 'Black and minority ethnic employment in housing organisations', in Somerville, P. and Steele, A. (eds) *Race, Housing and Social Exclusion*, London: Jessica Kingsley.

Sweet, R. (2005) 'Redressing the balance', *Construction Manager*, July/August, 12–13.

Watts, J. (2003) Women in Civil Engineering: Continuity and Change. Unpublished PhD thesis, London, Middlesex University.

Wilkinson, S.J. (1996) 'The Factors Affecting the Career Choice of Male and Female Civil Engineering Students in the UK', *Career Development International*, 1(5): 45–50.

Wilson, F. (2003) *Gender and Organizations*. Ashgate, London.

Wittenbaum, G. and Strasser, G. (1996) 'Management of information in small groups', in Nye, J. and Brower, M. (eds) *What's Social About Cognition? Social Cognition in Small Groups*, Sage, CA: 3–28.

Part I

Diversity and the law

Chapter 2

Discrimination law and the construction industry

Anne Ruff

Introduction

This chapter examines discrimination law in the context of the construction industry. The introduction describes the legislative background and assesses the impact of the law on the construction industry. The style of referencing and footnotes in this chapter has been adopted to facilitate ease of reading and the accurate attribution of legal sources.

English law traditionally considered that all men are equal, and that the law is 'colour blind' if not 'gender blind'. This traditional approach failed to protect individuals who were treated unfairly because of their sex or race. Anti-discrimination legislation was first introduced in the United Kingdom in 1965 when the Race Relations Act was passed. This proved to be too limited in scope and the Race Relations Act 1968 was passed. The 1968 Act extended the anti-discrimination legislation to employment. In 1976 a third Race Relations Act was passed. This Act which remains in force enabled industrial tribunals to hear claims by individuals against employers, and established the Commission for Racial Equality (CRE) so that patterns of discrimination could be identified and investigated. The 1976 Act was amended in 2000[1] in response to the Macpherson report, and most public authorities are required to document and monitor the operation of their equal opportunities policies.[2] The Act was amended again in 2003 by regulations[3] passed to give effect to the European Union Race Directive.[4]

In 1976 sex discrimination became unlawful[5] and the right of women to receive equal pay to men was introduced,[6] in part, at least, because the United Kingdom became a member of the European Economic Community, now known as the European Union. The content of the sex discrimination legislation was very similar to that of the race relations legislation. A counterpart to the CRE was also established in the form of the Equal Opportunities Commission.

The United Kingdom is required to comply with European Union (EU) treaties, regulations and directives, as well as decisions of the European

Court of Justice in Luxembourg. EU law has had a considerable impact on sex discrimination and equal pay law in the United Kingdom. Two recent pieces of EU legislation are having a major impact on discrimination law. The EU legislation is attempting to introduce common European wide principles into race discrimination legislation and into anti-discrimination employment legislation.

In 2003 as a consequence of the Employment Directive[7] it became unlawful in the United Kingdom for employers to discriminate on grounds of religion or belief,[8] or sexual orientation,[9] and in 2006 it will become unlawful for UK employers to discriminate on grounds of age. In Northern Ireland it has been unlawful to discriminate on religious or political grounds since 1976.

However, it was not until the Disability Discrimination Act 1995 that it became unlawful for employers to discriminate against disabled applicants and employees. This legislation has some similarities to the race and sex discrimination legislation but requires employers to be more pro-active. The legislation was amended in 2003[10] to give effect to the Employment Directive,[11] and in 2005 by a new Disability Discrimination Act.

The aim of this chapter is to examine discrimination law in employment with particular reference to the construction industry. Case studies are included to illustrate the actual or potential impact of the legislation on employers in the construction industry. The key legal provisions relating to race, sex and disability discrimination are explained as well as those relating to discrimination on the grounds of sexual orientation and on grounds of religious or other belief. Sexual discrimination against pregnant women and transsexuals, and equal pay legislation is also considered. The chapter first briefly examines the evidence which suggests that unlawful discrimination is widespread in the construction industry. Second, the general principles underpinning unlawful discrimination are explained. Third, each form of unlawful discrimination is considered in more detail. Fourth, the role of the Equality Commissions is examined. Finally the chapter concludes by identifying the two main ways in which the legislation impacts on employers in the construction industry and the issues which are most likely to cause concern to such employers.

The construction industry

Despite the existence of race and sex discrimination legislation for more than a quarter of a century, its impact on the construction industry appears to have been limited. Employees in the construction industry are predominantly white males.[12] In 1995 CRE concluded in relation to the construction industry that 'what is needed is a better appreciation by employers of the value of good employment practice, and a greater commitment by employers to fairness and equality of opportunity' (CRE, 1995: 13). In

2002 a report commissioned by the CITB indicated that there is scope for increasing the recruitment of ethnic minorities into the [construction] industry. In addition, action to address elements of discrimination and exclusion would result in more ethnic minorities staying in construction and moving into more senior positions' (Royal Holloway, University of London 2002).

Gender segregation is the norm. In 1998 'Nine out of ten employees in the construction industry [in England] were men' (EOC, 1998). The EOC commented that: 'A similar pattern can be seen in the occupations of women and men. Those showing the greatest concentrations of men were craft and related occupations and plant and machine operatives. In addition two-thirds of managers and administrators were men'. Similar patterns of gender segregation arose in Scotland and Wales. The outlook remains bleak. Men account for 99 per cent of modern apprentices in construction.[13] In 2003 the EOC launched a formal investigation into occupational segregation which is focusing on modern apprenticeships in five sectors, including construction and plumbing.[14]

In order to attract the necessary numbers of skilled employees, the industry needs to look beyond this traditional source of workers and recruit from a broader spectrum of the population.[15] Thus not only is it illegal to discriminate against protected groups it is economically necessary for the construction industry to train and employ a more diverse employee population.

Anecdotal evidence suggests that individuals from ethnic minorities employed in the construction industry continue to suffer discrimination, for example, being paid less than other employees on the same grade, or not being offered a job on the grounds of insufficient experience despite working in the industry for 20 years. Perceptions of ethnic minorities continue to be stereotyped particularly in those parts of the country where there are few ethnic minority residents. Women employed in the construction industry even at a senior level report discrimination or prejudice by managers rather than by the manual labour force. In a white male-dominated industry the culture is seen as akin to a rugby club, with some male managers finding it difficult to work with women of a similar or higher status. Despite attempts to encourage women civil engineers, they often find it difficult to obtain employment on construction sites. Reasons given include difficulties that they would face managing the work on site. Yet there is evidence that among the public there is customer preference for 'women tradespersons', who are perceived as less threatening than men, likely to be more careful in their work, and tidier (Norman, 2002).

General principles

The Race Relations Act 1976 and the Sex Discrimination Act 1975 define race discrimination and sex discrimination in virtually identical terms. The Disability Discrimination Act 1995 adopts a similar but not identical approach to discrimination, whereas the Equal Pay Act 1970 stands on its own. This chapter examines discrimination law in relation to employment, with the focus being on English[16] and EU law.

The law imposes a negative obligation on employers in the sense that employers[17] should not discriminate against job applicants, employees, or former employees[18] from the protected groups rather than positively discriminate in their favour. There are limited exceptions to this principle. In certain circumstances employers may positively discriminate. Claims may be brought against an employer in an employment tribunal. Unlike a claim for unfair dismissal there is no requirement that the employee should have been employed for a minimum period of employment before bringing such a claim, and there is no limit on the amount of damages that may be awarded against the employer by the tribunal. Temporary workers employed through an employment agency are protected by the legislation,[19] but self-employed workers are not generally protected. The legislation applies to employment in the United Kingdom as well as to British employers employing employees in another country who are ordinarily resident in Great Britain.[20]

Discrimination in relation to employment is normally unlawful in relation to

- job advertisements
- offers of employment
- the terms on which employment is offered
- opportunities for promotion, transfer, training, or receiving other benefits
- dismissal or subjecting a person to any other detriment.[21]

Job advertisements which indicate an intention to discriminate on grounds of sex,[22] race,[23] or disability[24] are unlawful. For example, advertising for a salesgirl or a foreman may be construed as being in breach of the legislation, unless the advertisement makes it clear that the job is open to persons of either sex. It is lawful for employers to advertise for employees of a particular nationality to work outside Great Britain.[25]

There are four main types of discrimination: direct, indirect, harassment and victimisation. In the case of disability discrimination there is no liability for indirect discrimination, but there is instead a requirement on employers to make reasonable adjustments.

Typical cases of *direct discrimination* involve employers refusing a person a job, not promoting them, not paying them a bonus, or dismissing them for reasons related to their race or sex. Pregnancy discrimination, discrimination against homosexuals and transsexuals and discrimination on grounds of religion or belief come within the ambit of direct discrimination.

Indirect discrimination arises where an employer imposes a requirement which is considerably more difficult for a member of a racial group or a woman to meet. For example an employer may require applicants for the position of a labourer to have a minimum C grade in English GCSE, or that a painter should have a minimum height of 1.70 m. Both such requirements are likely to be unlawful unless the employer can justify them.

Harassment can amount to direct or indirect discrimination. In 2003 legislation introduced a new statutory definition of harassment which applies to all forms of unlawful discrimination apart from sex discrimination, which is subject to similar provisions introduced in 2005.

Victimisation arises where an employer treats an employee less favourably because they have relied on procedures or rights contained in the anti-discrimination legislation. An employee will also be able to recover compensation where they are treated less favourably by their employer because, in reliance on the anti-discrimination legislation[26] they have

- brought proceedings under the legislation; or
- have given evidence or information in relation to a claim brought by any person under the legislation; or
- done anything by reference to the legislation in relation to any person or
- made allegations that any person has acted in breach of the legislation.

An employer will be similarly liable where they treat an employee less favourably because they know or suspect that an employee intends to do one of these things. However, the employee will not be able to bring a claim of victimisation where any allegation made by him was false and not made in good faith.

Positive discrimination is normally unlawful in the United Kingdom. Where an employer practises positive sex or race discrimination in favour of women or members of an ethnic minority, the employer is likely to be liable for race or sex discrimination where male employees or members of a different ethnic group object. The disability discrimination legislation unlike the sex and race legislation does not make it unlawful for an employer to treat disabled employees more favourably than other employees.[27]

Positive discrimination should be distinguished from *positive action* which is permitted in limited circumstances under both UK and EC law. Positive action arises, for example, where an employer wishes to encourage under-represented groups to apply for training or employment. For example, job advertisements may invite applications from under-represented or disadvantaged groups. Where in the previous 12 months no persons of a particular racial group or sex (or a comparatively small number) were doing particular work, an employer may provide training or encouragement to the under-represented racial group or sex to undertake that work.[28] European Union law also permits measures to promote equal opportunity for men and women in particular by removing existing inequalities which affect women's opportunities.[29]

Therefore, for example, where a building firm is experiencing a skills shortage it would be lawful to provide a training course for an ethnic minority or for women applicants where none or a comparatively small number were employed on that task during the previous year.[30] Positive action in employment is also permitted to prevent or compensate for disadvantages linked to religion or belief,[31] or sexual orientation.[32]

The disability legislation requires employers to make reasonable adjustments for disabled employees (see later), which is a form of positive action, rather than positive or reverse discrimination.

Recent changes in race relations legislation require public authorities to consider taking positive action. It is unlawful for public authorities to do any act which constitutes racial discrimination.[33] Specified public authorities, including county, borough and district councils as well as the Construction Industry Training Board,[34] are under a number of legal duties. These include a duty to eliminate unlawful racial discrimination and promote equality of opportunity and good relations between persons of different racial groups,[35] as well as a duty to monitor the racial composition of their work force and publish annually the results of the monitoring exercise.[36] Local councils are also required to publish race equality schemes.[37]

Codes of Practice

Employers should have regard to relevant Codes of Practice[38] which give practical guidance on the implementation of legislation. Codes of Practice are not law, but should normally be followed. They provide value assistance where there is ambiguity, but are illustrative rather than prescriptive.[39] In addition there is an EU code of practice on sexual harassment[40] and a code on age discrimination.[41] The former is well established and should normally be complied with by employers, whereas the latter is a voluntary code and there are currently no penalties for non-compliance.

Employment tribunals

Discrimination claims are heard by employment tribunals, which are composed of a lawyer chairman and two lay members. An application to the tribunal must be made within three months from when the act complained of was done.[42] The tribunal has the discretion to hear a case where the application was made outside the time limit and it is just and equitable to do so.

Burden of proof

An employee normally has to prove their case on a balance of probabilities, and if so proved the employer has to establish a defence to avoid liability.

However, in sex and race discrimination cases, the applicant does not necessarily have to show that a particular act was intended to be discriminatory. Where an applicant or an employee can prove facts from which it can be inferred that direct[43] or indirect[44] discrimination has occurred, the employer is required to deny or justify the direct or indirect discrimination. This approach was developed in the English courts, but EU directives expressly required it to be adopted and the approach can now be found in UK legislation.[45] The Employment Appeal Tribunal (EAT) has given guidance as to how this new approach in the sex discrimination provisions should be interpreted.[46] The legislation sets out a two-stage test. First, the applicant has to prove on a balance of probabilities facts from which the tribunal could conclude that the respondent employer had committed an unlawful act of discrimination. Second, where the applicant has proved such facts the respondent employer has to prove on a balance of probabilities that the unfavourable treatment was not on the grounds of sex.

Remedies

Where unlawful discrimination has been proved an employment tribunal may order one or more of the following three remedies:[47]

- compensation
- a declaration as to the legal position of the parties to the case
- a recommendation.

There is no ceiling on the amount of compensation that may be awarded under the discrimination legislation. Where the case was one of indirect race discrimination[48] and the employer was able to prove that the discrimination was unintentional and was not foreseeable by the employer the tribunal did not have the power to award compensation until 2003.[49] This restriction no longer applies to claims brought under the new definition of indirect discrimination introduced in 2003.[50] In the case of unintentional indirect discrimination on the grounds of sex,[51] religion or belief[52] or sexual orientation[53] the tribunal has always had a discretion to award compensation.

Where a recommendation is ordered, for example, to put in place a policy and a procedure relating to harassment, and the employer does not comply with the recommendation, the employee may return to the tribunal and obtain compensation for the failure to comply.

In the case of an equal pay claim a tribunal may:

- make an order declaring the rights of the parties and
- award damages or arrears or remuneration.[54]

Liability of employers and principals

Employers are liable for acts of discrimination by their employees in the course of their employment, whether or not it was done with the employer's knowledge or approval.[55] The employer will have a defence where they are able to show that they had taken such steps as were reasonably practicable to prevent such discriminatory acts taking place. This means that employers should have an equal opportunities policy and an effective system for reporting, investigating and addressing concerns raised under the policy. Similarly an employer will be liable for discrimination by a person acting on the employer's behalf, such as a sub-contractor. Employers may also be liable where their employees discriminate against a contract worker as illustrated by the following case study.

Case study

Mr Bassi owned a lorry. He entered into a contract with Pioneer Concrete (UK) Ltd to deliver ready-mixed concrete to building sites. He visited a site at which C.J. O'Shea Construction was working. Mr Bassi stated that he was racially abused by O'Shea's banksman whose task it was to indicate where, when and in what quantities the concrete delivered by Mr Bassi was to be unloaded. Mr Bassi was subsequently banned from visiting the site and he brought a claim of racial discrimination against O'Shea. The EAT held on a preliminary issue that the employment tribunal was entitled to find that Mr Bassi was employed by Pioneer to personally carry out work on their behalf for O'Shea and could come within the definition of a contract worker.[56]

Race discrimination

Race discrimination may be the consequence of direct discrimination, indirect discrimination, harassment or victimisation (see earlier). In practice the majority of claims are now based on indirect discrimination because direct discrimination has become less common.

Direct discrimination

Direct discrimination is defined as treating a person less favourably on racial grounds.[57] 'Racial grounds' means on grounds of colour, race, nationality or ethnic or national origins.[58] The test in direct discrimination cases is often referred to as the 'but for' test. The question asked, for example, is: 'Would a person have been treated in the way complained of but for his colour?' It is an objective test. It is no defence for an employer to show that they did not deliberately treat a person less favourably and that they are not prejudiced against a particular race (or sex). There are few claims now based on blatant direct discrimination. Usually such discrimination will have to be inferred from the surrounding circumstances.[59] Unreasonable treatment by an employer of a member of an ethnic minority is unlikely to amount to discrimination where the employer is able to show that all employees were treated equally badly. However, such a defence is unlikely to enhance an employer's public image.

Direct racial discrimination also occurs when a white employee is treated less favourably because they have complained about the discriminatory treatment of a fellow Asian employee. If the complaint related to sex discrimination, the complainant would bring a claim of victimisation, rather than direct sex discrimination. This is because direct sex discrimination only arises where the complainant has personally been treated less favourably.

Racial segregation of the work force by an employer also amounts to direct discrimination.[60] However, where, for example, there are a high proportion of Irish workers performing a particular job, because existing employees introduce a friend or relative from the same racial group whenever a vacancy arises, does not amount to an act of segregation by the employer.[61] Where the employer introduces a fairer system of recruitment, but the workforce refuses to co-operate with it, the employer will be potentially liable to a complainant from a different racial group who is unable to obtain a job.[62] Segregation of the work force along sexual lines may amount to direct discrimination, but an employer is likely to be able to demonstrate that it is not less favourable treatment to have, for example, separate rest rooms or toilet facilities for women and men.

An employer is liable whenever an employee or applicant is treated less favourably, which means that when making a comparison between the complainant and an actual or hypothetical comparator that 'the relevant circumstances in the one case are the same or not materially different, in the other'.[63] Unlike equal pay cases the complainant in direct discrimination cases is not required to identify an actual 'comparator' that is, another employee or applicant who has been treated more favourably than the complainant.

The main *defence* available where direct discrimination has been established is where the employer is able to show that being a member of a particular racial group is a genuine occupational *qualification* for the job.[64] For example, where the holder of the job provide persons of a particular racial group with personal services promoting their welfare, such as a nursery worker for a nursery where 84 per cent of the children were of Afro-Caribbean origin.[65] This defence has been narrowed in 2003 as follows. An employer will have to show that there is a genuine and determining occupational *requirement* that a person belongs to a particular race or has specific ethnic or national origins, and that it is proportionate to apply that requirement in the particular case.[66] Neither of these defences is likely to be relevant to the construction industry.

Indirect discrimination

The definition of indirect discrimination was extended in July 2003. Under the 1976 Act indirect race discrimination arose where an employer

- applied an unjustifiable *requirement or condition* to a job applicant or an employee,
- which was such that the *proportion* of persons of the same racial group who could comply with it was *considerably smaller* than the proportion of persons not of that racial group who could comply with it,
- and which was to the *detriment* of the applicant or employee because they *could not comply* with it and
- which the employer could not show to be *justifiable* irrespective of the racial group to which the applicant or employee belongs.[67]

Racial group means a group of persons defined by reference to colour, race, nationality or ethnic or national origin.[68] A 'racial group' should have a long shared history and a cultural tradition of its own.[69] The courts have held that Sikhs[70] and Jews[71] are each a 'racial group' but that gypsies[72] and Rastafarians[73] are not.

The employer had a defence where they were able to show that the requirement or condition was justifiable. The test of justifiability is objective based on reasonableness rather than the employer's convenience or subjective perception.[74]

In relation to placements in the construction industry a CRE report found that the placement methods used by the CITB could lead to unlawful indirect discrimination on racial grounds (CRE, 1995), because of the CITB's dependence on 'DIY' and 'kith and kin' approaches to finding placements for trainees.

The new definition of indirect discrimination introduced in 2003 only applies where the discrimination is on grounds of race or ethnic or national origins. The previous definition continues to apply where there is discrimination on grounds of colour or nationality; whether this makes any difference in practice remains to be seen.[75] However, there is in theory at least the possibility of different definitions being relied upon depending upon the reason for the discriminatory act. The new definition provides that indirect race discrimination arises where an employer

- applies to a job applicant or an employee a *provision, criterion or practice*,
- which he applies equally to persons not of the same race or ethnic or national origins as that other,

- which puts the job applicant or employee *at a particular disadvantage when compared* with those other persons,
- which puts the applicant or the employee *at that disadvantage* and
- which the employer cannot show to be a *proportionate* means of achieving a *legitimate aim*.[76]

The new definition is less stringent than the original one. 'Provision, criterion or practice' will cover desirable as well as essential requirements. The need for statistical evidence relating to comparators is likely to reduce if not disappear. The defence of justification is replaced by proportionality and the need to demonstrate a legitimate aim, which may make little difference in practice.

Racial harassment

Racial harassment by fellow employees was a form of direct or indirect discrimination under the Race Relations Act 1976, depending upon whether the harassment amounts to less favourable treatment or to a detriment. Stating that an employee is, for example, 'typically Irish' is likely to amount to racial harassment.[77] It is not necessary for the claimant to show that the harassing employee had an intention or a motive to discriminate. An employer will normally be liable where an employee is racially abused by a fellow employee[78] (see earlier). As shown by the following case study an employer may also be liable where an employee racially abuses an employee of one of their sub-contractors.

Case study

Mr Essa is Welsh, black, and of Somali ethnicity. He was a successful amateur boxer and in order to maintain himself he worked as a labourer and a construction worker. In 1999 he obtained work with Roy Rogers trading as R & R Construction, who was subcontracted by Laing Ltd to carry out work on the Millennium stadium in Cardiff. The foreman was noted to have an extremely abrasive style and on one occasion referred to Mr Essa as 'that black cunt'. Mr Essa complained to Laing but considered that his complaint was not taken seriously. Thereafter he was taunted by other employees and eight days after the incident he left the site and did not return having been reprimanded for not clocking off properly. His claims of constructive dismissal and race discrimination against Roy Rogers were dismissed, but his claim for direct race discrimination against Laing was successful. This was because Laing was responsible for the foreman and was on the facts unable to rely on the defence in s 32(3) RRA 1976 (see earlier) and show that it had taken such steps as were reasonably practicable to prevent the foreman acting in the abusive manner he did. The employment tribunal awarded Mr Essa £5,000 for injury to his feelings together with £519.76 in respect of financial loss. Mr Essa appealed to the EAT arguing that he should also have been compensated for the psychological condition said to have been caused by the discrimination. The majority of the Court of Appeal accepted Mr Essa's argument and held that he was entitled to be compensated for the psychological injury and illness directly caused by the foreman's abuse even though such injury was not reasonably foreseeable.[79]

The law on racial harassment was amended in 2003, and a new statutory definition of racial harassment was introduced by regulations[80] giving effect to EU directives. Harassment in employment arises under the new provisions where 'on grounds of race or ethnic or national origins' (but not colour or nationality which continue to be covered by the previous legislation and case law), an employee engages in unwanted conduct which violates another employee's dignity or creates an intimidating, hostile, degrading, humiliating or offensive environment for him or her.[81] Employers will also be liable where they racially harass an applicant, an employee or a former employee.[82]

Duties of public authorities

In 2001 local councils became subject to statutory duties to eliminate unlawful racial discrimination and promote equality of opportunity.[83] The legislation requires most local councils to publish a race equality scheme and to monitor by reference to racial groups the number of employees, who received training, and who were dismissed.[84] The results of such monitoring are to be published annually. These provisions are likely to have an impact on construction companies that undertake work for local authorities (see e.g. Godwin 2004).[85]

For example the City of Nottingham introduced a code of practice (City of Nottingham 2002)[86] which expressly refers to the duties imposed by the Race Relations (Amendment) Act 2000. The code includes a section on equal opportunities which states that Nottingham expects organisations that participate in its projects to be equal opportunities employers and sets out specific steps which it requires such organisations to take in order to achieve 'genuine equality in employment and service delivery'.

Religious discrimination

Religious groups in Great Britain were not protected by the Race Relations Act 1975 unless they could also show that they were a 'racial group' with a long shared history and a cultural tradition of their own.[87] Although the courts have held that both Sikhs and Jews are a 'racial group', Muslims and Christians normally are not. This is in part because Islam and Christianity transcend ethnic and national boundaries. However, in one case a tribunal held on the facts that direct discrimination on religious grounds amounted to indirect race discrimination where most of the workers who were Muslim were also Asian.[88] The Muslim workers had been disciplined for taking time off for Eid, a religious festival. Leave had been banned during the summer months. The workers had offered to make up the time missed but their request had been refused. In Northern Ireland religious discrimination is unlawful as is discrimination on the grounds of political belief.[89]

In 2003 the United Kingdom implemented EU legislation outlawing discrimination in employment on the grounds of religion or belief.[90] 'Religion or belief' is defined as 'any religion, religious belief, or similar philosophical belief'.[91] It is arguably unlawful to discriminate against an atheist, an agnostic or a humanist so long as their views amount to 'a similar philosophical belief'.

The types of discrimination made unlawful are the same as those under the 2003 race relations provisions (see earlier), that is: direct,[92] indirect,[93] victimisation[94] and harassment.[95] The regulations expressly protect Sikhs from discrimination in connection with requirements as to the wearing of safety helmets on construction sites.[96] Where an employer expects a Sikh to wear a safety helmet while on a construction site and there are no reasonable grounds for believing that the Sikh would not wear a turban at all times while on the site, this will amount to indirect discrimination. The employer will not have a defence that that the wearing of the safety helmet is 'a proportionate means of achieving a legitimate aim'. The regulations also provide that fellow workers will not be able to argue that they are discriminated against because of the special treatment afforded to Sikhs.[97]

Sex discrimination

The legislation, which was passed primarily to ensure that women should not suffer discrimination when compared to men, has been extended to encompass sexual harassment, pregnancy discrimination, discrimination against homosexuals[98] and transsexuals,[99] as well as discrimination on the basis of marital status.[100] Under EC law the principle of equal treatment means that there shall be no discrimination whatsoever on grounds of sex either directly or indirectly by reference in particular to marital or family status.[101] In addition to direct and indirect discrimination, an employee may also bring a claim of victimisation under the legislation (see earlier).

Direct discrimination

The sex discrimination legislation is phrased in similar terms to the race relations legislation, and has normally been interpreted in a similar way by the courts. Direct discrimination is defined as treating a person less favourably on grounds of her or his sex or marital status.[102] In other words the complainant would not have been treated that way but for their sex. The complainant is not required to identify an actual comparator, but if there is no such comparator a hypothetical comparator will be used. The House of Lords has stated that it is necessary to compare 'like with like'.[103] Lord Nicholls considered that it was often helpful to concentrate on why a complainant had been treated in the way she had been before considering whether it amounted to less favourable treatment. Examples of direct discrimination include dress codes which for example do not permit men to wear T-shirts at work, but allow women to do so.

Where there is a material difference between the applicant and the comparator this will be a defence to an allegation of direct discrimination. It is no defence for an employer to argue that a female employee was not offered the opportunity of going on a training scheme because the employer was concerned that as she was likely to be the only female trainee she would suffer verbal or physical harassment by the other trainees.[104] Similarly it is no defence for an employer to argue that women are automatically excluded from jobs requiring physical strength, because women are not normally as strong as men.[105]

Where direct sex discrimination is proved, an employer will have a defence where they can show that being a man is a genuine occupational qualification.[106] This would include where the nature of the job requires the person to be of particular sex to preserve another's decency or privacy. An example would be working in a care home looking after elderly male on female patients.

Direct discrimination has been interpreted by the courts to include sexual harassment, as well as discrimination against pregnant employees and transsexuals. Legislation has made it unlawful to directly discriminate against transsexuals,[107] and homosexuals.[108]

Indirect discrimination

Indirect discrimination under the Sex Discrimination Act 1975 was initially defined in the same way as in the original provisions in the Race Relations Act 1976.[109] The definition was changed in 2001 to give effect to a European Union Directive. An employer will now be liable for indirect sex discrimination where the employer

- applies a provision, criterion or practice (which has replaced 'requirement or condition') to a female job applicant or employee, which he applies or would apply to a man but,
- which is such that it would be to the detriment of a considerably larger proportion of women than men (which has replaced 'the proportion of women who can comply with it is considerably smaller than the proportion of men who can comply with it'),
- which he cannot show to be justifiable irrespective of the sex of the person to whom it is applied and
- which is to her detriment (the requirement 'because she cannot comply with it' is omitted).[110]

Therefore much of the previous case law is no longer binding because the definition of indirect discrimination has been changed. All aspects of the new definition will have to be satisfied and are likely to give rise to litigation. In particular employers should note that whereas 'requirement or condition' was interpreted as essential in order to be for example appointed, the new wording of 'provision, criterion or practice' is likely to cover most aspects of a job specification. There is still an element of comparison between the numbers of men and women who can satisfy the 'provision, criterion or practice', but the government's view is that there should be less need to rely on statistics. The defence of justification is still available.

Examples of indirect discrimination under the previous definition included introducing a flexible shift system which has a disproportionate impact on female employees;[111] requiring management training or supervisory experience when women were mainly employed in basic grade posts;[112] and requiring employees to work extra hours.

One contentious area relates to discrimination against part-time workers. There were conflicting decisions about whether refusing to permit a woman to transfer from full-time to part-time work because of child care responsibilities amounted to indirect sex discrimination. UK legislation which required part-time employees to be employed for a longer period than full-time employees before they could bring a claim for unfair dismissal was held by the European Court of Justice (ECJ) to have a disparate impact on women because the great majority of part-time workers were women.[113] However, the House of Lords considered that that requirement was objectively justifiable.[114] Part-time workers who are discriminated against may also have rights under the Equal Pay Act 1970 (see below) and the Part-time Workers (Prevention of Less Favourable Treatment) Regulations 2000.[115]

Pregnancy

The law is to be found in both UK and EU legislation and case law, with EU law making a considerable impact in this area. Dismissal of a pregnant woman is automatically unfair if the principal reason for the dismissal is connected with the pregnancy or the birth.[116] In addition a pregnant woman has a right to paid time off for ante-natal care during pregnancy[117] as well as certain rights in relation to suspension from work because of pregnancy or breast-feeding under statutory provisions including health and safety regulations.[118]

Refusal to appoint a woman because she is pregnant is direct discrimination, as is treating her less favourably in relation to training and other benefits including bonuses. As can be seen from the following case study dismissal of a woman because she is pregnant also amounts to sex discrimination. However, dismissal of an employee who is absent from work after giving birth and taking maternity leave should be treated in the same way as dismissal for absence from work for any other reason.[119]

Case study

Ms Webb was recruited by EMO Air Cargo(UK) Ltd to a permanent post primarily as a replacement for an employee who was going on maternity leave in about six months time. She found that she was pregnant and due to give birth at much the same time as the employee she was replacing. The employers dismissed her. She claimed direct sex discrimination. The House of Lords dismissed her claim stating that she was dismissed because she would be absent from work at the required time rather than because of her pregnancy. The ECJ held that the dismissal was contrary to the Equal Treatment Directive (ETD). The House of Lords considered the case again in the light of the ECJ's judgment and interpreted the Sex Discrimination Act 1975 in accordance with the ETD, holding that the dismissal amounted to unlawful direct sex discrimination.[120]

Sexual orientation

The English courts and the European Court of Justice have consistently held that employment discrimination against male or female homosexuals is not unlawful under the Sex Discrimination Act 1975[121] or the Equal Treatment Directive.[122] However, the European Court of Human Rights held that such discrimination was in breach of the European Convention on Human Rights 1950.[123] In 2003 discrimination in employment against homosexuals became unlawful when the UK government implemented[124] the EU Employment Directive.[125]

The legislation makes it unlawful to discriminate in employment[126] against another person on grounds of sexual orientation whether they are homosexual or heterosexual or bisexual.[127] Direct and indirect discrimination[128] in employment are unlawful as are victimisation[129] and harassment.[130] The regulations use similar definitions to those in the Race Relations Act 1976 as amended in 2003 (see earlier). Again there is provision for an actual or hypothetical comparator and the circumstances between the complainant and any comparator must not be materially different. Where indirect discrimination is alleged the perpetrator will have a defence where they are able to show that the discriminatory act is a proportionate means of achieving a legitimate aim. An employer will have a defence where they can show the employee should be of a particular sexual orientation, for example heterosexual, because it is a genuine occupational requirement or where the employment is for the purposes of an organized religion which, for example, does not tolerate homosexuality.[131]

Gender reassignment

In 1996 the ECJ held that it was in breach of EC law to treat a transsexual less favourably than a person of the sex to which he or she was deemed to belong before undergoing gender reassignment.[132] This decision was followed by the English courts.[133] As a consequence of the ECJ decision the UK government introduced regulations[134] which amended the Sex Discrimination Act 1975 and make direct but not indirect discrimination against transsexuals unlawful. Indirect discrimination is in any case probably unlawful under EC law and the Human Rights Act 1998.

The regulations make it unlawful to treat a person less favourably on the ground that they intend to undergo, are undergoing, or have undergone gender reassignment.[135] In particular a person who is absent from work because they are undergoing gender reassignment should not be treated less favourably than a person who is absent due to sickness or injury or for some other reason.[136] An employer has a defence where they can show that being a man or a woman is a genuine occupational qualification for the job.[137] There is also a defence of a 'supplementary genuine occupational qualification'.[138] This arises, for example, where the job involves a close degree of physical or social contact in a private home, or requires the employee to share residential accommodation with other employees. It is not a defence in the second example where the employee has already undergone gender reassignment.

'Gender reassignment' means a process which is undertaken under medical supervision for the purpose of reassigning a person's sex by changing physiological or other characteristics of sex, and includes any part of such a process'.[139] Therefore the legislation only applies once a person starts the process under medical supervision, and not necessarily when a person decides to start dressing or living as a person of the sex to which they aspire to belong.[140] Once a transsexual person has undergone gender reassignment surgery and has lived as a member of the reassigned gender they are for the purposes of employment to be treated equally with non-transsexual members of that gender.[141]

Sexual harassment

Sexual harassment was not expressly referred to in the Sex Discrimination Act 1975. However, it could amount to direct discrimination where an employee was treated less favourably, or alternatively it may have amounted to indirect discrimination where the employee suffered a detriment, for example, decided to transfer to a lower paid job. The European Union has been influential in developing protection from harassment in the workplace. The EU Code of Practice *Protecting the Dignity of Women and Men at Work* (1991) defines sexual harassment as 'unwanted conduct of a sexual nature, or other conduct based on sex affecting the dignity of women and men at work'. Although a single incident may amount to sexual harassment, tribunals should consider the whole picture, rather than look at each incident in isolation. Whether words or conduct amounts to harassment is in part objective. The conduct should be 'unwanted' and there may be issues about whether the complainant has consented to or provoked the conduct.

In extreme cases an employee may be liable to pay compensation where they repeatedly harass a fellow employee and they may also be charged with a criminal offence.[142]

Whereas changes in 2003 expressly included harassment within the race, religion or belief, sexual orientation and disability discrimination legislation, there was no such amendment to the Sex Discrimination Act 1975. The 1975 Act was only amended in 2005 once European-wide agreement was reached on the legislative provisions. The new legislation[143] introduced two types of sexual harrassment: harrassment on grounds of sex; and unwanted verbal or physical conduct of a sexual nature.

As can be seen from the following case study an employer may be liable where one employee sexually harasses a colleague.

Case study

Ms Stedman was employed as a secretary by Reed and Bull Information Systems Limited. She complained to her mother and colleagues, but not to her own line manager at work that the manager to whom she reported made comments with sexual innuendos to her, on one occasion tried to look up her skirt, and frequently stood behind her when telling dirty jokes to colleagues which stopped when she complained to him. Eventually she left the company because of the harassment and the effect it was having on her health. The employment tribunal held that the employer had discriminated against her by subjecting her to a detriment and in effect dismissing her. The company had failed to investigate the cause of her illness and the complaints she had made to other members of staff had not been taken seriously. The employer was responsible for these omissions and Ms Stedman was entitled to regard herself as constructively dismissed. The EAT dismissed the employer's appeal.[144]

Equal pay

The Equal Pay Act was passed in 1970 but came into force in 1976 at the same time as the Sex Discrimination Act 1975. The Equal Pay Act is primarily concerned with contractual arrangements as to pay and pensions, whereas the employment provisions of the Sex Discrimination Act are mainly concerned with other contract terms, employment procedures and practices as well as non-contractual fringe benefits. Where, for example, there is a dispute about a woman's entitlement to fringe benefits she may choose to bring a claim under either Act. The Equal Pay Act 1970 must be interpreted by the domestic courts in accordance with EU law[145] and decisions of the ECJ. The purpose of the equal pay legislation is not to achieve fair wages, it is to eliminate inequality in pay insofar as it arises from sex discrimination. A man may also rely on the provisions of the Act.

The Equal Pay Act inserts an equality clause into a woman's contract of employment, where one does not exist.[146] An equality clause relates to the terms of the contract and modifies the terms in three situations.

The first situation is where the woman is doing 'like work' with a man.[147] 'Like work' is defined as where the work 'is of the same or a broadly similar nature, and the differences (if any) between the things she does and the things they [the men] do are not of practical importance in relation to terms and conditions of employment'.[148] When setting pay rates employers may take into account differences such as the degree of responsibility, whether the person works at night, or alone, or without supervision.

The second situation arises where a woman is doing 'work rated as equivalent'.[149] This provision only applies where a formal job evaluation study has been undertaken by the employer.[150] Such a study must be analytical, and capable of objective and impartial application. The study should preferably focus on the specific demands made and skills required of a worker, rather than on a whole job basis.

The third situation arises where neither of the above situations applies, but a woman is employed on 'work of equal value' to a man in terms of the demands made on her (for instance, under such headings as effort, skill, and decision).[151] The employer is likely to be required to objectively justify any difference in the woman and the man's terms and conditions.

In all three situations any less favourable terms in the women's contract are modified so that they are no longer less favourable, or any term in the man's contract which benefits him is included in the woman's contract. The tribunal should look at individual terms rather than the contract terms as a whole.[152]

An employer has a defence where they can establish that there is a genuine material factor which is not the difference of sex and which amounts to a material difference between the woman's and the man's position.[153]

In order to succeed in an equal pay claim the woman must identify a male comparator. A comparison is made not just of their respective terms and conditions, but also of the actual work they each undertake. The comparator may be currently employed, or previously employed, or subsequently employed by the same employer, an associated employer, or in certain circumstances in the same establishment or service. However, as highlighted by the case study below a woman may not compare herself with male workers doing work of equal value who are still employed by the woman's previous employer. This is likely to disadvantage women employed in public sector jobs which were subsequently privatised.

Case study

School dinner ladies employed by North Yorkshire County Council (NYCC) were paid the same rates of pay as male manual workers, the jobs having been given equal ratings under a job evaluation study. The rates for dinner ladies employed by private contractors were much lower, because the work was predominantly done by women. NYCC reduced the pay of its dinner ladies so that it could compete with private contractors in the compulsory competitive tendering (CCT) process. NYCC could not rely on the material factor defence because the dinner ladies' wages were reduced because they were women and this amounted to direct discrimination.[154] The CCT process led to private contractors winning the school meals contract in some areas within NYCC. Some of the dinner ladies became employees of the private contractors. They were paid less than the council employees. They brought equal pay claims against their new employers relying on EU law, and used the male council employees whose work had been rated as equivalent as comparators. The Court of Appeal referred the case to the ECJ.[155] The ECJ held that the dinner ladies could not rely on EU law in order to compare their pay with the male council workers because there was no single body which was responsible for the inequality and which could restore equal treatment.[156]

Disability discrimination

The Disability Discrimination Act 1995 makes it unlawful for employers to discriminate against disabled applicants and employees on the grounds of their disability.[157] The legislation is similar but not identical to the race and sex legislation. In particular indirect discrimination was not unlawful, but there is a duty on employers to make reasonable adjustments. In 2004 the employment provisions in the 1995 were amended to give effect to the EU Employment Directive.[158] Small businesses employing fewer than 15 employees exempt from the disability discrimination legislation until 2004. This exemption no longer applies from 1 October 2004.[159]

'Disability' is defined as a physical or mental impairment which has a substantial and long-term adverse effect on a person's ability to carry out normal day-to-day activities.[160] Addiction to alcohol, nicotine or any other substance is not an 'impairment' for the purposes of this legislation.[161] Certain other conditions are not treated as impairments. These include hay fever, a tendency to physical or sexual abuse of other persons, or a tendency to steal.[162] A 'mental impairment' should result from a clinically well-recognised illness,[163] such as schizophrenia; this requirement has been removed by the Disability Discrimination Act 2005.

Case study

Andrew Watkiss, a Chartered Secretary was offered a job as a Company Secretary for a leading construction company. However, the offer was withdrawn after he told the company that he had been diagnosed 20 years ago as suffering from schizophrenia and was on medication. He had successfully worked as a company secretary during much of that 20 year period, although there had been periods of illness involving hospital admissions on three occasions. He successfully brought a claim before an employment tribunal under the DDA and won damages for loss of earnings and injury to feelings (DRC 2003).

Difficulties may arise where a person claims that they suffer pain and are 'disabled' following an accident at work but there is no evidence of any physical injury. The issue arises whether such pain amounts to a mental or a physical impairment. The medical description sometimes used is a functional or psychological 'overlay'. As the case study below illustrates the courts do not consider that this condition amounts to a 'disability' under the Disability Discrimination Act 1995.

Case study

Mr McNicol was employed as a trackman on the railways. He claimed that the vehicle that he was driving in the course of his employment in October 1995 went over a pothole, causing him to be jolted upwards towards the roof. He claimed that he suffered an injury to his back and his lower neck and that it amounted to a disability. He brought a claim of disability discrimination against his employers arguing that they had failed to make any reasonable adjustments to ensure that he was not substantially disadvantaged. The issue before the tribunal was whether he was suffering from a disability. There was no medical evidence of any physical injury to his back, despite his claim that he suffered pain. There was a suggestion by a spinal surgeon that Mr McNicol was suffering from a psychological or psychiatric impairment rather than a physical impairment. An employment tribunal rejected the argument that a mental condition which causes pain is a physical impairment, and held that there was no clear evidence that Mr McNicol was suffering from a clinically well-recognized mental illness. A video of Mr McNicol secretly made the day before the hearing was also inconsistent with his claim that he was disabled within the meaning of the 1995 Act. The EAT and the Court of Appeal upheld the decision of the employment tribunal.[164]

Medical treatment or aids should not normally be taken into account when deciding whether a person is disabled.[165] However, when considering eyesight, disability is assessed when wearing glasses or contact lens.

'Long-term' means has lasted or is likely to last at least 12 months.[166] 'Normal-day-to-day activities' are defined and include mobility, manual dexterity, physical co-ordination, ability to lift, carry or otherwise move everyday objects, and perception of the risk of physical danger.[167] Account should be taken of how far an activity is normal for most people and carried out regularly by most people. Any indirect effects of a change of routine, fatigue or duration should also be taken into account.[168] Where a person has a progressive condition such as cancer, multiple sclerosis, muscular dystrophy or HIV infection, he is taken to have an impairment which has a substantial adverse effect whether it does so.[169] However, the Guidance states that a medical diagnosis alone is not enough; there must be some effect on the individual's ability to carry out normal day-to-day activities, but the effect need not be substantial or continuous.[170] This is no longer required by the Disability Discrimination Act 2005.

There are four types of unlawful disability discrimination: direct; failure to make reasonable adjustments, as well as victimisation[171] (see earlier) and harassment.[172] However, a non-disabled person cannot rely on the legislation to claim that they have been discriminated against because a disabled person has been treated more favourably than them.

Direct discrimination

The definition of direct discrimination is virtually identical to that under the sex and race legislation (see earlier). An employer discriminates against a person if for a reason which relates to that person's disability, he treats him less favourably than he treats others to whom that reason does not apply.[173] The focus should be on the reason (and the underlying facts) for the treatment. Where a disabled employee is treated less favourably than those employees able to perform the job, this is likely to amount to discrimination, unless the employer can show that the difference in treatment is justifiable.[174] From 2004 direct discrimination will also occur where a disabled person is treated less favourably than a person without their particular disability and whose circumstances are the same or not materially different.[175] As the following case study illustrates employers in the construction industry cannot use health and safety concerns as a blanket excuse for refusing to employ a person with a disability.[176]

Case study

Richard Cleaver was employed by Persimmon plc to oversee the construction of houses. He was suspended for seven months from his job because of his diabetes. He had worked for six weeks without difficulty before going on a two week holiday. On his return to work he was suspended on full pay because of health and safety concerns. This had not been previously raised as a concern. His employer did not refer him to an occupational health consultant claiming that they needed to obtain copies of his medical records which they failed to do. Mr Cleaver's rheumatology consultant sent Persimmons plc a report stating that Mr Cleaver was able to perform the job. In particular he was able to walk, climb ladders and work on the upper floors of buildings. The company took no action. Eventually a second medical report was obtained which confirmed that Mr Cleaver could continue in his job, yet he remained suspended. Mr Cleaver won £20,000 from his employer in an out of court settlement. Mr Cleaver, supported by the DRC, successfully argued that he was less favourably treated on the ground of his disability. However, in return for the compensation Mr Cleaver agreed to resign (DRC, 2003).

Duty to make reasonable adjustments

An employer also discriminates against a person if he fails to comply with the duty to make reasonable adjustments[177] to prevent employment arrangements or the employer's premises putting the employee at a substantial disadvantage.[178] This obligation relating to employment arrangements initially only applied in relation to

- arrangements for determining to whom employment should be offered;
- any term, condition or arrangements on which employment, promotion, a transfer, training or other benefit is offered or afforded.[179]

The obligation did not originally apply to dismissal or other detriment, but does so from October 2004.

The duty to make reasonable adjustments underwent a major revision with effect from 1 October 2004. The duty was extended to include a form of indirect discrimination similar to that found in other discrimination legislation.

The new definition no longer refers to 'arrangements made by or on behalf of an employer'. It provides that where

- a provision, criterion or practice applied by or on behalf of an employer, or
- any physical feature of premises occupied by the employer places the disabled person at a substantial disadvantage it is the duty of the employer to take reasonable steps to prevent the provision, criterion or practice having that effect.[180]

The 1995 Act gives examples of reasonable steps which an employer may have to take in relation to a disabled person in order to comply with this duty. Similar provisions are contained in the 2004 amendments. They include

- making adjustments to premises
- allocating some of the disabled person's duties to another person
- altering his working hours
- allowing him to be absent during working hours for rehabilitation, assessment or treatment
- giving him, or arranging for him to be given training
- acquiring or modifying equipment
- modifying instructions or reference manuals
- providing supervision.[181]

The legislation also sets out the factors to be taken into account when deciding whether a particular step should be considered reasonable for a particular employer. The factors include

- the extent to which taking the step would prevent the effect in question
- the extent to which it is practicable for the employer to take the step
- the financial and other costs which would be incurred by the employer in taking the step and the extent to which taking it would disrupt any of his activities
- the extent of the employer's financial and other resources
- the availability to the employer of financial or other assistance with respect to taking the step.[182]

Defence of justification

The defence of justification is available to an employer where an applicant claims either form of disability discrimination. In either case the employer has to show that the reason for the less favourable treatment is both *material* to the circumstances of the particular case and *substantial*.[183] However, from 1 October 2004 the defence of justification to a claim of direct discrimination is no longer available where there is no material difference between the disabled person, who is treated less favourably, and another employee who does not have the particular disability.[184]

Where the employer is under a duty to make reasonable adjustments and has not done so the employer will not be able to plead justification to a claim of direct discrimination unless they can also establish justification for a breach of the duty to make reasonable adjustments.[185]

Role of the equality commissions

The Commission for Racial Equality, The Equal Opportunities Commission, and the Disability Rights Commission publish statutory Codes of Practice on Employment[186] which give practical guidance to employers on how to implement the legal requirements under the anti-discrimination legislation. The Commissions may also support individual litigants and bring cases against employers who authorise discriminatory advertisements.

In addition the Commissions have power to carry out formal investigations in order to work towards the elimination of discrimination and promote equality of opportunity.[187] The EOC and the CRE may carry out either general or named person investigations. The Commissions have limited powers when undertaking general investigations. In 2003 the EOC announced that it is undertaking a general investigation into gender segregation in employment, focusing on modern apprenticeships in five sectors, including construction and plumbing (EOC, 2003). In 2004 the EOC published two reports arising out of the investigation (EOC, 2004).

Named person investigations take place where a Commission has a 'belief' based on evidence that a named employer is discriminating in breach of the sex or race legislation.[188] The Commissions have power to obtain written and oral evidence from the employer. Where an investigation confirms that there have been breaches of the legislation by the employer the Commission may issue a non-discrimination notice which is enforceable in the courts.

In 1995 the Commission for Racial Equality carried out a formal investigation into the Construction Industry Training Board (CRE 1995), which was concerned with the testing, interview and placement processes of the Youth Training Scheme (YTS) at selected centres in London and the West Midlands in 1991. The CRE concluded that the selection tests conducted by the CITB did not discriminate against ethnic minority candidates. However, the CRE was concerned that three aspects of the CITB's placement process was potentially indirectly discriminatory. Two aspects are particularly relevant to individual employers in the construction industry. First is the reliance by many small and medium-sized businesses on recruiting trainees through kith and kin. Second, such business sometimes required applicants to live locally. Both practices could place ethnic minority candidates at a disadvantage and amount to indirect discrimination.

The case study below illustrates the type of formal recommendations which may be made by a Commission following a formal investigation.

Case study

The CRE report of a formal investigation into the Construction Industry Training Board recommended that the CITB selection tests and procedures should be properly validated for ethnic minority candidates, that the placement process should be more systematic, and that the CITB should improve its ethnic record-keeping and monitoring systems (CRE 1995: 12). In particular, test results should continue to be analysed annually by ethnic origins and sex, and where that analysis shows differences between different ethnic groups, such differences should be investigated and the reasons for them addressed. Staff and candidates should know and understand the selection criteria. In relation to placements the CRE report found that the placement methods used by the CITB could lead to unlawful indirect discrimination on racial grounds. The CRE recommended that the CITB should, for example, maintain 'an effective, central role in the placement process: by initiating placements, monitoring and managing them directly rather than leaving employers and trainees to get on with it themselves'. In addition the CRE recommended that the CITB 'should introduce a special support service for ethnic minority applicants, especially young women'.

Conclusions

Discrimination law affects employers in the construction industry in two main ways. First, employers who do not ensure that their employment policies, procedures and practices comply with the legislation risk being taken to employment tribunals by job applicants, employees, and former employees. Second, under changes introduced by the Race Relations (Amendment) Act 2000 employers who do not practise as well as preach equal opportunities may find that they are less likely to procure contracts from public authorities.

Specific issues which are likely to predominate in the construction industry are: recruitment practices, bullying and harassment at work, the 'canteen' and 'rugby club' culture, as well as the need to balance health and safety requirements with the obligation not to discriminate on grounds of disability.

The construction industry is being encouraged to move forward and widen its recruitment and training strategies so that more women and members of ethnic minorities become valued and long-term members of the workforce. Economic necessity as well as compliance with the law should ensure that this happens.

Discussion questions

1   Consider how trainees are recruited by a particular construction operation with which you are familiar. Is any aspect of the recruitment process potentially discriminatory? What mechanisms should an employer have in place to assess whether or not the process might be discriminatory? How can an employer minimise the risk of discrimination in this context?

2   Draft an equal opportunities policy in relation to discriminatory harassment which could be used in a small or medium-sized construction business. What should the employer do to try and ensure that such harassment is not tolerated?

3   Discuss the practical actions that an employer should take in determining whether a diabetic applicant or employee should or should not be employed because they may pose a health and safety risk.

Notes

1 Race Relations (Amendment) Act 2000.
2 Race Relations Act 1976 (Statutory Duties) Order 2001 SI 2001/3458 arts 2, 5, as amended by the Race Relations Act 1976 (Statutory Duties) Order 2003 SI 2003/3006 arts 2, 4, 5.
3 Race Relations Act 1976 (Amendment) Regulations 2003 SI 2003/1626 which came into force on 19 July 2003.

4 The Race Directive (Council Directive 2000/43/EC of 29 June 2000).

5 Sex Discrimination Act 1975.

6 Equal Pay Act 1970.

7 The Employment Directive (Council Directive 2000/78/EC of 27 November 2000).

8 Employment Equality (Religion or Belief) Regulations 2003 SI 2003/1660 (EE (RoB) R 2003) came into force on 2 December 2003, as amended by the Employment Equality (Religion or Belief) (Amendment) Regulations 2003 SI 2004/437.

9 Employment Equality (Sexual Orientation) Regulations 2003 SI 2003/1661 (EE(SO) R 2003), which came into force on 1 December 2003.

10 Disability Discrimination Act 1995 (Amendment) Regulations 2003 SI 2003/1673. The amendments come into force on 1 October 2004.

11 The Employment Directive (Council Directive 2000/78/EC of 27 November 2000).

12 In 1995 the CRE stated that 'people from visible ethnic minorities are poorly represented in the construction industry. In the 1991 census 8 per cent of 16–17 year olds were from ethnic minority groups whereas only 1.1 per cent of the Construction Industry Training Board's trainees in 1991 were from these groups', *Building Equality*, Report of a formal investigation into the Construction Industry Training Board.

13 EOC Press Release 25 June 2003.

14 In May 2004 the EOC published its first report arising out of the investigation (*Plugging Britain's Skills Gap*). The report found a direct link between skills shortages and the under-representation of women in key parts of the labour market such as construction. The EOC published a second report in November 2004.

15 There is a shortfall from formal construction training schemes of 17,000 each year (EOC 2003).

16 The law is the same in Wales and very similar in Scotland and Northern Ireland.

17 Similar obligations are also imposed on, for example, partnerships, trade organisations, providers of vocational training and employment agencies.

18 SDA 1975 s 20A as inserted by the Sex Discrimination Act 1975 (Amendment) Regulations 2003 SI 2003/1657 reg 3; RRA 1976 s 27A as inserted by the Race Relations Act 1976 (Amendment) Regulations 2003 SI 2003/1626; DDA 1995 s 16A as inserted by the Disability Discrimination Act 1995 (Amendment) Regulations 2003 SI 2003/1673 reg 15; EE(RoB) R 2003 SI 2003/1660 reg 21; EE(SO) R 2003 SI 2003/1661 reg 21; Relaxion Group plc v Rhys Harper [2003] IRLR 484.

19 Contract workers: RRA 1976 s 7; SDA 1975 s 9; DDA 1995 s 12 (replaced from 1 October 2004 by s4B as substituted by the Disability Discrimination Act 1995 (Amendment) Regulations 2003 SI 2003/1673 reg 5); EE(RoB) R 2003 SI 2003/1660 reg 8; EE(SO) R 2003 SI 2003/1661 reg 8.

20 RRA1976 s 8 as substituted by the Race Relations Act 1976 (Amendment) Regulations 2003 SI 2003/1626 reg 11; DDA 1995 s 68(2) as substituted by the Disability Discrimination Act 1995 (Amendment) Regulations 2003 (DDA1995 (A)(R) 2003 SI 2003/1673 reg 27; EE(RoB) R 2003 SI 2003/1660 reg 9; EE(SO) R 2003 SI 2003/1661 reg 9. SDA 1975 does not apply where the employee works wholly outside Great Britain: s 10 as amended.

21 RRA 1976 s 4; SDA 1975 s 6; DDA 1995 s 4 as substituted from 1 October 2004 by DDA1995(A)R 2003 SI 2003/1673 reg 5; EE(SO) R 2003 SI

2003/1661 reg 6; EE(RoB) R 2003 SI 2003/1660 reg 6. 'Dismissal' includes constructive dismissal.

22 SDA 1975 s 38.

23 RRA 1976 s 29.

24 DDA 1995 s 11; from 1 October 2004 s 16B as inserted by DDA(A)R 2003 SI 2003/1673 reg 15.

25 RRA 1976 s 29(3).

26 Race Relations Act 1976 s 2, Sex Discrimination Act 1975 s 4 (including claims under the Equal Pay Act 1970, and the Pensions Act 1995 ss 62–65), Disability Discrimination Act 1995 s55 (from 1 October 2004 as amended by DDA1995(A)R 2003 SI 2003/1673 reg 21), EE(SO) R 2003 SI 2003/1661 reg 4; EE(RoB) R 2003 SI 2003/1660 reg 4.

27 See also in relation to supported employment DDA 1995 s 10 (from 1 October 2004 is renumbered s 18C by DDA1995(A)R 2003 SI 2003/1673 reg 11).

28 Race Relations Act 1976 s 38; Sex Discrimination Act 1975 s 48.

29 Equal Treatment Directive Article 2(4) (Council Directive 76/207/EEC of 9 February 1976).

30 See also RRA 1976 ss 35, 36 in relation to certain types of training.

31 EE(RoB) R 2003 SI 2003/1660 reg 25.

32 EE(SO) R 2003 SI 2003/1661 reg 26.

33 RRA 1976 s 19B as amended by the Race Relations (Amendment) Act 2000.

34 RRA 1976 sch 1A as inserted by RR(A)A 2000 sch 1 and amended by the Race Relations Act 1976 (General Statutory Duty) Order 2001 SI 2001/3457.

35 RRA 1976 s 71 as substituted by RR(A)A 2000 s 2(1) and sch 1A as inserted by RR(A)A 2000 sch 1.

36 Race Relations Act 1976 (Statutory Duties) Order 2001 SI 2001/3458 article 5.

37 Race Relations Act 1976 (Statutory Duties) Order 2001 SI 2001/3458 article 2.

38 CRE Code of Practice for the Elimination of Racial Discrimination and the Promotion of Equality of Opportunity in Employment (1983), a revised draft code was issued in June 2004; EOC Code of Practice for the Elimination of Discrimination on the grounds of Sex and Marriage and the Promotion of Equality of Opportunity in Employment (1985); DfEE Code of Practice for the Elimination of Discrimination in the field of Employment against Disabled Persons or Persons who have had a Disability (1996), likely to be replaced in October 2004; EOC Code of Practice on Equal Pay (2003). Code of Practice on Immigration and Asylum Act 1999 – section 222 – on the avoidance of race discrimination in recruitment practice while seeking to prevent illegal working.

39 Vicary v British Telecom [1999] IRLR 680 in relation to DfEE Guidance on matters to be taken into account in determining questions relating to the definition of disability.

40 EU Code Protecting the Dignity of Women and Men at Work (1991).

41 Age Diversity in Employment: A Code of Practice (1999).

42 RRA 1976 s 68, SDA 1975 s 76, DDA 1995 para 3 Sch 3, EE(RoB) R 2003 SI 2003/1660 reg 34; EE(SO) R 2003 SI 2003/1661 reg 34.

43 King v Great-Britain China Centre [1991] IRLR 513.

44 Enderby v Frenchay Health Authority [1994] ICR 112.

45 RRA 1976 s 54A as inserted by the Race Relations Act 1976 (Amendment) Regulations 2003 SI 2003/1626 reg 41; Sex Discrimination Act 1975 s 63A as inserted by Sex Discrimination (Indirect Discrimination and Burden of Proof Regulations) 2000 SI 2000/2660 reg 5; EE(SO) R 2003 SI 2003/1661 reg 29; EE(RoB) R 2003 SI 2003/1660 reg 29.

46 Barton v Investec Henderson Crosthwaite Securities Ltd [2003] IRLR 332; see also Igen v Wong [2005] IRLR 258.

47 RRA 1976 s 56, SDA 1975 s 65, DDA 1995 s 8; EE(RoB) R 2003 SI 2003/1660 reg 30; EE(SO) R SI 2003/1661 reg 30.
48 RRA 1976 s 1(1)(b).
49 RRA 1976 s 57(3).
50 RRA 1976 s 1(1A).
51 SDA 1975 s 66(3).
52 EE(RoB) R 2003 SI 2003/1660 reg 30(2).
53 EE(SO) R 2003 SI 2003/1661 reg 30(2).
54 Equal Pay Act 1970 s 2.
55 SDA 1975 s 41; RRA 1976 s 32 – see Essa v Laing 2003; DDA 1995 s 58; EE(RoB) R 2003 SI 2003/1660 reg 22; EE(SO) R 2003 SI 2003/1661 reg 22.
56 C.J. O'Shea Construction Ltd v Bassi [1998] ICR 1130.
57 Race Relations Act (RRA)1976 s 1(1)(a).
58 RRA 1976 s 3(1).
59 Nagarajan v London Regional Transport [1999] IRLR 572.
60 RRA 1976 s 1(2).
61 See, for example TATU v Modgill, Pel Ltd v Modgill [1980] IRLR 142.
62 See, for example R v Commission for Racial Equality ex p Westminster City Council [1985] ICR 827; [1985] IRLR 426.
63 RRA 1976 s 3(4).
64 RRA 1976 s 5.
65 Tottenham Green Under Fives v Marshall [1989] IRLR 147.
66 RRA 1976 s 4A as inserted by the Race Relations Act 1976 (Amendment) Regulations 2003 SI 2003/1626 reg 7.
67 RRA 1976 s 1(1)(b).
68 RRA 1976 s 3(1).
69 Mandla v Dowell Lee [1983] IRLR 209.
70 For example Mandla v Dowell Lee [1983] IRLR 209.
71 Seide v Gillette Industries [1980] IRLR 1980.
72 CRE v Dutton [1989] IRLR 8, although a county court is reported to have held otherwise O'Leary v Allied Domecq Inns Ltd (July 2000).
73 Crown Suppliers (PSA) v Dawkins [1993] IRLR 284.
74 Hampson v DES [1989] IRLR 69.
75 One difference was the availability of damages for unintentional indirect discrimination (see above).
76 RRA 1976 s 1A as inserted by Race Relations Act 1976 (Amendment) Regulations 2003 SI 2003/1626 reg 3.
77 McCauley v Auto Alloys Foundry Ltd and Taylor (1994) DCLD 21.
78 Jones v Tower Boot Co [1997] ICR 254.
79 Essa v Laing Ltd [2004] IRLR 313.
80 Race Relations Act 1976 (Amendment) Regulations 2003 SI 2003/1626 reg 5, which came into force on 19 July 2003.
81 RRA 1976 s 3A as inserted by the Race Relations Act 1976 (Amendment) Regulations 2003 SI 2003/1626 reg 5.
82 RRA 1976 s 4(2) and s 27A as amended and inserted by the Race Relations Act 1976 (Amendment) Regulations 2003 SI 2003/1626 regs 6,29.
83 RRA 1976 s 71, sch 1A as substituted by the Race Relations (Amendment) Act 2000 s 2 and sch 1.
84 Race Relations Act 1976 (Statutory Duties) Order 2001 SI 2001/3458 regs 2,5.
85 See, for example Godwin 2004.
86 See also West Midlands Forum Common Standards for Council Contracts (July 1998).
87 Mandla v Dowell Lee [1983] IRLR 209.

88 In JH Walker v Hussain [1996] IRLR 11.
89 Fair Employment and Treatment Order 1998 SI 1998/3162 (NI 21).
90 EE(RoB) R 2003 SI 2003/1660.
91 EE(RoB) R 2003 SI 2003/1660 reg 2.
92 EE(RoB) R 2003 SI 2003/1660 reg 3(1)(a).
93 EE(RoB) R 2003 SI 2003/1660 reg 3(1)(b).
94 EE(RoB) R 2003 SI 2003/1660 reg 4.
95 EE(RoB) R 2003 SI 2003/1660 reg 5.
96 EE(RoB) R 2003 SI 2003/1660 reg 26.
97 EE(RoB) R 2003 SI 2003/1660 reg 26(2).
98 EE(SO) R 2003 SI 2003/1661.
99 Sex Discrimination (Gender Reassignment) Regulations 1999 SI 1999/1102.
100 Sex Discrimination Act (SDA) 1975 s 3, as substituted by the Sex Discrimination (Indirect Discrimination and Burden of Proof) Regulations 2001 SI 2001/2660 reg 3.
101 Article 2(1) Equal Treatment Directive (ETD).
102 s 1(2)(a) SDA as substituted by the Sex Discrimination (Indirect Discrimination and Burden of Proof) Regulations 2001 SI 2001/2660.
103 Shamoon v Chief Constable of the Royal Ulster Constabulary [2003] 2 All ER 26.
104 Greig v Community Industry [1979] ICR 356, 1979 IRLR 158.
105 FM Thorn v Meggitt Engineering Ltd [1976] IRLR 241.
106 SDA 1975 ss 7, 9(3B); Equal Treatment Directive Article 2(2).
107 SDA 1975 s2A(1) as inserted by the Sex Discrimination (Gender Reassignment) Regulations 1999 SI 1999/1102 reg 2.
108 EE(SO) R 2003 SI 2003/1661 reg 3(1)(a).
109 s 1(1)(b). In 2003 an additional definition of indirect race discrimination was introduced (RDA 1975 s 1(1A)) which is in similar terms to the SDA definition introduced in 2001 (see above).
110 s 1(2)(b) Sex Discrimination Act (SDA) as substituted by the Sex Discrimination (Indirect Discrimination and Burden of Proof) Regulations 2001 SI 2001/2660.
111 London Underground v Edwards (No 2) [1998] IRLR 364.
112 Falkirk Council v Whyte [1997] IRLR 560.
113 R v Secretary of State for Employment ex p Seymour-Smith and Perenz [1999] IRLR 253.
114 R v Secretary of State for Employment ex p Seymour-Smith (No 2) [2000] 1 WLR 435.
115 SI 2000/1551.
116 Employment Rights Act 1996 (ERA 1996) s 99.
117 ERA 1996 ss 55–57.
118 ERA 1996 ss 66–71.
119 Brown v Rentokil Ltd [1998] IRLR 445.
120 Webb v EMO Air Cargo (UK) Ltd [1995] IRLR 645.
121 Smith v Gardner Merchant Ltd 1[998] IRLR 510; MacDonald v Secretary of State for Defence [2003] IRLR 512; Pearce v Governing Body of Mayfield School [2003] IRLR 512.
122 Grant v South Western Trains [1998] IRLR 206.
123 Smith and Grady v UK [1999] IRLR 734; Lustig-Prean v UK (2000) EHRR 548.
124 EE(SO) R 2003 SI 2003/1661 which came into force on 1 December 2003.
125 The Employment Directive: Council Directive 2000/78/EC.
126 EE(SO) R 2003 SI 2003/1661 reg 6.
127 EE(SO) R 2003 SI 2003/1661 reg 2.
128 EE(SO) R 2003 SI 2003/1661 reg 3.

129 EE(SO) R 2003 SI 2003/1661 reg 4.
130 EE(SO) R 2003 SI 2003/1661 reg 5.
131 EE(RoB) R 2003 SI 2003/1660 reg 7; R(ota Amicus–MSF section and Others) v Secretary of State for Trade, Christian Action Research Education, Evangelical Alliance, Christian Schools Alliance [2004] EWHC 860 (Admin).
132 P v S and Cornwall County Council [1996] IRLR 347.
133 Chessington World of Adventures Ltd v Reed [1998] IRLR 56.
134 Sex Discrimination (Gender Reassignment) Regulations 1999 SI 1999/1102.
135 SDA 1975 s 2A(1).
136 SDA 1975 s 2A(3).
137 SDA 1975 s 7A.
138 SDA 1975 s 7B.
139 SDA 1975 s 82 as inserted by the Sex Discrimination (Gender Reassignment) Regulations 1999 SI 1999/1102 reg 2(3).
140 See e.g. Croft v Royal Mail Group plc [2003] IRLR 592.
141 Chief Constable of West Yorkshire and Another v A (HL) [2004] IRLR 573.
142 Protection from Harassment Act 1997.
143 Employment Equality (Sex Discrimination) Regulations 2005 SI 2005/2467. See also Ruff 2005.
144 Reed and Bull Information Systems Ltd v Stedman [1999] IRLR 299.
145 Article 141 European Union Treaty; Equal Pay Directive – Council Directive 75/117/EEC.
146 Equal Pay Act 1970 s 1(1).
147 Equal Pay Act 1970 s 1(2)(a).
148 Equal Pay Act 1970 s 1(4).
149 Equal Pay Act 1970 s 1(2)(b).
150 Equal Pay Act 1970 s 1(5); Bromley v H & J Quick Ltd [1988] IRLR 456. The evaluation of the applicant's job and the comparator's job must be carried out in the same study: Douglas and Others v Islington London Borough Council and Others, The Times, 27 May, 2004.
151 Equal Pay Act 1970 s 1(2)(c).
152 Hayward v Cammell Laird Shipbuilders Limited [1988] IRLR 257.
153 Equal Pay Act 1970 s 1(3).
154 Ratcliffe v North Yorkshire County Council [1995] IRLR 439.
155 Lawrence v Regent Office Care [2000] IRLR 608.
156 Lawrence and Others v Regent Office Care Ltd and Others [2002] IRLR 822.
157 Disability Discrimination Act 1995 (DDA 1995) s 4. From 1 October 2004 the definition of discrimination was replaced by that contained in Disability Discrimination Act 1995 (Amendment) Regulations 2003 SI 2003/1673 DDA(A)R 2003 SI 2003/1673 reg 5.
158 DDA(A)R 2003 SI 2003/1673 which came into force on 1 October 2004.
159 DDA(A)R 2003 SI 2003/1673 reg 7.
160 Disability Discrimination Act 1995 s 1(1); Disability Discrimination (Meaning of Disability) Regulations 1996 SI 1996/1455; DfEE Guidance on matters to be taken into account in determining questions relating to the definition of disability.
161 Disability Discrimination (Meaning of Disability) Regulations 1996 SI 1996/1455 reg 3.
162 Disability Discrimination (Meaning of Disability) Regulations 1996 SI 1996/1455 reg 4.
163 DDA 1995 para 1(1) of Sch 1.
164 McNicol v Balfour Beatty Rail Maintenance Ltd [2002] IRLR 711.
165 DDA 1995 para 6 of Sch 1.

166 DDA 1995 para 2(1) of Sch 1.
167 DDA 1995 para 4(1) of Sch 1.
168 See DfEE Guidance on matters to be taken into account in determining questions relating to the definition of disability.
169 Disability Discrimination Act 1995 para 8(1) of Sch 1.
170 See DfEE Guidance on matters to be taken into account in determining questions relating to the definition of disability. Para A 15.
171 DDA 1995 s 55 as amended from 1 October 2004 by DDA 1995(A)R 2003 SI 2003/1673 reg 21.
172 From 1 October 2004 a statutory definition of harassment similar to that for racial harassment applies: DDA 1995 s 3B as inserted by DDA 1995(A)R 2003 SI 2003/1673 reg 4.
173 DDA 1995 s 5(1)(a). Replaced from 1 October 2004 by DDA 1995 s 3A as inserted by DDA 1995(A)R 2003 SI 2003/1673 reg 4.
174 Clark v TDG t/a Novacold [1999] IRLR 318.
175 From 1 October 2004 DDA 1995 s 3A(5) as inserted by DDA 1995(A)R 2003 SI 2003/1673 reg 4. The concept of material difference is also used under the SDA 1975 (see above).
176 See also Mallon v Corus Constructions & Industrial 2003 EAT (unreported).
177 DDA 1995 s 6; from 1 October 2004 s 3A(2) as inserted by DDA 1995(A)R 2003 SI 2003/1673 reg 4.
178 DDA 1995 s 5(2); from 1 October 2004 s 4A(1) as inserted by DDA 1995(A)R 2003 SI 2003/1673 reg 5.
179 DDA 1995 s 6(2); from 1 October 2004 s 4(1)(2) as inserted by DDA 1995(A)R 2003 SI 2003/1673 reg 5.
180 DDA 1995 s 4A as inserted by DDA 1995(A)R 2003 SI 2003/1673 reg 5.
181 DDA 1995 s 6(3); from 1 October 2004 s 18B(2) as inserted by DDA 1995(A)R 2003 SI 2003/1673 reg 17.
182 DDA 1995 s 6(4); from 1 October 2004 s 18B(1) as inserted by DDA 1995(A)R 2003 SI 2003/1673 reg 17.
183 DDA 1995 s 5(3)(4); from 1 October 2004 s 3A(1)(b) (3)(4)as inserted by DDA 1995(A)R 2003 SI 2003/1673 reg 4. See, for example Mallon v Corus Constructions & Industrial 2003 EAT (unreported) where the employers terminated the interview for the post of occupational health nurse on discovering that the applicant was an insulin dependent diabetic. The employers successfully established the defence of justification.
184 DDA 1995 s 3A((4)(5)) as inserted by DDA 1995(A)R 2003 SI 2003/1673 reg 4.
185 DDA 1995 s 5(5); from 1 October 2004 s 3A(6) as inserted by DDA 1995(A)R 2003 SI 2003/1673 reg 4.
186 CRE Code of Practice for the Elimination of Racial Discrimination and the Promotion of Equality of Opportunity in Employment (1983), which was replaced by a new code published on 24 November 2005. EOC Code of Practice for the Elimination of Discrimination on the grounds of Sex and Marriage and the Promotion of Equality of Opportunity in Employment (1985); DFEE Code of Practice for the Elimination of Discrimination in the field of Employment against Disabled Persons or Persons who have had a Disability (1996); EOC Code of Practice on Equal Pay (2003) The DRC published a new Code of Practice on Employment and Occupation in 2004.
187 SDA 1975 ss 57–61; RRA 1976 ss 48–52; Disability Rights Commission Act 1999 ss 3–6.
188 CRE v Prestige Group plc [1984] IRLR 166.

References

City of Nottingham, *Code of Practice of Employment and Training Construction Sector 2002.*

CRE 1995, *Building Equality*, Report of a formal investigation into the Construction Industry Training Board.

DRC 2003 website (www.drc-gb.org).

EOC, *Women and Men in Britain, The Labour Market* 1998.

EOC Press Release 25 June 2003.

EOC 2004 (May) *Plugging Britain's Skills Gap.*

EOC 2004 (November) *Britain's Competitive Edge: Women, Unlocking the Potential.*

Godwin, Kate, Contracting for Equality, *Equal Opportunities Review*, No. 130, June 2004, pp. 8–14.

Norman, Irene (Rhyl College), Keynote Speech to the Equality Exchange Annual Conference, 2002.

Royal Holloway, University of London, *Retention and Career Progression of Black and Asian People in the Construction Industry*, Centre for Ethnic Minority Studies, January 2002.

Ruff, Anne, Harrassment in the Workplace – What's New?, *Business Law Review*, 26(10), October 2005, pp. 226–230.

Chapter 3

Diversity in the employment relationship

Malcolm Sargeant

Introduction

This chapter considers the issue of employment status. The traditional model of a contractual relationship between an employer and a full-time, permanent, employee is one that does not match the reality of relationships between employers and workers. A diverse workforce, such as that which exists within the Construction sector, requires a variety of contractual relationships, so that some types of relationship do not remain as 'inferior' to this traditional model. Both the EU and the UK Government have accepted that to have a flexible workforce requires that different types of working arrangements, such as part-time, fixed-term and temporary work, need to be encouraged. One way of helping this flexibility is to extend employment protection from those who work on the traditional model to those who work on these different patterns.

The style of referencing and footnotes in this chapter is adopted to facilitate ease of reading and accurate attribution of legal sources.

Dependent labour

One of the features of employment law in the United Kingdom is the distinction between employees and workers. The latter tends to have a wider meaning. Section 230(1) Employment Rights Act 1996 (ERA 1996) defines an employee as 'an individual who has entered into or works under (or, where the employment has ceased, worked under) a contract of employment'. Section 230(2) then defines a contract of employment, for the purposes of the Act, as meaning 'a contract of service or apprenticeship, whether express or implied, and (if it is express) whether oral or in writing'. The meaning of worker can be the same, but it can also have a wider meaning, that is, an individual who has entered into, or works under, a contract of employment or

> any other contract, whether express or implied and (if it is express) whether oral or in writing, whereby the individual undertakes to do or

perform personally any work or services for another party to the contract whose status is not by virtue of the contract that of a client or customer of any profession or business undertaking carried on by the individual.[1]

Thus there are some individuals who will not be under a contract of employment to a particular employer, but are under a contract to perform personally any work or services for an employer. Often these latter will be treated as self-employed, which means, for example, that they would not receive the benefits of employment protection measures attributable to employees. Nevertheless, these workers may be as dependent on one employer as that same employer's employees.

The numbers of self-employed workers has grown significantly in the last 20 years and, in 2004, amounted to approximately 3.4 million people, compared to about 24.5 million employees. Many more male workers (over 20 per cent of male workers) than female workers (less than 4 per cent of female workers) are self-employed (Labour Force Survey 2004). Over two thirds of the self-employed have no employees themselves and are dependent upon using their own skills and labour (Bevan). For some workers, self-employment is an illusion. They will be dependent upon one employer for their supply of work and income, but may be lacking in certain employment rights because of their self-employed status. One study of freelancers in the publishing industry (Stanworth and Stanworth), for example, concluded that

> Freelancers in publishing are essentially casualised employees, rather than independent self-employed ... in objective terms they are disguised wage labour.
>
> (1995)

There is then a real difficulty in distinguishing between those who are genuine employees and those who are self-employed, especially if they have the same dependence on one employer as do employees.

Differences in employment protection

To some extent this difficulty is recognized by the Government when certain employment protection measures are applied to workers and others to employees only. The Working Time Regulations 1998,[2] for example, refer, in regulation 4(1), to a 'worker's working time'; whilst the Maternity and Parental Leave etc Regulations 1999 apply only to employees.[3] The limited extent to which those who are not classified as employees receive employment protection is illustrated in Table 3.1.

An example of the protection offered to employees is contained in *Costain Building & Civil Engineering Ltd v Smith*,[4] where a self-employed

Table 3.1 Employment protection rights of employees and workers

Employment right	Employees only	All workers (including employees)
Written statement of employment particulars	x	
Itemised pay statement	x	
Protection against unlawful deductions from wages		x
Guarantee payments	x	
Time off – for public duties, to look for work or arrange training in redundancy, ante-natal care, dependants, pension trustees, employee reps, for young person to study or train, for members of EWCs	x	
Ordinary or additional maternity leave	x	
Parental leave, paternity leave, adoption leave	x	
Right to notice	x	
Written statement of reasons for dismissal	x	
Unfair dismissal	x	
Right to be accompanied to disciplinary/grievance meetings		x
Right to a redundancy payment	x	
Right to an insolvency payment	x	
Protection by the Transfer Regulations	x	
Protection by the Fixed-term Work Regs	x	
Right to be informed and consulted about collective redundancies	x	x
Right to national minimum wage		x
Right to rest beaks, paid annual leave and maximum working time		x
Protection by Part-time Workers Regs		x
Rights connected with belonging to a trade union or time off for trade union duties and activities	x	
Right to new dispute resolution procedures	x	

Source: Department of Trade and Industry (2002) *Discussion Document on Employment Status in Relation to Statutory Employment Rights* URN 02/1058 pp. 15–16, London.

contractor was appointed by the trade union as a safety representative on a particular site. The individual had been placed as a temporary worker with the company through an employment agency. After a number of critical reports on health and safety he was dismissed by the agency at the request of the company. He complained that he had been dismissed, contrary to section 100(1)(b) ERA 1996, for performing the duties of a health and safety representative. In the course of the proceedings he failed to show that he was other than self-employed. This proved fatal to the complaint, as the relevant Regulations only allowed trade unions to appoint safety representatives from amongst its members who were employees.[5] As the Employment Appeal Tribunal concluded that there was not a contract of employment in existence between the agency and the individual, and that he did not come within the protection offered to trade union health and safety representatives.

Vicarious liability

Employers are vicariously liable for the actions of their employees, rather than for independent contractors. Lord Thankerton summed up the test for vicarious liability:

> It is clear that the master is responsible for acts actually authorised by him; for liability would exist in this case even if the relation between the parties was merely one of agency, and not one of service at all. But a master, as opposed to the employer of an independent contractor, is liable even for acts which he has not authorised, provided they are so connected with acts which he has authorised that they may rightly be regarded as modes – although improper modes – of doing them.[6]

An example of the duty of care to employees was shown in *Lane v Shire Roofing*[7] where the claimant was held to be an employee rather than a self-employed contractor. As a result of this damages were awarded, after a work related accident, in excess of £100,000, which would not have been awarded if the claimant had been carrying out work as an independent contractor. In *Makepeace v Evans Brothers (Reading) Ltd*[8] the Court of Appeal dismissed a claim by a sub contractor's employee against the main site contractor. In this case the employee had been seriously injured whilst using a tower scaffold lent by the main contractor to the sub-contractor. The court held that no duty of care was owed by the main contractor as it was not their duty to ensure that the sub-contractor's employees knew how to erect and use the equipment. This was the responsibility of the individual's employer.

In *Waters v Commissioner of Police of the Metropolis*[9] a police constable complained that the Commissioner of Police had acted negligently in failing to deal with her complaint of sexual assault by a colleague and the harassment and victimization resulting from making the complaint. The House of Lords held that:

> If an employer knows that acts being done by employees during their employment may cause physical or mental harm to a particular fellow employee and he does nothing to supervise or prevent such acts, when it is in his power to do so, it is clearly arguable that he may be in breach of his duty to that employee. It seems that he may also be in breach of that duty if he can foresee that such acts will happen.

Thus the employer owes a duty of care to employees who may be at physical or mental risk or for whom it is reasonably foreseeable that there may be some such harm.[10]

Identifying the employee

The common law has developed, at different times, a number of tests for distinguishing those that have a contract of employment from those that are self-employed contractors. In *Carmichael*,[11] the House of Lords approved the conclusion of an employment tribunal, which had held that the applicant's case 'founders on the rock of the absence of mutuality'. The case concerned the question as to whether two tour guides were employees under contracts of employment and therefore entitled under section 1 of the ERA 1996 to a written statement of particulars of the terms of their employment. The House of Lords accepted that they worked on a casual 'as and when required' basis. An important issue was that there was no requirement for the employer to provide work and for the individual to carry out that work. Indeed the Court heard that there were a number of occasions when the applicants had declined offers of work. There was an 'irreducible minimum of mutual obligation' that was necessary to create a contract of service. There needed to be an obligation to provide work and an obligation to perform that work in return for a wage or some form of remuneration. Part-time home workers, for example, who had been provided with work, and had performed it, over a number of years, could be held to have created this mutual obligation.[12]

Byrne Brothers (Formwork) Ltd v Baird[13] concerned the employment status of a 'labour-only subcontractor'. The individual, with others, claimed that they were workers in terms of the Working Time Regulations 1998, and therefore were entitled to be paid for the Christmas/New Year holiday period. The Employment Appeal Tribunal held that they were entitled to the protection of the Regulations and that they were neither employees nor in business on their own account. They occupied an intermediate category:

> But typically labour-only subcontractors will, though nominally free to move from contractor to contractor, in practice work for long periods for a single employer as an integrated part of his workforce; their specialist skills may be limited, they may supply little or nothing by way of equipment and undertake little or no economic risk. They have long been regarded as being near the border between employment and self-employment.

Thus there was sufficient dependency and mutuality of obligation to ensure that these individuals should not be regarded as running their own business.[14]

What is the effect of the parties to the contract deciding that, for whatever reason, it should be a contract for services, rather than a contract of service? This clearly happens in different occupations, where there is an acceptance that individuals are to be treated as self-employed contractors, rather than employees. One study of the construction industry (Harvey, 1995) concluded that some 58 per cent of the workforce, excluding local government, was treated as self-employed. This was some 45 per cent of the total workforce. The author concluded that there 'is the strongest indication that self-employment, as an employment status, is an economic fiction'. In *Ferguson v Dawson & Partners*[15] the court considered an individual who worked on a building site as a self-employed contractor. He did not have an express contract of any kind, although the Court came to accept that implied terms of a contract did exist. Although the label of self-employment, as agreed by the parties, was a factor to be considered, it could not be decisive if the evidence pointed towards a contract of employment.[16]

Other types of employment relationship

There are a number of other types of employment relationship which are worth considering within the context of the construction sector with its fragmented and diverse workforce. These are the use of temporary agency staff, part-time workers and those on fixed term contracts.

Temporary agency staff

The employment agency industry is an important part of the UK economy. It grew from an industry that merely supplied domestic staff to the current day, one that supplies individuals on a wide range of skills and levels. During 1997–1998 it was estimated, for example, that the industry placed 379,000 individuals into permanent positions with employers and in one sample week in November 1997 it placed 879,000 temporary and contract workers with employers.

In March 2002 the European Commission published its latest proposal for a Directive on the working conditions for temporary agency workers.[17] This is the latest stage of a long saga beginning when the Commission published its first proposal over 20 years ago in 1982. One of the purposes of the initiative is to raise the status of temporary work. It is regarded as an important element in the new flexible economy which the EU hopes will create more and better jobs in the future.

The purpose of the draft Directive, as described, is to be twofold. First, in Article 2(1), the principle of non-discrimination is to be applied to temporary workers in order to improve the quality and status of such work. Interestingly nowhere in the draft Directive is the word 'temporary' defined, so it would presumably apply equally to a very short term posting of a few hours and a longer term placement which could be over a number of years. The one exception to this that is proposed is that the need to pay temporary workers on a par with similar permanent workers may not apply to those employed for less than six weeks (Article 5(4)). It is an interesting thought that part of the objective of the Fixed-term Work Directive was concerned to stop the abuse of fixed-term contracts by limiting the number of such contracts before a person would be assumed to be a permanent employee. In comparison there is no such issue with temporary agency work. Indeed temporary work is to be encouraged and its status improved. Why is there this difference? The second purpose contained in Article 2 is that the Directive aims to establish a suitable framework for the use of temporary work 'to creating jobs and the smooth functioning of the labour market'.

Article 1(1) specifies that it applies to the contract of employment or the employment relationship which exists between a temporary agency and the worker who is posted to a user undertaking to work under its supervision. The final definition as to who will be covered is, as usual, left to the Member State as a worker is defined as someone who is protected as such under national law. It is perhaps unlikely that some agency temporary workers will be covered. It may be difficult to show an employment relationship with those that work under the guise of a limited liability company. An example of such a complex, but not unusual, employment situation is found in *Hewlett Packard Ltd v O'Murphy*.[18] Here the individual concerned formed a private limited company which then entered into a contract with the agency which in turn had a contract with Hewlett Packard. This case concerned the relationship between the worker and the user company. He failed to show that there was an employment relationship between the two, but the relationship with the agency must also be unclear.

Article 3(2) does specify that people may not be excluded solely on the basis that they are part-time workers or on fixed-term contracts within the meaning of the Directives on part-time work and fixed-term contracts.

The basic working and employment conditions of the temporary worker will need to be at least as good as those that would have applied if the individual had been recruited directly by the employer rather than via an agency (Article 5(1)). Basic working and employment conditions are those relating to working time, rest periods, night work, paid holidays and public holidays and pay. They also relate to work done by pregnant women, nursing mothers, children and young people, as well as any action taken to combat discrimination on other grounds. This seems an important list of terms, but there are some significant omissions. There is no mention of any notice period, so temporary staff can still, subject to any other issues, be removed at short notice. Nor is there any opportunity for any disciplinary or grievance appeals procedure. Presumably this is an issue that is assumed to be between the worker and the Agency employer, even though any disciplinary or grievance matters are likely to be between the individual and the user enterprise. Last, there is no mention of pension arrangements. The exclusion of these would make any attempt to lift the status of temporary agency workers to the same status as permanent workers meaningless.

There are three categories of temporary worker mentioned in the Directive. They may each receive differing levels of protection. The first is the worker who has a permanent contract of employment with an Agency, so that they continue to be paid between assignments. Member States may exclude such workers from the principle of non-discrimination. This is regardless of the pay of such people. Permanent employees of the Agency will not therefore have the opportunity to compare themselves with comparable workers from client businesses when they are working on a posting. The second category comprises those who are on a posting, or postings with the same employer, of less than 6 weeks. Member States are to be permitted to exclude these temporary workers from protection against discrimination on the grounds that they are temporary workers. This will exclude large numbers of people. In France and Spain, for example, some 80 per cent of temporary staff work on assignments of less than one month and all could be excluded. In a survey done by the UK Recruitment and Employment Confederation in early 2002, the responding agencies estimated that 64.4 per cent of their temporary agency workers would be able to accumulate six weeks or more employment with one employer. This would leave over one third of all such staff in the United Kingdom excluded in this category. The final category appears to be everyone else, that is, those posted on an assignment, or assignments, lasting for more than six weeks and who do not have a permanent contract of employment with the Agency through which they work.

Employers' organisations were opposed to the idea that a temporary agency worker should be treated as favourably as employees in the host undertaking. It was even suggested that the comparison should be with permanent members of staff of the employment agency, the effect of which would be to neuter the aims of the Directive. This issue is of importance in the construction sector where there will inevitably be considerable amounts of temporary work. The issue of treating temporary employees equally with permanent employees may be somewhat academic if the temporary employees turn out to really be the employees of the host employer and not the agency. In *Brook Street Bureau v Patricia Dacas*[19] an individual was employed as a cleaner in a mental health hostel. She was supplied to the employer, the Council, by an employment agency and stayed in the job for some 5 years. The question for the Court of Appeal was what her employment status was. Was she employed by the local authority or the agency or neither? The Court pointed out that the statutory definition of a contract of service includes an implied contract and that, in such cases, employment tribunals should consider the possibility of an implied contract of service. In this case a contract could be implied because of the levels of control and direction which the end user, the Council, had over the individual. She was therefore able to bring a claim for unfair dismissal against the Council as her true employer. The implications of such decisions for the construction industry, as with any employer, using temporary staff hired through an agency, are important. It shows the need to clarify the contractual arrangements between all the parties and to ensure that the agency has a role in the control and direction of the temporary staff.

Like the Directive and Regulations concerning part-time work, temporary agency workers are to be notified of permanent vacancies in the user enterprise to give them the opportunity to find permanent employment (Article 6). In addition measures will need to be taken to improve access to training both in the Agency employer and in the user enterprise.

Fixed-term contracts

The number of employees on fixed-term contracts increased by over 100,000 between 1994 and 2001 and the Labour Force Survey for Spring 2001 estimated that the total was then 1,396,000 individuals. This figure includes significant numbers of seasonal and casual workers. The majority of those working on fixed-term contracts are women (some 55 per cent), which is reflected in the rest of the European Community. There are also some discrepancies related to ethnic origin: some 5.5 per cent of white employees are on such contracts, compared to 7 per cent of black employees and 8–9 per cent of those employees of an Indian/Pakistani/Bengali ethnic origin. Of no surprise to those that work in the public sector is the fact that over half of all employees on fixed-term contracts are in that sector. More disappointingly is that some 70 per cent of all those who have been on fixed-term contracts for more than 2 years are in the public sector. Perhaps reflecting the nature of charity funding, there are also some 60,000 employees of charities who are on these type of contracts.

Fixed-term work makes up a significant amount of work in the European Union, with approximately 12.2 per cent of its working population being employed on such contracts. On 18 March 1999 the Social Partners at European Community level concluded a framework agreement on fixed-term work. This in turn became Directive 1999/70/EC on fixed-term work.[20] This Directive comes after a lengthy period of the European Commission attempting to obtain agreement amongst the Member States. Proposals were first introduced in 1990 and, until 1999, only one measure had been adopted.[21]

The purposes of the Directive, which apply in the construction industry as well as elsewhere, are to, first, improve the quality of fixed-term work by ensuring the application of the principle of non-discrimination; and, second, to establish a framework to prevent abuse arising from the use of successive fixed-term employment contracts or relationships.[22]

A fixed-term worker is defined by clause 3 of the framework agreement as

> A person having an employment contract or relationship entered into directly by an employer and a worker where the end of the employment contract or relationship is determined by objective conditions such as reaching a specific date, completing a specific task, or a occurrence of specific event.

The agreement introduces a principle of non-discrimination against fixed-term workers,[23] with stricter controls over the renewal of such contracts.

That such workers need to be protected is illustrated in *Booth v United States of America*.[24] The appellants had been employed by the US Army on fixed term contracts with a break of around two weeks between each contract. At the end of each contract the appellants had been informed in writing of the termination and were paid outstanding holiday pay and benefits. They had to fill in a new application form before each new contract began. On termination of the contract all the appellants claimed redundancy payments and one claimed unfair dismissal. Despite arguing that the arrangement for two week breaks between contracts was designed to defeat the legislation, the applicants were unsuccessful in their claims even though, apart from three two week breaks, they had some 5 years service. One of the concerns of the Fixed-Term Work Directive is to encourage some objective justification for continuing fixed-term contracts of employment.

Section 45 of the Employment Act 2002 provided the authority for the introduction of the Fixed-term Employees (Prevention of Less Favourable Treatment) Regulations.[25] They define a fixed-term contract as either a contract of employment which is made for a specific term, or a contract that terminates automatically on the completion of a particular task, or the occurrence or non-occurrence of any specific event except one resulting from the employee reaching normal retirement age or such conduct of the employee that might entitle an employer to summarily dismiss that employee.[26]

The Government decided to apply the regulations to employees only. Those who are not treated as employees are to be excluded. Whilst recognising the problems associated with including non-employees, the decision does have the result of excluding significant numbers of individuals who work on fixed-term contracts and who, apart from their employment status, are indistinguishable from permanent employees or employees on fixed-term contracts. This might especially be true in the construction industry where there is a high proportion of self-employment.

The definition of the comparator uses the same approach as that used by the Part-time Workers Regulations. The individual with whom a fixed-term worker is to be compared is someone who, at the time when the alleged treatment takes place, is employed by the same employer and is engaged on the same or broadly similar work, having regard for whether they have similar skills and qualifications, if this is relevant. The comparable permanent employee must work or be based at the same establishment, although other locations will be considered if there is no one appropriate at the same establishment.[27]

The Regulations provide that a fixed-term employee has the right not to be treated by the employer less favourably than a comparable permanent employee with regard to the terms of the contract or by being subject to any other detriment related to being a fixed-term employee. This includes less favourable treatment in relation, first, to any period of service qualification related to a condition of service, second, to training opportunities and, third, to the opportunity to secure permanent employment in the establishment.

Importantly, however, the Government decided to include less favourable treatment in relation to pay and pensions. As a result, the rules on statutory sick pay, rights to guarantee payments and payments on medical suspension ensure that fixed-term employees and comparable permanent employees are treated in the same way. Similarly where there are qualifying rules for membership of pension schemes, then these rules should be the same for fixed-term and comparable permanent employees, unless the different treatment can be objectively justified. The Government believes that this will help reduce pay inequalities because the majority of fixed-term employees are women, so the inclusion of pay and pensions will help the reduction of inequalities between the sexes.

There is a defence of objective justification to the provisions of the Regulations.[28] The Government has opted to allow the 'package' approach, as an alternative to the 'item by item' approach, when deciding whether an individual has been treated less favourably on the grounds of being a fixed-contract employee. It will not be necessary to compare each part of the terms of employment and ensure that each individual part is comparable to the permanent employee, unless the employer so wishes. Such treatment is objectively justifiable if the terms of the fixed-term employee's contract of employment, as a whole, are at least as favourable as those of the permanent comparator. This presumably means that it will be permissible to, say, pay a higher salary in compensation for other benefits such as holidays and pensions, so long as the value of the 'package' overall is equivalent or better than the permanent employee.

If an employee considers that he or she has been treated less favourably on the grounds of being a fixed-term employee, which is not objectively justifiable, then the employee is entitled to request a written statement giving particulars of the reasons for the treatment. This must be provided by the employer within 21 days of the request. Such a statement will be admissible in any future employment tribunal proceedings.[29] A dismissal connected to enforcing an employee's rights under the Regulations will be treated as an unfair dismissal.[30]

Unfortunately, the Regulations seem unlikely to stop the repeated use of fixed-term contracts, which seems a rather strange outcome. Where there is a fixed-term contract or a succession of such contracts resulting in the employee being continuously employed for 4 years or more, then the contract will automatically be deemed a permanent contract, unless there is objective justification suggesting otherwise.[31] The problem comes in the definition of continuous employment contained in the Employment Rights Act 1996. Any week during which a contract of employment exists will count towards continuity of employment (section 212(1) Employment Rights Act 1996).[32] Thus any sort of substantial break not covered by section 212(3) Employment Rights Act is likely to break continuity and make the Regulations ineffective. If an employee considers that he or she has become a permanent employee because of these rules, then the employee may request a written statement from the employer stating that he or she is now a permanent employee or, if not, the reasons why the individual is to remain a fixed-term employee. This statement must be given within 21 days and is admissible in future employment tribunal proceedings.[33]

Provision is made for some flexibility as the maximum period can be varied by collective or workforce agreements. This opportunity for flexibility may, however, be of limited benefit. The agreement will be reached with employee representatives, the majority of whom are likely to be permanent employees. This may not be a problem for situations where the employees are represented by a trade union and reach a collective agreement on the issue. It might be a problem where employees elect their own representatives, the majority of whom will not be affected. It is more likely to be a problem if the workforce agreement is reached by a majority vote of the workforce (this can be done where there are less than twenty employees). It is perhaps questionable whether employees in such situations will resist management demands for a more flexible approach if the majority are unaffected by the proposals.

Employees will be able to claim unfair dismissal if they are dismissed with regard to their rights under the Regulations. They may take their claim for less favourable treatment to an employment tribunal which will have the right to award compensation to the claimant and recommend action for the employer to take, within a specific period, to obviate or reduce the adverse effect complained of. The compensation is to be limited, however, as the tribunal is specifically forbidden from awarding damages for injury to feelings (although such claims will still be possible if the individual was dismissed for exercising their rights under the Regulations). The matters that will be taken into account will be the loss of benefit arising from the infringement and any reasonable expenses of the complainant as a result of the infringement.

Part-time employment

There has been a significant increase in part-time employment in the last two decades. Between 1989 and 1999 the number of part-time jobs as a proportion of all jobs has risen from 24 per cent to 29 per cent. In 2002 the total number of part-time workers had exceeded 6.9 million people. This growth has been fuelled by the growth of the service sector and by the increase in the female work force. Between 1987 and 1997 there was a 13 per cent increase in the number of women employees and a 14 per cent increase in the number of part-time women employees.[34] The number of men working part-time has also been increasing, but from a much lower base than women. Between March 1980 and March 1998 the proportion of employed men who were working part-time increased from 8 to 12 per cent, compared to 44 per cent of women employees. Over 80 per cent of part-time employees are women. This growth has taken place throughout the European Union. In the European Union as a whole part-time work accounts for 16.5 per cent of total employment, with women accounting for 79.3 per cent of all part-time workers.

The Government claimed that less-favourable treatment of part-time employees is not widespread in the United Kingdom,[35] although a House of Commons report[36] pointed out that 54 per cent of male part-timers and 42 per cent of women part-timers work for employers who do not have a pension scheme, compared to 25 per cent of full-time employees. Discrimination against part-time workers has often, however, been interpreted as indirect sex discrimination, as there are a much greater proportion of women working part-time than men.

The Part-Time Work Directive adopted a framework agreement reached by the Social Partners at EU level.[37] The purpose of the Directive is set out in clause 1 of the framework agreement. First, it aims to provide for the removal of discrimination against part-time workers and to improve the quality of part-time work; and, second, it aims to facilitate the development of part-time work on a voluntary basis and to contribute to the flexible organization of working time in a manner which takes into account the needs of employers and workers.

The Directive, in clause 4, introduces a principle of non-discrimination, so that part-timers, in respect of employment conditions, should be treated no less favourably than full-time workers, solely because they work part-time. There is the opportunity to 'objectively justify' differences. The Directive was transposed into national law by The Part-time Workers (Prevention of Less Favourable Treatment) Regulations 2000 (PTW Regulations),[38] which came into effect on 1 July 2000.

Regulation 2 of the PTW Regulations has a similar definition for both full-time and part-time workers. They are individuals who are paid wholly or in part by reference to the amount of time worked and who are, in relation to other workers employed under the same type of contract, defined as full-time or part-time. The definition for a full-time comparator follows closely the definition in the Directive, although in terms of defining where the comparator needs to be based, it does have a wider definition. Full-timers, in relation to the part-timers, need to be engaged in the same or broadly similar work having regard, where relevant, to whether they have similar levels of qualifications, skills and experience and to be based at the same establishment or, if there is no full-time comparator at the same establishment, at a different establishment.

This is a very demanding test for establishing whether an individual's job can be used as a comparator on which to base a claim for discrimination. The first issue, of course, is what happens if there is no full-time person who can meet the criteria. Where a workforce is made up entirely of part-time employees in a particular category, the Regulations will be of no assistance in enabling them to claim discrimination on the basis of being a part-time worker.[39] One example might be a contract cleaning operation. All the employees concerned with cleaning might be part-time and all the supervisory, management and administration employees might be full-time. The result is that there is no full-time comparator on whom the cleaning staff can base a claim. These employees may be low paid because they are part-time and, perhaps, because they are not organized collectively, but they are unable to base a claim using the Part-time Workers Regulations. Although there may be a *prima face* case for showing discrimination, it could not be a case based upon these Regulations. Even the Government's own figures suggest that 80 per cent of part-time workers will not have a full-time comparator available to them.

Regulation 5 establishes the principle of non-discrimination. A part-time worker has the right not to be treated less favourably than how the employer treats a comparable full-time worker as regards the terms of the contract or by being subject to detriment by any act, or failure to act, by the employer. The right only applies if the treatment is on the grounds that the worker is part-time and that the treatment cannot be justified on objective grounds. In determining whether a part-timer has been treated less favourably, the principle of *pro rata temporis* applies. The one exception to this concerns overtime. Not paying overtime rates to a part-time worker until they have at least worked hours comparable to the basic working hours of the comparable full-time worker is not to be treated as less favourable treatment.

In the Government's compliance guidance, accompanying the Regulations, the following examples are some of those given as arising from the principle of non-discrimination:

- previous or current part-time status should not of itself constitute a barrier to promotion

- part-time workers should receive the same hourly rate as full-time workers
- part-time workers should receive the same hourly rate of overtime pay as full-time workers, once they have worked more than the normal full-time hours
- part-time workers should be able to participate in profit sharing or share option schemes available for full-time workers
- employers should not discriminate between full-time and part-time workers over access to pension schemes
- employers should not exclude part-timers from training simply because they work part-time
- in selection for redundancy part-timer workers must not be treated less favourably than full-time workers.

These examples only apply, of course, if the employer cannot objectively justify a distinction in treatment or if there is no full-time comparator, meeting the criteria provided, with whom the employees can relate.

As with other discrimination measures, the employer or agent of the employer will be liable for anything done by an employee in the course of their employment.[40] There is a defence of having taken all reasonable steps to prevent the worker doing the act in the course of their employment.[41] There is the right not to be dismissed or suffer detriment as a result of exercising any rights under the Regulations.[42] Interestingly, before a complainant brings a case to an employment tribunal they may request in writing from the employer the reasons for the less favourable treatment. The worker is entitled to a reply within 21 days and that reply, or lack of it, is admissible in tribunal proceedings. The tribunal can award compensation, although not for injury to feelings, and make a recommendation for action by the employer to correct the fault.

Conclusion

A workforce, such as that in the construction industry, which reflects both the requirements of employers for flexibility and also recognizes the diversity of potential contractual relationships, requires consideration of the status of the individuals concerned. The European Union has recognised that in order to make diversity attractive, there is a need to move away from a traditional model of full-time employer/employee relationships. This is partly to be done by increasing the employment protection offered to those with non-standard contractual relationships and, as a result, increasing the status and attractiveness of different ways of working.

The European Union and the Government, however, can only provide a framework. It is for employers to realise the importance of equality in diversity. It is only employers who can ensure that flexible forms of working, whether it be through self-employment or one of the other types of contractual relationship that have been considered in this chapter, increase in status and attractiveness to the benefit of themselves as well as to the individuals concerned.

Discussion questions

1   Do you agree with the premise that increasing employment protection for those working on non-standard contracts will improve the status and attractiveness of those types of contract, thus giving employers in the construction sector a greater choice in the quality of workers employed?

2   What are the consequences of such a large part of the workforce in the construction industry being self-employed, rather than employees? Do you think the legal distinction between the employed and the self-employed is a useful one as far as the construction industry is concerned?

3   Do you agree that a principle of non-discrimination should apply to those working under 'non-standard' contracts of employment?

Notes

1 Section 230(3) ERA 1996.
2 SI 1998/1833.
3 See regulation 13(1) where only employees with one year's continuous service, and responsibility for a child, are entitled to parental leave.
4 [2000] ICR 215.
5 Regulation 3(1) of the Safety Representatives and Safety Committees Regulations 1977, SI 1977/500; employee is defined by reference to section 53(1) HASAWA 1974, which defines employee as a person who works under a contract of employment.
6 *Canadian Pacific Railway Company v Lockhart* [1942] AC 591 at p599 PC.
7 *Lane v Shire Roofing Company (Oxford) Ltd* [1995] IRLR 493 CA.
8 [2001] ICR 241 CA.
9 [2000] IRLR 720 HL.
10 See also, for example, *Spring v Guardian Assurance plc* [1994] IRLR 460 HL on the duty of care owed on employment references and *Wigan Borough Council v Davies* [1979] IRLR 127 on bullying and harassment by fellow employees.
11 *Carmichael v National Power plc* [2000] IRLR 43 HL.
12 *Nethermere (St. Neots) Ltd v Gardiner* [1984] ICR 612 CA; see also *Clark v Oxfordshire Health Authority* [1998] IRLR 125 CA where the position of a nurse in the staff bank was considered and held that there was an absence of mutuality of obligation.
13 [2002] IRLR 96.

14 See also *Lee v Chung* [1990] IRLR 236 and *Ready Mixed Concrete (South-East) Ltd v Minister of Pensions* [1968] 2 QB 497 HC.
15 [1976] 1 WLR 1213 CA.
16 See also *Young & Woods Ltd v West* [1980] IRLR 201 CA and *Lane v Shire Roofing Company (Oxford) Ltd* [1995] IRLR 493 where the courts relied upon the test in *Market Investigations Ltd* [1968] 3 All ER to decide that the applicants in both cases were employees, even though treated as self-employed for tax purposes.
17 COM (2002) 701.
18 [2002] IRLR 4.
19 [2004] EWCA Civ 217.
20 Council Directive 1999/70/EC of 28 June 1999 concerning the framework agreement on fixed-term work OJ L175/43 10.7.99.
21 Council Directive 91/383/EEC supplementing the measures to encourage improvements in the safety and health at work of workers with a fixed-duration employment relationship or a temporary employment relationship OJ L206 29.07.1991.
22 Clause 1 of the framework agreement.
23 Clause 4 of the framework agreement.
24 [1999] IRLR 16.
25 The Fixed-term Employees (Prevention of Less Favourable Treatment) Regulations 2002 SI 2002/2034.
26 Regulation 1(2).
27 Regulation 2.
28 Regulation 4.
29 Regulation 5.
30 Regulation 6.
31 Regulation 8.
32 See *Sweeney v J & S Henderson Ltd* [1999] IRLR 306.
33 Regulation 9.
34 All these statistics and those that follow are taken from Part-time Working Second Report of the House of Commons Education and Employment Committee, session 1998–1999; HC 346-1.
35 See Consultation on Part-time Work DTI 26/11/99.
36 Supra note 34.
37 Council Directive 97/81/EC concerning the Framework Agreement on part-time work OJ L14/9 20.1.1998.
38 SI 2000/1551.
39 See evidence of Business Services Association to the Education and Employment Committee, House of Commons Hansard HC 346 vol. 2 session 1998–1999 p. 138.
40 Regulation 11(1) PTW Regulations 2000.
41 Regulation 11(3) PTW Regulations 2000.
42 Regulation 7 PTW Regulations 2000.

References

Bevan, J. *Barriers to Business Start Up: A Study of the Flow into and out of Self Employment*, Department of Employment Research Paper No. 71, HMSO, London.

Harvey, M. (1995) *Towards The Insecurity Society: The Tax Trap of Self-employment*, Institute of Employment Rights, October, London.

Labour Force Survey Quarterly Supplement, April 2004, No. 25, Office for National Statistics, London.

Part-time Working, Second Report of the House of Commons Education and Employment Committee, session 1998–1999; HC 346-1.

Stanworth, C. and Stanworth, J. (1995) 'The self employed without employees – autonomous or atypical?', *Industrial Relations Journal*, Vol. 26, No. 3, September, pp. 43–55.

Part II

Gender and equality

Chapter 4

Social exclusion

Women in construction

Clara Greed

Introduction

One of the objectives of the diversity movement in construction is to increase the number of women in the industry. This is generally seen as a good thing to do, presumably based on the assumption that 'more' will mean 'better' (for women and for the industry) (Greed, 1988, 2000). It is assumed that greater numbers of women will be a clear sign of increased equal opportunities (EO) policy working, and that the women will magically soften and improve the worst aspects of the construction industry culture. Attracting more women will, it is often assumed, lead to more humane forms of management and thus greater productivity and less of a confrontational, conflict-ridden 'macho pack culture' and thus more efficiency and cost-effectiveness. It has often been assumed that 'what is good for women is good for the industry as a whole', that is, the business case but one must add that 'what is good for the industry is not necessarily good for women' (Rhys Jones *et al.*, 1996; Wall and Clarke, 1996; Greed, 1999a,b; Clarke *et al.*, 2004). Expecting a small minority of women to be the change agents to turn around an entire industry is putting a tremendous burden and responsibility upon women entrants. This stance ignores the need for major cultural and organisational change upon the part of the men who comprise 95 per cent of this sector.

In this chapter it is argued that more [women] does not necessarily mean better for a range of complex reasons. It is important to question whether an increase in the number of women entering the construction professions will result in changes in the culture of the construction industry, and thus in the organisation and conditions of employment within professional occupations in construction. This article is based on a range of research (including Greed, 1997a, 1999a, 2000; De Graft-Johnson *et al.*, 2003) which has shown that little commensurate change has occurred so far. Likewise a perusal of the professional construction and built environment journals will reveal that there is still a lack of female faces particularly in construction journal articles on 'top men' in the industry (e.g. peruse any copy of current journals such as *Property Week, Construction Manager, Building inter alia*). Only two of the eighteen members of the current Egan Committee, established to advise on the skills needed by the built environment professions to create sustainable communities, were women, and there were no other visible minorities to be seen (Egan, 2004). Meanwhile student research dissertations and women graduates' experiences still recount the same old problems, albeit now encountered by a new generation of women, a greater number of whom, significantly, may now be drawn from ethnic minority backgrounds too (Uguris, 2001).

First the 'problem' is presented in terms of the quantitative lack of women in construction, whilst the qualitative aspects of the situation, not least the lack of women at senior levels and the discriminatory and macho culture of the industry, are referred to throughout the chapter. Next, a theoretical perspective on the situation is given, based on the concepts of social exclusion and closure that might help explain the situation. References to a range of academic and professional literature that helps cast light on the problem is included at appropriate points throughout. There is no doubt that the industry is trying to improve itself and therefore in the penultimate section initiatives and programmes are identified that might yet result in change. It is culturally significant that there is such a range of these but they seem to be missing the mark and having little effect, presumably because they are aimed at the wrong level or people, and are not culturally appropriate or acceptable to those suffering from the problems. This section is linked to a list of relevant web sites provided at the end. The concluding section sets out some guiding principles for the future. Readers should note that a new raft of generic EO initiatives are currently being introduced at EU and UK level that might help in the future, and these are dealt with in their own right in the chapter on women in planning.

Setting the scene

The Construction industry employs around 1.4 million people, and is responsible for around 10 per cent of the Gross National Product (Fielden *et al.*, 2000). It employs around 10 per cent of the total male workforce, making it one of the largest employment sectors in the United Kingdom (UK) (Clarke *et al.*, 2004). Of these, women constitute less than 6 per cent of the 15 per cent of construction staff who are in the professional and managerial levels of the industry (Greed, 2000), and around 2 per cent of those in the manual trades (Clarke *et al.*, 2000; CITB, 2003). Although these figures are very low, they are a gradual improvement from 10 years earlier (Greed, 1991). But very little progress has been made, compared with other erstwhile professional areas such as medicine and law that started from an equally low base but now manifest more women than men graduating in these specialisms (De Graft Johnson *et al.*, 2003).

Precise figures are difficult to obtain and verify for the construction industry (Greed, 1997b), as there is a general ethos of secrecy swirling around the construction sector. There is often a lack of clear written evidence and criteria, too, on matters such as recruitment, promotion, career development, and qualification. Much of the system still seems to run on the *'he's a good chap, I knew his father'* principle, with informal associations, 'pub culture' fraternities, masonic and sporting enclaves playing a key role in the management of the industry, and the chance of work, at all levels (Greed, 1999b).

It is important to investigate 'why' there remains such a low proportion of women in this sector as a significant proportion of all jobs nationally, at least 10 per cent, are effectively assumed to be earmarked as male territory, because they are within the bounds of construction. This restricts women's chances of EO within society and the economy as a whole. Paradoxically, the industry is suffering a crisis of under-recruitment. It was stated in the building press that if every 16-year-old boy currently at school were to go into construction, this would still not be enough to meet future demands (Dainty and Geens, 1993) and this appears still to be the situation (*Construction Manager*, 2002). Applications to construction courses dropped by over 40 per cent between 1992 and 1997 and this trend has continued resulting in several departments being in terminal decline as students opt for computing, management and media studies in preference to built environment degrees.

Because of the '*man*power shortage' the industry wants to know *'how to get more women and other minorities into construction'* and quickly. This objective is by no means the same as seeking to make the industry, and the occupations therein, more welcoming and ('workable-in-able') to women, in terms of overall ethos and organisation. The assumption still appears to be that women will enter the industry on its terms and will fit in and be grateful, with little change or adjustment being required on the part of the industry itself. But women have their own agendas, lives and particular requirements, not least because many are carers of children, and increasingly elderly relatives, as well as being workers outside the home.

Once in the industry, relatively few women are to be found in senior posts. However, there are now more young professional women in construction, and some large firms have positive recruitment programmes. Attempts to track individual women down have often resulted in responses such as, *'oh there was one, but she left'*. The *'there was one'* phenomenon proved common and there seems to be high levels of minority staff turnover. Whereas in town planning and surveying it was found that there are log jams, women at middle management level seeking to be promoted, and feeling overtaken by younger men, in construction, the women have difficulty even getting a job, let alone promotion, in spite of the much stated 'shortage' of qualified professionals. Log jams exist in construction too but are composed of technician level men seeking to better themselves by getting into management, rather than women professionals.

Of those women who are employed in construction, the majority are in administrative, secretarial and other support service roles, and significantly, are frequently referred to (by mainstream professionals in the industry) as 'not contributing to', or 'not being part of' the [sacred] 'construction process' (Clarke et al., 2000). Many of them also work on site, but in office terrapins with their own toilets and amenities. They appear to exist in quite different 'social space' within the 'same' physical space that women professionals might find demanding and alienating. So there are many women in construction, but they 'don't count' because they are 'office ladies' (Greed, 1999a). However, they may be included in the total workforce when firms are boasting as to how many women, or indeed people, they employ. However, the CITB has set up initiatives to attract 'office ladies' to construction careers (and, for example, Martin Selman and Cathy Higgs at the University of the West of England are course leaders in this initiative in the South West of England).

The industry is not only highly gendered it is also strongly classed, that is vertically divided into distinct socio-economic groups (cf. Evetts, 1996: 27), quite feudal, and military in structure. Roles are strongly differentiated and skills are fragmented in an overwhelming range of specialisms and areas of expertise. Subtle vertical divisions among men with different skills, and between those with different levels of responsibility, are to be found within the industry. Every man knows his place, and 'who' is above and below him. This differential levels are captured in the 'map' of the construction industry produced for National Vocational Qualifications (NVQ) purposes by the Construction Industry Standing Conference (CISC, 1994) upon which the NVQ system, currently in force, is based. Relatively upper class white female professionals, who might have traditionally been officers' wives staying at home (cf. Building, 14.10.94: 210–213), may find there is no place for them in a male world divided into officers and men and remarkably this trend continues up to the present (De Graft Johnson et al., 2003).

The construction industry is one of few remaining areas where a large 'male manual working class' still exists, but unlike factory workers or miners (the darlings of traditional marxist analysis and fordist management theory) this 'working class' is mobile, independent and often self-employed (partly for taxation reasons, see Druker and White, 1996) and culturally entrepreneurial and not necessarily deferential to managers (the alternative to site work is to become a 'man with a van' running your own building business). When looking at 'women in management' it is important to consider 'who' or 'what' is being managed, and at what human cost. The culture and workforce is very different from what a senior woman manager might encounter in a bank or retail organisation which has a more 'captive' and female workforce.

Many women have commented upon the strange contrast between the 'clean' image of 'management' found in the literature of the construction tribe (e.g. Langford *et al.*, 1994), with its polite discussions of conflict management, and what are seen as the harsh realities of labour relations, characterised by bullying (Brown, 1997), conflict, exploitation, poor conditions, and pressure. Some women have talked of seeing the 'slave culture' of the building site as the 'guilty secret' which professional men are hiding from women seeking to work in construction. Several women professionals commented that the hidden agenda behind CDM in Britain (1995 Construction (Design and Management) Regulations), was to reduce the pressure and bullying, under the guise of health and safety legislation. Indeed H&S (health and safety) appears to be a default container for 'equal opportunities' issues, in the absence of other provision. But, within the construction industry H&S agenda disability (and indeed pregnancy) may be equated with injury and incapacity, and seen as a reason for leaving, not entering, the industry. In fact, the whole industry seems to work by each level putting pressure on the next person down. There is a ladder of command stretching from senior management down to the 'overseer' (sergeant, clerk of works) type levels. For such posts, ability to shout and command instant obedience of subordinates are essential attributes, but unquestioning personal obedience to a woman superior may go against the grain.

Women in construction are increasingly likely to work in smaller firms or as self-employed professionals. Many 'prefer' off-site consultancy work, to predominantly on-site roles. The 'big boys' in the 'big firms' are only the tip of the iceberg, as the majority of all firms in construction are small, even though greater attention is often given to the large 'big name' contractors in Britain (Druker and White, 1996). Black (especially Afro-Caribbean) women professionals may feel even more out of place, particularly if working for contractors which are part of international construction organisations where colonial and 'ex-pat' attitudes prevail. Asian women may be the subject of racial stereotyping that typifies them as '*businesslike, obedient, and hardworking*' especially the Chinese (such comments were given to the researcher in a friendly, paternalistic manner by several senior men) (see Chapter 10 for more in-depth discussion of this phenomenon). Moslem construction professionals are increasingly looked upon with suspicion, their religion being seen as a potential barrier, '*always wanting time off to pray*' whilst a Moslem woman's apparel may be seen as unsuitable for a building site (Ismail, 1998). So those few women construction professionals are seen as not fitting easily into pre-existing organisational structures and male pecking orders, and may be far less likely to be given '*the benefit of the doubt*' if they are seen to be '*too ethnic*'. Indeed it may take them quite a while to 'read' the situation and suss out the workplace dynamics in order to survive.

Conceptualising the problem

In seeking to understand the apparent complacency of the industry, and the wariness of would-be female entrants, it was useful to view the construction industry as a tribe, which is itself divided into competitive, aggressive sub-tribes, corresponding to the different professional bodies and specialisms within construction. These all live on 'Planet Construction' – a hostile world which has very little contact with other inhabited planets (such as Planet Modern Management, or Planet Equal Opportunities, or for that matter Planet Careers Guidance) (Greed, 1999b). The author uses the word 'tribe' advisedly, because her research methodology in earlier research (Greed, 1991, 1994) has been primarily ethnographic in approach (Greed, 1997, Greed, 2000). Ethnography may be defined, etymologically, as the process of writing about a particular 'tribe' or people (ethnos). Thus the methodology has its antecedents in anthropology (Hammersley and Atkinson, 1995). The research approach was based upon studying the way the different construction tribes 'see' the world, especially the members of the different professional bodies in construction. The research investigated tribal cultural taboos and rituals, and sought to identify the values and attitudes prominent within each professional tribal group. Particular attention was given to how women were 'seen' as potential fellow members of their chosen tribe (i.e. as professionals and employees or perhaps just as desirable tribal artefacts).

'Subculture' is taken to mean the cultural traits, beliefs, and lifestyle peculiar to the construction tribe (Greed, 1991: 5–6). One of the most important factors seems to be the need for a person to fit in to the sub-culture. It is argued that the values and attitudes held by its members have a major influence on their professional decision-making, and therefore ultimately influence the nature of 'what is built'. The need for the identification with the values of the subculture would seem to block out the entrance of both alternative ideas and people that are seen as 'different' or 'unsettling', especially 'women'. But these are the very elements, and potential components, which may, in fact, be more reflective of the needs and composition of wider society. Such processes, of exclusion and closure, thus contribute the low percentage of women in construction (Greed, 2003a).

The concept of 'closure' is a key factor in understanding the composition of the professions (Greed, 1991: 6). It is important to investigate the powers of the various subcultural groups to control who is included in, or out of their specialism. This is worked out on a day-to-day basis at an interpersonal level with some people being made to feel awkward, unwelcome and 'wrong' and others being welcomed into the subculture, and made to feel comfortable and part of the team. Many a woman in construction has commented that at various times in their careers they were told, *'who do you think you are? It's not for people like you'*.

Within this highly controlled culture, some women are more acceptable to the construction industry than others. The glass ceiling is still definitely there but for some individual women it would appear that the sky's the limit. If they have to appoint a woman because of EO policy less progressive firms are more likely to appoint one who is 'acceptable' and not known for her feminist views or who is 'different' but one who will support the existing male power base and not rock the boat. Such a woman, therefore, is not necessarily going to have a helpful or encouraging relationship with other women seeking to establish themselves in the construction industry. Such are the powers of professional socialisation, social class position, and personal perspective that it should never be assumed that any woman professional is going to hold substantially different or more enlightened views from her male counterparts. Those who 'fit in' are the most likely to gain seniority. Such women contrasted sharply with some of the women graduates, especially those from ethnic minorities and/or inner city backgrounds who were seen as too 'different'. Such women, who were mainly from the new universities, had entered lower status branches of the construction professions. Different experiences of class and 'life' further distance women from each other and prevent the build-up of critical mass. Significantly those women who are from a higher social class, with the resources and contacts to weather the storm of uncertain employment prospects were found in current research to have very different life experiences, and more ambiguous attitudes to gender issues, than those who had come up the hard way via access to courses and on to new university courses (De Graft Johnson *et al.*, 2003).

Conceptualising change: critical mass

Within this apparently fortress-like setting outsiders seeking entrance appear to be either 'socialised' to conform, or are marginalised, discouraged or ejected. A crucial question, therefore, was *'how can change be generated and transmitted within professional subcultural groups?'*. In previous research, emphasis was put upon identifying potential agents and 'mapping' pathways of change. Concepts which informed the investigation of change included critical mass theory as to 'how many people are needed to change an organizational culture', (cf. Morley, 1994: 195, who refers to Bagilhole's work (1993), and see subsequent work by Bagilhole *et al.*, 1996 and Dainty *et al.*, 2000). Kanter (1977) suggests that 15–20 per cent (minority composition) is needed to change the culture of an organisation. Gale (1995) suggests 35 per cent is necessary in the construction industry, but research respondents in the construction industry have suggested that the percentage should be much higher. But originally, in physics, from which the theory is derived, critical mass was defined as an amount, not a percentage (like the minimum size snowball that holds together without melting), which would trigger a chain reaction. Only 20 pounds of Uranium 235 was needed to create critical mass in an atomic bomb weighing 9000 pounds which is 0.2 per cent of total matter (Larsen, 1958: 35, 50, 55 and 73) – comparable to the percentage of minority individuals in construction! Thus it is important to recognise and mobilise the powers of 'prime movers' in detonating critical mass explosion (Kanter, 1983: 296). These might be dynamic powerful individuals or the powers of EU and UK regulations that overarch the industry (as discussed in Chapter 7 in respect of the effects of the EU Amsterdam Treaty requirements for equality mainstreaming in the United Kingdom). But, the process is not guaranteed, for as a male physicist colleague pointed out, just one drop of the 'wrong' (negative) ingredient can contaminate the whole mixture, and stop the whole process from taking place.

Actor network theory was also of interest in the research. This is concerned with the ways in which change is transmitted (or blocked) by means of the activities of influential social groups within an organisation who maintain power and control (cf. Callon *et al.*, 1986). In order to generate change, actor network theory argues there is a need to create new groupings and networks that transcend existing divisions and alliances (Panelli, 2004: 189). Much depends upon creating new pathways and alliances so that, for example, minority groups might pool their strength in the construction industry to generate change (Murdock, 1997; Law and Hassard, 1999). However, it was found from previous research on the town planning profession that division and fragmentation among minority groups was commonplace (Greed, 1988, 1994, 2000) and that this continues particularly in respect of gender and ethnicity being seen as two competing factions, not as two manifestations of the same problem (Onuoha and Greed, 2003). Indeed the emphasis upon separating out the needs of different minority groups ostensibly for equalities-proofing purposes may even cynically be interpreted as part of a wider 'divide and rule' strategy on the part of controllers of the status quo who do not want change (Greed (ed.), 2003a,b). So the chances of progress for minorities in the construction industry is to a considerable extent dependent on generating such institutional change, which in turn is reliant upon cultivating cultural change, and thus linking with change agents within and outside the industry (Greed, 1999b: 246–254).

Thankfully, it has been found from this research and from the author's experience in the built environment professions, that a resilient and influential minority of women does exist within the industry, including key charismatic individuals and leaders, who *are* likely to hold alternative viewpoints. This group is likely to increase (God willing) whilst the present young cohorts of women professionals grow older and become more 'cynical', thus contributing to the build-up of critical mass. But, more negatively, one still hesitates to use the phrase 'critical mass'. Although there is some truth in the concept that once a certain proportion of women is achieved the culture will shift, it should not be assumed that it will be for the better. In parallel with the situation in physics, it is easier to create an uncontrolled, than a controlled, chain reaction resulting from detonation, particularly when the setting is unstable (Larsen, 1958: 50). But, critical mass is one of the most frequently used terms in the industry when discussing (EO). It fits well with the scientific and quantitative bent of the construction subculture (Larsen, 1958), but is highly optimistic and over-simplistic if used as a predictive social concept without acknowledging the immense cultural and structural obstacles present.

Clearly the situation is highly complex and cannot easily be solved by a few directives from central government or from senior management alone (Greed, 2000). Far from 'critical mass' being achieved in construction areas where women are more numerous, such as in housing specialisms, one can observe a 'tilting' or tipping down of this sector's status. Once a specialism is seen as being suitable for women it seems to lose much of its power and status and *post hoc* reasons, structures and rules are created to legitimate the situation.

In spite of all the problems, some young women construction university students interviewed by the author still come across as being full of excitement about their future career, unaware of what awaits them. Indeed the apparent innocence of such young women was commented upon by many older and wiser women construction academics in the research (De Graft Johnson *et al.*, 2003). With the growth of higher education intake more and more young people are potentially drawn into this web of lies and they may lack the cultural capital and social background to see what they are really up against. Indeed many women built environment students simply do not seem to be interested in women's issues (WDS, 1998). One bright young woman kept saying at a college meeting, *'but women students are all getting better degrees than the men, so it's obvious they will do better'*. An older woman responded *'but that's got nothing to do with it, it won't make any difference'*, and the meeting nearly came to blows. Clearly the age gap further militates against critical mass being achieved readily. In contrast one woman of about 25 years of age, spoke out on women's issues at a professional mainstream meeting with great awareness, and such individuals might generate more change than say 20 senior women who have been neutralised by the processes of professional socialisation. Reading EO 'guidelines' as 'actions' imminently to be implemented can generate misplaced optimism (cf. Sharpe, 1995).

Relevant literature: casting light on the situation

A range of academic literature is useful in making comparisons with other parallel professions, such as architecture, and also other countries where progress is greater, such as Sweden (Mellström, 1995), enables one to identify the peculiarities of the British construction professions, and to 'fight familiarity'. Also it was helpful to make comparisons with other industries altogether, and draw upon previous comparative studies – although even these proved to be gendered as to choice of comparators. For example studies in which women researchers were prime movers (such as Druker *et al.*, 1996; and Bagilhole *et al.*, 1996 (which refers to Kanter (1977) on critical mass) tended to draw comparisons with feminised areas such as personnel management, banking and retail.

In contrast, predominantly male research teams within the construction industry, when investigating 'what is wrong with the industry', typically chose to make comparisons with equally male-dominated sectors such as the petro-chemical and motorcar industries. They also tend to concentrate upon comparing industrial processes rather than investigating the composition and culture of the workforce (Egan Report, 1998). A few years ago the 'famous' Latham Report (Latham, 1996) was produced which addressed a whole range of criticisms about the industry, and sought to offer policy solutions. In a nutshell Latham saw the solution as one of more effective 'team-building'. Indeed both construction education and practice seemed to be obsessed with 'working in teams' and woe betide the individual who does not fit readily into a team. Many a young woman construction professional told of the difficulties she has had working out, whether it is wiser to *'be one of the lads'* and *'prop up the bar to all hours'* or whether to stay away as *'you know when they don't really want you there don't you?'*.

Whilst Egan has failed to address gender (Egan, 1998 and again in Egan, 2004), it is significant that Working Group 8 of Latham (CIB, 1996) did address gender issues. It was headed by Sandi Rhys Jones who has a respected track record on researching women's as well as men's issues in construction (Rhys Jones *et al.*, 1996). But it was found that Group 8's work had not even been mentioned at major national liaison conferences which were meant to be giving feedback on the draft recommendations from the Latham report. Gender was crowded out by discussions on environmental sustainability, productivity and multi-tasking *inter alia* (cf. CORE, 1995). This was in spite of the fact that Rhys Jones (CIB, 1996) had argued, and continues to argue, for the need to increase women's participation primarily from the 'business case' and 'manpower resources' perspective, rather than from what might be seen as a more aggressive 'feminist' perspective (see Chapter 7). Although the *content* of Working Group 8 was quickly forgotten by the construction industry, nevertheless astute male construction professionals obviously thought it would show them in a good light if they could remember to mention the name of the report: a useful shield to protect them from likely accusations of lack of gender awareness. Some male professionals, feeling both threatened and somewhat politically superior commented, *'oh we've done women, you should be concerned about the environment'*.

In the author's research in 1997–1998 innocent enquiries about the topic were often met with yawns, and boredom by senior male construction professionals, and with the response, *'You ought to look at the Latham Report Working Group 8'*. It was as if everyone had had their jabs, they had all been inoculated against the EO virus, and there was no need for them to further consider the matter of EO. A Latham 'script' now exists which male construction professionals can confidently recite: the topic has been 'done'. One gets the same sound bites again and again. This script still exists nearly ten years later, although the word 'Latham' has long since been forgotten, and male managers have become even more adept at spinning women and other minorities the correct spiel on the importance of EO whilst successfully making the individual in question feel it is her/his fault that she is not getting on as well as she should. Indeed the construction industry may have to spin the EO script to ensure that they get contracts in areas subject to urban regeneration and EU funding, for, as is explained in the chapter on women and planning, 'gender-proofing' that is contract compliance is now a requirement for many public/private partnership schemes. But male managers proving to other male managers that they are complying is more to do with peer approval than with making sure that women are really getting a fair deal in their respective firms.

Several years on after Latham, two Egan reports and a spate of other committees on similar lines, little has changed, except for the way in which initiatives and proposals are now presented. Indeed another generation is discovering the problems afresh and seldom does a month go by without receiving an e-mail from yet another keen woman student doing an undergraduate dissertation on 'women in construction'. Many of these individuals give the impression they are the first to have thought of the subject and to whom all pre-existing literature and research is invisible, yet to be discovered territory. Clearly this material has never been mentioned on their course. Young women graduates entering the built environment professions, especially architecture, were often quite shocked at finding a macho, confrontational and long hours work culture which their university course had not prepared them for (De Graft Johnson *et al.*, 2003).

But within the industry there is a plethora of web pages, e-mail groups, and electronically transmitted documents that seek to address the problem of 'how to get more women into construction'. Committees seem to assume that creating a web page is a crucial step in creating change. But whether anyone, let alone the right people, ever read this mass of material is another matter altogether, but at least everyone feels something has been done. So let us now look at the range of potential change agents seeking to ameliorate the situation. A web page list is included at the end of the chapter.

Change agents

Generic or gendered change?

As can be seen, and is readily acknowledged by many construction professionals, *'the industry is in a mess'*. There are demands for change within the industry from a range of mainstream and minority sources. Government initiatives such as the Foresight programme and Partners in Technology, and a range of working groups under the auspices of CIC (Construction Industry Council), CIB (Construction Industry Board), CIOB (Chartered Institute of Building), ICE (Institution of Civil Engineers) and CITB (Construction Industry Training Board) *inter alia*; and a range of research projects have all highlighted the need for cultural change within construction.

Reasons for change

The reasons cited for change variously include

- 'the business case'
- increased efficiency
- health and safety considerations
- greater competitiveness
- recruitment crises
- European harmonisation
- down-sizing
- multi-tasking
- greater flexibility
- environmental, economic and social sustainability
- improved human resource management
- qualification rationalisation
- educational reform
- creating a climate of technological innovation and progress
- pressures from the European Union
- joined up thinking
- social inclusion.

But, as can be seen from the above list, arguments for diversity, or indeed gender, awareness are not strongly promoted. More likely a generic emphasis on equality, or an enthusiasm for the creating of 'community' (often framed in very vague terms as to which groups actually comprise this mystic phenomenon) tend to be promoted (Egan, 2004).

The leaders of the construction industry are the first to admit there are major problems in the industry. The organisation of frequent deprecatory rituals, which include the production of a major report and follow-up committee meetings, appears to be an expected part of tribal life. Perhaps these are intended, in part, to mislead outsiders, and to give the impression that something is being done about 'bad practices' which, in reality, may well be to the benefit of the maintainers of the status quo on Planet Construction. Egan (1998) was one of a long series of much heralded reports on how to solve the problems of the industry, preceded, as described earlier by the Latham Report (CIB, 1996), which even included a sub-report on gender. Yet again Egan prepares another report, albeit this time with out gender even being mentioned (Egan, 2004).

Thus religious penance has been done, sacrifice had been made to the gods of the construction industry. In other words extremely thick, expensive, heavy reports had been produced and the gods had been satisfied. No further action is required except perhaps for producing a web page in the hope that someone might read it and change the world!

Many of the above initiatives adopt a generic approach. They do not disaggregate the needs of different groups within the construction workforce upon the basis of gender, race, age or other social differences. Indeed it seems that many women-led initiatives, groups and key individuals remain 'invisible' to mainstream committees and networks when it comes to looking for new members, or to seeking advice on equal opportunity policies. Likewise little of the 'feminist research' in this field is being used in shaping policy. Much of the sterling work that women have done over the years appears to be invisible, although it might offer better understanding and realistic solutions. A colleague noted following a meeting in May 2002 at the ICE, at which a male engineer commented, *'I haven't heard anyone complain about these issues before'* and commented that once men take up the issues they 'reinvent the wheel' with extreme self-confidence, congratulating each other for their efforts, as if they are the first to think of the issue and are the champions and leaders of change. But they seem completely unaware that all this has been done already. Subsequent incidences in 2003 and 2004 have been noted by the author and her research colleagues of this mentality still continuing to the present day. Clearly there is a need to involve women and other minorities in high-level initiatives, whether they are government led or professional body led. Also there is a strong need for male politicians and policy makers to engage with community and minority groups who are already tackling construction issues in terms of employment, training, design and policy issues that affect them directly.

Bottom-up and top-down change agents

Bottom-up change agents

Far from altering the culture of construction, one result of increased access by women and other minorities, has been the development of a series of

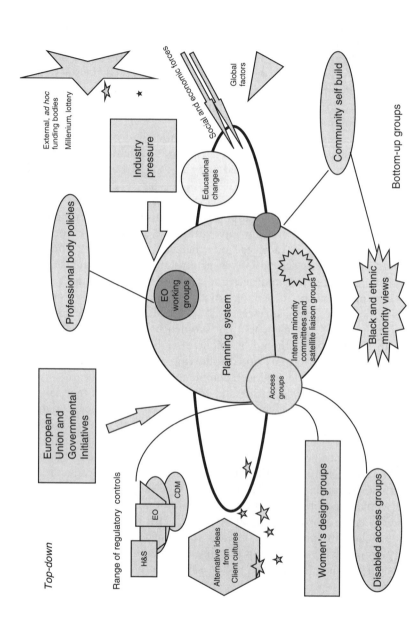

Figure 4.1 Diagram of change agents impinging on construction.

new satellites circling 'Planet Construction' with their own subcultures and organisational structures (Figure 4.1).

Significant groups containing women professionals include

- Architects for Change (incorporating 'Women in Architecture')
- Centre for Accessible Environments
- CIC EO Taskforce
- Planning Aid for London
- SOBA (Society of Black Architects)
- WIP (Women in Property)
- Women and Manual Trades (and London Women and the Manual Trades)
- Women's Design Service.

The establishment of ICEFLOE, the Equality Forum of ICE (civil engineers) has been an extremely significant development, and several of the other chartered bodies now have their own 'women' and groups which in some cases are no longer marginal, 'bottom up' groups but are becoming established committees within their professional body liasing and advising top-down policy makers. Several of the other main built environment professional bodies have also developed new initiatives, not least the planners, architects and surveyors as is explained further in an other chapter of this book.

Those actively involved are admittedly quite small in number, but they may be seen as a force for change. They form part of a powerful network of alternative groups, they are often highly productive in publication, research and campaigning. Also the organisations they run offer a model of alternative management structures. They are often based on a more co-operative, inclusive attitude towards employees at all levels, and a greater level of the communication with 'society' particularly when their 'client' is the community, or a disadvantaged or under-represented minority group. These organisations are not commercial enterprises, and therefore some may say irrelevant as models for the construction industry. Yet their ability to achieve a great deal with limited resources shows they have a respect for the 'cost factor' and for efficiency, economy, and flat management structures, providing examples of working practices which might be usefully emulated within the mainstream industry. One often finds among women's groups a complete lack of the social division between manual trades and professional levels found in the mainstream industry.

Top-down change agents

The above satellite, minority and community groups may be seen as potential 'bottom up' grassroots change agents (see Figure 4.1). A range of 'top-down' change agents could be identified, exerting influence over and above the industry, including governmental initiatives, regulatory and funding bodies. Some of these have already been discussed at the start of this section. Others that may be considered to be potentially more significant, but perhaps more unexpected as potential sources of change, are also highlighted. For example, the National Lottery, Millennium, Sports Council, and Arts Council all have higher powers to require higher accessibility and design standards than are found under 'normal' legislation (Arts Council, 1996) in respect of 'what is built' and 'who ' is building it (drawing on contract compliance principles found in North America). Voluntary bodies, representing minority groups are likely to be among the beneficiaries of grants, thus enabling them to produce exemplar schemes in terms of both design and employment practice. However some 'women and planning' groups are highly suspicious of the *ad hoc* lottery approach to funding which is based upon competition, rather than comprehensive long-term policy, and which may yet prove temporary. Also, black groups may still feel excluded even when other community groups are getting recognition, particularly when it comes to 'competitions' for new schemes. Many are critical of the perceived racism of some housing associations who put black women professionals on their management committee *'because it looks good'* (as two black women architects separately explained to me), but never actually use black professionals in construction projects (Harrison and Davies, 1995). Black built environment professionals have set up a range of network groups, to raise the 'visibility' of all black practices and individual practitioners, to counter further 'assimilation' or 'exclusion'.

Local government still has a role to play, in spite of the decline in DLO (Direct Labour Organisations) which used to be popular employers for women and other minorities providing support, training, and often the first opportunity for 'real' employment (Wall and Clarke, 1996). During the late 1990s Local Labour in Construction Initiatives, such as GLIiC (Greenwich) provided employment and information (LWMT, 1996), in association with major contractors and developments in the area, whilst providing liaison with training schemes and local small business initiatives. Also, many London borough town planning departments make it a requirement of any major planning permission that under a 'Section 106 Agreement' certain EO measures and design features must be integrated in the development, although such measures have been subject to legal challenge, as *ultra vires*. Single Regeneration Budget programmes, and Housing Association schemes, and HATs (Kelly, 1997), have created new possibilities for involvement of people and professionals on specific building projects.

However more recent information on the situation from groups such as Women's Design Service, the London Boroughs Women and Planning Forum, and a range of rural groups too, suggest that women are often marginalised in the consultation processes, which are dominated by local businessmen, because of the emphasis upon economic rather than community considerations (Little, 2002). Also major clients with more 'feminised' cultures such as retail developers, and health authorities, offer alternative business culture role models. A range of controversial 'top-down' urban renewal schemes in inner London have created tremendous levels of community involvement and 'bottom-up' response. In fact there is now such a plethora of urban regeneration programmes under New Labour that many local inner city communities are having to confront planners and developers to defend their territory, and in the process some are getting interested in construction.

Conclusions

Whilst there has been a growth in initiatives and committees, people want deeds not words. Much of the output does not seem to reach the right level of people, or it is still seen as being only to do with women and not the men who rule the construction industry. Therefore publicity material advertising new initiatives is passed on to the 'personnel lady' in the organisation rather than being mainstreamed into decision-making by senior male managers. Bottom-up groups tend to be more successful, in spite of less resources, than top-down groups. Overall, many women exhibit a weariness and frustration at the lack of progress, bearing in mind that it now over 15 years, at least, Since the arguments and reasons for attracting more women have been presented to the industry. Yet one still comes across even young men who will innocently comment, *'well I never thought of that before, I wonder if anyone else has done any work on it'*. But, as many women have commented in respect of the attitudes of older senior managers, *'you have to keep on reminding them of the issues or they will forget all about it'*.

As to what should be done, the CIC EO Taskforce in Construction has produced '10 commandments for equal opportunities' (*Building* 6.11.98: 73) which might be applied within professional bodies and large construction firms. Gale (Fielden *et al.*, 2000) have been working on the Building Equality in Construction Project at UMIST,[1] Manchester, and set out another list of key points on 'How to encourage more women into construction'. These two lists comprise the objectives the industry should be working towards to achieve EO.

Ten commandments for equal opportunities in construction

- Set goals for EO
- Have a clear vision of what you are trying to achieve with EO
- Have a long-term plan to achieve EO
- Have a board member responsible for the policy

- Ensure women are well represented at all levels, and in all areas
- Monitor the progress of women and the reasons why they leave
- Identify and remove barriers to women
- Accommodate women's non-linear career paths
- Establish mentoring systems for women.

(Equal Opportunities Taskforce in Construction)

How to encourage more women into construction

- Good and continuous training
- Flexible working hours and help with childcare
- Equal pay
- School visits and bringing girls on site
- Provide female role models and networking systems.

(Building Equality in Construction Project
at UMIST, Manchester)

In all these activities it is very important to have timescales, targets, monitoring, and proper resources in terms of time, money and person power to implement change as is discussed further in Chapter 7 on women in planning and surveying.

Discussion questions

1 In what ways does the construction industry need to change in order to attract more women, ethnic minorities and other minority groups?
2 How can such change be generated? By whom?
3 The system of having 'chartered professional bodies' to cover each of the main fields of construction expertise is often seen as uniquely British and rather old-fashioned compared with the rest of the world. Discuss, with reference to the situation in one other country.

Note

1 UMIST is referred to in various chapters in this book and has now merged with Victoria University of Manchester to form The University of Manchester.

References

Arts Council (1996) *Equal Opportunities: Additional Guide*, London: National Lottery.

Bagilhole, B. (1993) How to keep a good woman down: an investigation of institutional factors in the process of discrimination against women academics, *British Journal of Sociology*, Vol. 14, No. 3, 262–274.

Bagilhole, B., Dainty, A. and Neale, R. (1996) 'Women in construction: a view of contemporary initiatives in the United Kingdom, and a proposal for international collaborative research', in *Proceedings of the GASAT Conference*, Ahmedabad, India (Gender and Science and Technology).

Brown, N. (1997) 'Bullying and Harassment in the construction industry: a cause for concern, or just part of the Job?', Unpublished dissertation for BSc (Hons) in Health and Community Studies at University College, Chester.

Callon, M., Law, J. and Rip, A. (1986) *Mapping the Dynamics of Science and Technology*, London: Macmillan.

CIB (1996) *Tomorrow's Team: Women and Men in Construction*, Report of Working Group 8 of Latham Committee, *Constructing the Team*, London: Department of the Environment, and Construction Industry Board (CIB).

CIOB (1995) *Balancing the Building Team: Gender Issues in the Building Professions*, Institute of Employment Studies, Report No. 284, Commissioned by CIOB and the Department of the Environment written by G. Court and J. Moralee, London: CIOB.

CISC (1994) *Occupational Standards for Professional, Managerial, and Technical Occupations in Planning, Construction, Property and Related Engineering Services*, London: CISC (Construction Industry Standing Conference).

CITB (2003) *Construction Industry: Key Labour Market Statistics*, CITB (Construction Industry Training Board), Kings Lynn, Norfolk.

Clarke, L., Michielsens, E. and Wall, C. (2000) 'Diverse equality: the example of the construction section', in M. Noon and E. Ogbonno (eds) *Equality, Diversity and Disadvantage in Employment*, London: Macmillan.

Clarke, L., Michielsens, E., Pederson, E.F., Susman, B. and Wall, C. (eds) (2004) *Women in Construction*, Brussels: Centre for Construction Labour Research.

Construction Manager (2002) *Ex-tearaways and Drop-outs Could Be Construction's Best Hope*, May Issue: p. 14 (unattributed article).

CORE (1995) *Building your Future: Will the Latham Report Change our Lives?*, Conference held at Ashton Court, Bristol, in *Proceedings of CORE Conference*, Bristol: CORE (Centre for Organisations Related to the Environment).

Dainty, A. and Geens, A. (1993), *Crossing the Sexual Divide*, Chartered Builder, September edition, p. 8.

Dainty, A., Bagilhole, B. and Neale, R. (2000) *A Grounded Theory of Women's Career Underachievement in Large UK Construction Companies*, in Construction Management and Economics, Vol. 18, pp. 239–250.

De Graft-Johnson, A., Manley, S. and Greed, C. (2003) *Why do Women Leave Architecture?*, RIBA commissioned study, London: Royal Institution of British Architects. http://www.riba.org.uk

Druker, J. and White, G. (1996) *Managing People in Construction*, London: Institute of Personnel and Development.

Druker, J., White, G., Hegewisch, A. and Mayne, L. (1996) 'Between hard and soft HRM: human resource management in the construction industry', *Construction Management and Economics*, Vol. 14, pp. 405–416.

Egan Report (1998) *Rethinking Construction: The Report of the Construction Task Force*, London: HMSO. (The Egan Report, Construction Industry Council.)

Egan (2004) *The Egan Review: Skills for Sustainable Communities*, London: ODPM (Office of the Deputy Prime Minister) in association with RIBA Enterprises Ltd.

Evetts, J. (1996) *Gender and Career in Science and Engineering*, London: Taylor and Francis.

Fielden, S., Davidson, M.J., Gale, A.W. and Davey, C. (2000) 'Women in construction: the untapped resource', *Construction Management and Economics*, Vol. 18, pp. 113–121.

Gale, A.W. (1995) Women in Construction, in D. Langford, M.R. Hancock, R. Fellows and A.W. Gale (eds) (1994) *Human Resources in the Management of Construction*, Longmans, Chapter 9, pp. 161–187.

Greed, C. (1988) 'Is more better?: with reference to the position of women chartered surveyors in Britain', *Women's Studies International Forum*, Vol. 11, No. 3, pp. 187–197.

Greed, C. (1991) *Surveying Sisters: Women in a Traditional Male Profession*, London: Routledge.

Greed, C. (1994) *Women and Planning: Creating Gendered Realities*, London: Routledge.

Greed, C. (1997a) *Social Integration and Exclusion in Professional Subcultures in Construction*, ESRC funded research, (Reference Number R000 22 1916).

Greed, C. (1997b) 'Cultural change in construction', in *Proceedings of the ARCOM Conference*, Cambridge, 15–17.9.97, Association of Researchers in Construction Management.

Greed, C. (1999a) *The Changing Composition of the Construction Professions*, Faculty of the Built Environment, Occasional Paper No. 5, Bristol: University of the West of England.

Greed, C. (1999b) *Changing Cultures*, in Greed, C. (ed.) Social Town Planning, London: Routledge.

Greed, C. (2000) 'Women in the Construction Professions', *Gender Work and Organisation*, Vol. 7, No. 3, July 2000, pp. 181–196.

Greed, C. (ed.) (2003a) *Report on Gender Auditing and Mainstreaming: Incorporating Case Studies and Pilots*, Research Report edited by Clara Greed, with research contributions by Linda Davies, Caroline Brown and Stephanie Duhr, London: Royal Town Planning Institute and at http://www.rtpi.org.uk

Greed, C. (ed.) (2003b) *The Rocky Path from Women and Planning to Gender Mainstreaming*, Clara Greed (ed.) with Linda Davies, Caroline Brown and Stephanie Dühr, Faculty Occasional Paper, Faculty of the Built Environment, University of the West of England, Bristol, Occasional Paper No. 14, May 2003, Bristol: University of the West of England.

Hammersley, M. and Atkinson, P. (1995) *Ethnography, Principles in Practice*, London: Routledge, Second edition.

Harrison, M. and Davies, J. (1995) *Constructing Equality: Housing Associations and Minority Ethnic Contractors*, London: Joseph Rowntree Trust.

Ismail, A. (1998) 'An Investigation of the reasons for the low representation of black and ethnic minority professionals in contracting', unpublished special research project report, available at University of the West of England, Bristol.

Kanter, R.M. (1977) *Men and Women of the Corporation*, New York: Basic Books.

Kanter, R.M. (1983) *The Change Masters: Corporate Entrepreneurs at Work*, Counterpoint, London: Unwin, p. 296.

Kelly, M. (1997), *The Good Practice Manual on Tenant Participation*, WDS (Women's Design Service), in association with DoE Special Grant Programme.

Langford, D., Hancock, M.R., Fellows, R. and Gale, A.W. (1994) *Human Resources in the Management of Construction*, Harlow: Longmans.

Larsen, E. (1958) *Atomic Energy: The Layman's Guide to the Nuclear Age*, London: Pan.

Latham (1996) *Tomorrow's Team: Women and Men in Construction*, Report of Working Group 8 of Latham Committee, *Constructing the Team*, London: Department of the Environment, and Construction Industry Board (CIB).

Law, J. and Hassard, J. (1999) *Actor Network Theory: and After*, Oxford: Blackwells.

Little, J. (2002) *Gender and Rural Policy: Identity, Sexuality and Power in the Countryside*, Harlow: Pearsons.

LWMT (1996) *Building Careers: Training Routes for Women*, London: London Women and Manual Trades. See WMT (renamed) on web list.

Mellström, U. (1995) *Engineering Lives: Technology, Time and Space in a Male-Centred World*, Occasional paper, Linköping Studies in Arts and Sciences No. 128, Institute of Tema Research, Linköping University, Sweden, pp. 168–172.

Morley, L. (1994) 'Glass ceiling or iron cage: women in UK academia', *Gender, Work and Organisation*, Vol. 1, No. 4: 194–204, October.

Murdock, J. (1997) 'Inhuman/nonhuman/human: actor network theory and the prospects of nondualistic and symmetrical perspective on nature and society', *Planning and Environment D*, Vol. 15, No. 6: 731–756.

ONS (2003) *Gender Statistics Review*, London: Office of National Statistics, and see National Statistics Online service at http://www.statistics.gov.uk/CCI/nugget.asp?ID=441

Onuoha, C. and Greed, C. (2003) *Racial Discrimination in Local Planning Authority Development Control Procedures in London Boroughs*, Faculty of the Built Environment, Occasional Paper No. 15, Bristol: University of the West of England.

Panelli, R. (2004) *Social Geographies: From Difference to Action*, London: Sage.

Rhys Jones, S., Dainty, A.W., Neale, R. and Bagilhole, B. (1996) 'Building on fair footings: improving equal opportunities in the construction industry for women', in *Proceedings of CIB (Construction Industry Board) Conference*, Glasgow.

Sharpe, S. (1995) *Great expectations: young women today think the battles have been won*, in Everywoman, December 1994/January 1995, No. 110, pp 14–16.

Uguris, Tijen (2001), 'Ethnic and Gender divisions in Tenant Participation in Public Housing', upublished PhD, Woolwich: University of Greenwich.

Wall, C. and Clarke, L. (1996) *Staying Power: Women Direct Labour Building Teams*, London: London Women and the Manual Trades.

WDS (Women's Design Service) (1998) 'Gender issues within planning education', in *Proceedings of the London Women and Planning Forum Symposium on 'Planning education' at Faculty of the Environment*, University of Westminster, London, WDS Broadsheet No. 28, March, London: WDS.

Useful web links for women and minorities in construction

Sources: The Strategic Forum's People Group, 2002, from CIC Cascade and additional web links from C. Greed's ongoing research 2004.

Access mainstreaming

Disability and construction
http://www.cityoflondon.gov.uk

Architects for change

Equal Opportunities forum at the RIBA incorporating 'Women in Architecture'
  Contact: Sumita Sinha,
  wia@bma.dircon.co.uk
Also see http://www.riba.org.uk, for Report entitled, 'Why are women leaving
  architecture?'

Audit and equalities

Audit Commission Best Value Performance Plans toolkit see:
  http://www.bvpps.aud-commission.gov.uk/toolkit/default.htm
Benchmarking Club see: http://www.thehousingforum.demon.co.uk
Best Value: http://www.lg-employers.gov.uk/mainstream.index.htm   and   see
  http://www.bvpps.aud-commission.gov.uk/toolkit/default.htm

Building work for women (changing the face of construction)

A constructive partnership between employers, training providers and support
organisations in the public, private and voluntary sectors. Focuses on increasing
opportunities for newly trained tradeswomen, reducing construction skills shortage
in London, lowering barriers faced by women entrants to construction, responding
to new NVQ requirements for more site work. Contributes to work of National Skills
Task Force in offering extra help for employers trying to recruit where there are skills
shortages. Partners include CITB and London Development Agency. Formed in 1999,
the success of Building Work for Women (BWW) has been confirmed with the news
that the partnership has secured funding to embark on Phase II of its work. BWW
will now be able to develop the new Build Up programme, as well as sustaining the
innovative work placement scheme matching trainee tradeswomen with employers
for highly supported on-site experience.
Contact: http://www.buildingworkforwomen.org.uk
Tel: 020 7637 8265

Cabinet office

see: http://www.womens-unit.gov.uk   and   see: http://www.cabinet-office.
  gov.uk/womens-unit/index.htm

CAE

Disability issues
Centre for Accessible Environments see: http://www.cae.org.uk

Changing the face of construction

This is a joint government/industry funded project aimed at helping to improve day to day working conditions through positive action. Initiatives include Building Equality in Construction and Building Work for Women. This aims to bring about a cultural change in the construction industry equality and greater respect for the workforce.
Contact: http://www.change-construction.org/building
Tel: 020 8305 2277

Construction Industry Trust for Youth (CITY)

A charity formed in 1961 which provides grants to disadvantaged students studying for construction qualifications. Originally set up to provide finance for youth club building projects but broadened its remit in 1993 to include grants to students.
Contact: http://www.charitynet/org-city
Tel: 020 7608 5184

Construction Modern Apprenticeship

This is a planned industry-designed training programme which should be completed by the Modern Apprentice's 25th birthday. This complements the CAS and can also be used as part of a business and learner's plan, Investors in People and on-the-job training and assessment.
Contact: http://www.citb.org.uk
Tel: 01485 577 577
Also see JIVE partnership Pat Turrell at Sheffield Hallam University.

Construction Industry Training Board (CITB)

CITB is a statutory body established to improve the quality and efficiency of training in the construction industry. It supports training in many ways, including the payment of grants and other financial support to firms who carry out training to its approved standards. It is also a national training organisation (NTO), ensuring that employers and individuals their sectors have access to accurate information about future learning and skill needs, and that there is sufficient provision to meet those needs as well as a managing agent (training provider).
Contact: http://www.citb.org.uk

Construction Industry Council Diversity Toolkit

http://www.cic.org.uk; http://www.rethinkingconstruction.org; http://www.m4i.
   org.uk
You can download the whole document pack at http://www.m4i.org.uk

Construction Industry Council Equal
Opportunities Task Force

Focuses on diversity and has concentrated its efforts on attracting more women into the industry. Also research into opportunities for disabled people.

Contact: CIC, 26 Store Street, London WC1E 7BT.
Tel: 020 7637 8692 contact Patricia Behal

DIALOG dialog@-employers.gov.uk

Equality Standard for Local Government in services and employment in England,
London: Employers Organisation for Local Government.

Equal opportunities

EOC (Equal Opportunities Commission) http://www.eoc.org.uk
Equal Opportunities Commission for Northern Ireland http://www.equalityni.
   org.uk

European comparisons

Gender and Urban Planning (genero urban) Spanish European Wide web site, see
http://www.angelfire.com/home/generourban and e-mail generourban-admin@
Listas.net

FIG

International Federation of Surveyors, FIG working group on under-represented
groups in surveying
http:// http://www.fig.net/figtree/pub/tf/unrep/200101/newsletter200001.htm
http:// http://www.fig.net/figtree/underrep/tfunrep.htm

GLA

Greater London Council
GLA http://www.london.gov.uk
LPAC (London Planning Advisory Committee) see: lpac@lpac.gov.uk
LRN (London Regeneration Network) see The Regenerator: Voice of the London.

GSUG

Statistics
(Gender Statistics Users' Group) see e-mail: gender@ons.gov.uk

ICEFLOE

The Equalities Forum for the Institution of Civil Engineers, Fair, Level, Open and
Equal Initiative. Aims at increasing diversity among ICE membership with emphasis upon gender, race, disability and other minority issues with publicity, careers
advice, videos and training http://www.ice.org.uk

Knowledge exchange

The Knowledge Exchange web site connects all organisations who are committed to Rethinking Construction and provides a one-stop, on-line search facility for anyone seeking to gain or share information on practical methods for implementing change and best practice.
Contact: http://www.knowledgeexchange.co.uk

Local collaborative partnership

Will run in all CITB's ten area offices, working with the local community and education specialists, employers and regeneration agencies who support CITB's objectives for diversity and equal opportunities.
Contact: http://www.citb.org.uk

M4I

Movement for Innovation
The M4I Board works to achieve high level support for Rethinking Construction from the industry's most forward thinking companies and has also delivered a number of significant practical outputs to help establish the business case for industry and client engagement.
Contact: http://www.m4i.org.uk

New deal for construction

Aims to improve future job prospects for the young unemployed by helping them to come off benefit and into work.
Contact: http://www.dti.gov.uk
Northern Ireland see e-mail: info.gender2@equalityni.org.uk

NOW INSET

This is a project developed by Business in the Community to enhance and promote the image of science, engineering, construction and technology (SECT) to young women. Funded by the European Social Fund's Employment NOW (New Opportunities for Women) initiative.
Contact: NOW INSET, Business in the Community, 80 Bournville Lane, Birmingham B30 2HP.
Tel: 0121 451 2227
And check SET in general.

OXFAM

OXFAM UK poverty programme and gender mainstreaming
http://www.oxfam.org.uk and see e-mail: ssmith@oxfam.org.uk

Respect for People

The DTI aims to help improve the industry's performance on 'Respect for People' issues by promoting diversity, health, safety, good site conditions, welfare and training, and helping to demonstrate that problems of recruiting and retaining the right people with the right skills requires a radical improvement in the way the industry treats its workforce.

The report A Commitment to People 'Our Biggest Asset' and the associated toolkits aim to provide the industry with practical help.

Contact for toolkit details. http://www.cic.org.uk and http://www.dtlr. gov.uk

Rethinking Construction

Rethinking Construction is the banner under which the construction industry, its clients and the government are working together to improve UK construction performance.

Contact: http://www.rethinkingconstruction.org.uk

Royal Institution of British Architects

RIBA Report with recommendations at http://www.riba.org.uk based on research on 'Why are women leaving architecture?'

Royal Institution of Chartered Surveyors

RICS Report 2003 entitled 'Raising the Ratio' Investigation of composition of the surveying profession (led by Louise Ellison and Sarah Sayce) London: University of Kingston on Thames and see http://www.rics.org.uk

Royal Town Planning Institute

RTPI see: http://www.rtpi.org.uk

RTPI Report 2003 on Gender Mainstreaming in local planning departments plus a Gender-Mainstreaming Toolkit, available at http://www.rtpi.org.uk

Scottish Executive http://www.scotland.gov.uk/government/devolution/ meo-00.asp

Scottish parliament: http://www.wsep.co.uk

SOBA

Society of Black Architects http://www.riba.org.uk

USA

Affirmative Action for Surveying and Mapping particularly concerned with black and women professionals.

Contact: wendy@netsync.net

Also provides on line PDF Acrobat Format quarterly newsletter.

WAMT

Women and Manual Trades ww (previously London Women and the Manual Trades)
52–54 Featherstone Road,
London EC1Y BRT
Tel: 020 7251 9192
info@wamt.org.uk

WDS

Women's Design Service
http://www.wds.org.uk
(wider group that share building with WAMT)
52–54 Featherstone Road,
London EC1Y BRT
Tel: 020 7251 9192
info@wds.org.uk, cheath@wds.org.uk

Wales

Welsh Assembly Equality Unit http://www.assembly.wales.gov.uk

Women

Women and Transport http://www.uel.ac.uk/womenandtransport
Women's National Commission http://www.thewnc.org.uk
Women's Unit http://www.cabinet-office.gov.uk/womens-unit/2001/htm
WRC (Women's Resources Centre) http://www.wrc.org.uk

Women's and men's careers in the UK construction industry: a comparative analysis

Andrew R.J. Dainty and Barbara M. Bagilhole

Introduction

This chapter examines the factors and interrelated decisions that influence the career dynamics of female and male construction professionals in the United Kingdom. It draws on research by the authors which compared the determinants and resultant career patterns of construction professionals through detailed career history profiles of matched 'pairs' of male and female informants. These data were used to establish any disparity between men's and women's career progression and to explore the determinants of women's organizational and occupational mobility patterns. Initially, the chapter outlines the nature of employment within the UK labour market and the impact that male domination has on the sector. Next, the literature surrounding the nature of careers in organization is reviewed as a precursor to explaining how women's and men's careers were examined in this study. The chapter then presents and discusses the findings of the research, which reveal the importance of addressing gender inequalities in career development if the industry is to benefit from the advantages that a diverse workforce can bring.

Background

Although the construction industry is one of the UK's largest employers, recruitment to the sector is homogeneous, being dominated by white males in both craft and professional positions. Women comprise a very small proportion of total employment within the sector, currently around 9.2 per cent of the workforce (DTI, 2003) compared to an economically active female population of 45 per cent (Labour Market Trends, 2001). In addition there is also evidence of horizontal segregation within the sector, as only 1 per cent of craft positions are occupied by women (CITB, 2002) and they form less than 4 per cent of the professional membership of the UK's construction related professional bodies (Davey *et al.*, 1998). Thus, women remain significantly under-employed in all of the main construction occupations.

In recent years, some construction companies have begun to recognize that workforce homogeneity is detrimental to their long-term growth. Proponents of diversification argue that it leads to a more broadly informed, more adaptable organization, which is closer to customers, more responsive, and more able to attract better quality employees (Ross and Schneider, 1992: 99; Greenhaus and Callanan, 1994). Accordingly, there have been signs that the UK construction industry has begun to seek to redress the historical gender imbalance. For example, governmental task forces have called on the sector to address these issues (Latham, 1994; Construction Industry Board, 1996; Rethinking Construction, 2000), and now individual organizations, professional bodies, and national training organizations have all undertaken initiatives to improve women's representation and the level of their involvement. These efforts have resulted in women's employment increasing in construction in recent years. There has been an upward trend in women studying for construction degrees (20 per cent) and they now make up 11.6 per cent of all those employed as professionals and managers within the industry (CITB CS, 2004).

Despite the increase in female representation, concerns remain that barriers to women's career progression may threaten their continued presence in the future. There is anecdotal evidence to show that women face discrimination and harassment, and have found developing their careers problematic (Building, 1995). As such, it has been acknowledged that, as women enter the construction work force in increasing numbers, employers will have to promote equality of opportunity in order to retain them (Khazenet, 1996; Yates, 1992). This need has added significance within the UK industry, where a tight labour market has begun to lead to increased salary levels in both craft and professional positions (Cargill, 1996; Knutt, 1997). Consequently, there is now a real need to both attract women to and *retain* women in the sector, particularly as they may act as role models and mentors for women considering construction careers in the future.

Developing strategies to retain women first requires an in-depth understanding of women's careers and the determinants of their progression. This knowledge will allow the industry to make informed judgements of how to develop human resources management (HRM) policies to develop a more fair and equitable workplace environment. Whilst previous research on women in construction in the United Kingdom has made considerable strides in understanding how women can be attracted to the industry (Gale, 1994), how their experiences of construction education can influence their career outlook (Srivastatva, 1996) and the identified barriers that they confront during the transition into paid work (Wilkinson, 1993), little work has explored the nature of women's careers. Accordingly, the research referred to within this chapter explored women's career progression dynamics and compared their career paths and progression characteristics with those of their male peers. By taking account of these factors within the context of modern HRM practice, initiatives are suggested as to how the sector can address equality issues.

Understanding organizational career dynamics

Career development at work is defined through 'mobility', a term used by labour economists to imply labour turnover. This can be defined as inter-firm movement (between organizations); intra-firm movement (within an organization); job mobility (within and between organizations); geographic mobility (movement of geographical location); or occupational mobility (changes of industry or sector) (see Young, 1988). All may contribute towards an individual's career profile and how they progress. 'Career dynamics' defines *how* such career progression takes place and, in particular, the way in which careers develop within organizations or over a working life (Armstrong, 2001: 453). Understanding career dynamics allows organizations to be sympathetic to the decisions and dilemmas facing their employees (Greenhaus and Callanan, 1994). Moreover, it allows firms to identify where inequalities may exist which have the potential to affect the success of a particular group.

Isolating or focusing on particular influences when trying to describe career dynamics is overly simplistic. In reality, even single-organization careers are influenced by a complex interaction of structures, cultures, individual actions and outcomes from both within and outside of the organization (Evetts, 1994). Moreover, structures and cultures are influenced by the decisions and actions of the individual, whilst at the same time helping to determine these actions and decisions (Evetts, 1992, 1996: 24). This perspective sees the individual as defining their growth throughout their life of work and not as moving along pre-determined career paths (Sonnenfeld and Kotter, 1982). By taking this perspective in this research, it allowed the interaction between the individual and their work environment to be studied and provided insights into the decision-making processes that define women's careers. Furthermore, it highlighted the compatibility and conflicts between women's personal actions and their employer's HRM policies and the attitudinal barriers that women perceive that they confront within the construction workplace. These are the barriers which need to be removed if a more equitable workplace is to be created in which women can progress in parity with their male colleagues and hence, be retained by the industry in the future (Agapiou and Dainty, 2004).

Research design and methodology

In order to understand the gender specific factors affecting careers, a comparison of women's development with that of their male peers was necessary. Such a comparison required an assessment of both the physical nature of their career dynamics and of the determinants which defined their progression patterns. In order to gather the range of insights required to make an assessment of careers, in-depth ethnographic interviews were used to compile detailed career history profiles of a range of matched pairs of male and female informants. This approach was adapted from a similar study which analysed women's careers in the Norwegian engineering professions (see Kvande and Rasmussen, 1994). Each female informant was matched as closely as possible with an equivalent male colleague according to their chosen vocational path, experience, length of service in their organization, and qualifications.

Five of the largest UK contracting organizations provided a total of 41 pairs (82) informants. The principal collaborating organization (Company A) provided a sample of 25 pairs, with the remaining informants being drawn from the other organizations. The sample within Company A represented all of the women employed in professional positions that were willing to participate in the study (86 per cent of the total number employed at the time of data collection). The sample from the other companies represented a random stratified sample of female employees matched with equivalent male peers. The majority of those interviewed worked in site-based positions, although a sample was also included from those working in office-based positions including those which supported the main operational functions (i.e. IT support, design work, and administrative roles). This allowed women's careers to be examined in a variety of career paths.

Initially, each informant's career was mapped against a seven-point career scale which represented all key positions from graduate trainee (level 1) to director (level 7). This scale was used due to an incompatibility of job titles between the different participating organizations. The informants' positions on the scale were established through questions about the level of their responsibilities. Next, each informant was asked to describe the major determinants of their career progression and to explain their strategies, actions, and personal resolutions on their careers. From these data, the gender-specific determinants of their progression dynamics were ascertained, and the underlying reasons for any disparity in progression discerned. A leading qualitative analysis software package, QSR NUD.IST™, was used to analyse and categorize the emerging career determinants and to extract the factors leading to disparate development patterns at particular career stages. The categorized data were then transformed into causal models in order to understand the underlying processes which impinged upon women's careers.

Findings

It was immediately noticeable that a disparity existed between women's and men's careers in terms of their vertical progression characteristics. Notably, after an initial period of graduate training, women's careers accelerated less rapidly than men's in the early years of their development. However, it was also noticeable that after 11 years, women with longer service in their respective organization experienced a more rapid overall career progression rate than their male pairs. This typically equated to an employee reaching 32–34 years of age. There was little variance in the promotion rates of the women taking part in the study through this period, and so it does not appear that the level of achievement of women has been distorted by some having achieved an exceptionally rapid rate of vertical progression. Women spent longer than men in virtually all positions before gaining promotion, particularly at level 5 (project manager or equivalent). Thus, for those women following a traditional operational (site-based) career path, the transition to office-based senior management positions appeared particularly problematic.

In order to explore reasons for this disparity in career progression characteristics, the determinants that had produced the informants' career progression characteristics were explored through their subjective career accounts. Eight generic analytical categories emerged from this analysis which were used to categorize and conceptualize the explanations of the under-achievement of women.

Entrance to the industry

Men tended to have chosen a construction career in response to a family member or someone else they knew who worked in the sector. In contrast, younger women tended to have embarked upon a construction career as a result of targeted recruitment campaigns aimed at attracting them into the industry. The consequences of very few women having been advised to join the industry by their friends and family was that they had a poor initial understanding of the culture of the industry and the other inherent difficulties of working in such a male-dominated environment. Also, it emerged that women tended to be ambitious, academic high-achievers and so a lack of progression opportunity soon led to dissatisfaction.

> My initial interest stemmed from a Women Into Science & Engineering course in my last year of A levels. They said that there were more women joining than at any time before. ... I think that once I realised that an engineer wasn't the bloke with his head under the bonnet of the car it became a lot more appealing. It appealed to my desire to want to get on, to progress rapidly and make my mark.
>
> (Female site engineer – 24 yrs)

Women's higher education experience was found to provide a sheltered environment and an unrealistic interface between career choice and working life within the sector, and it did not prepare them for what they would encounter. Women were surprised to find that they confronted barriers to their progression such as sexist behaviour, harassment, deliberate attempts to undermine their workplace contribution and work/life balance conflict.

> I think there is more picking on women than there is picking on men. I have been shocked at just how blatant the discrimination has been that has been directed towards me ... they didn't warn us about that in the careers talks!
>
> (Female assistant construction manager – 26 yrs)

In contrast, most men, having entered the industry as a result of advice from family or friends, had a good understanding of the nature of the work environment and of the likely career development opportunities. Few men expressed regrets over their career choice, as most had chosen construction because of the nature of the work and careers that it offered. The net result of the routes to entry taken by most women was that many had become disillusioned with their career choice and were seeking to leave the industry at early stages of their careers. Many felt that career opportunities had been over-sold to them by the targeted campaigns and that very little effort had been put into explaining the realities of working within such a male-dominated workplace. This provides a context for many of the other findings presented later.

Entrance to companies and initial career experiences

Women found the process of entering companies problematic in comparison to men, both in terms of their initial entry to employment and in their subsequent attempts to move between different companies. A key factor influencing this was that responsibility for HRM had been largely devolved to operational line managers. Such managers had set opinions about the types of people that they wished to recruit. These tended to be men who conformed to traditional male organizational norms or to their own work ethics. Therefore, women were disadvantaged in this process through stereotyped expectations of their career and personal priorities which were assumed to be different to men's. In addition, the informal recruitment and selection processes used by operational managers favoured applicants with existing contacts within the companies to which they were applying. Accordingly, men who tended to have industry contacts used them to secure positions and good remunerative packages.

> I rang all the people I had worked with in London and I asked a couple for a reference. The next thing I knew I had a phone call from the

personnel department offering me a job. I found out that this guy I had
asked for a reference from had phoned them up and sold me to
them. ... it's 'who you know' in this industry that counts.

(Male assistant project manager – 26 yrs)

The implications of this were that entry to employment and subsequent
promotions were more difficult for women, particularly for those in their
late 1920s to early 1930s where they were disadvantaged by manager's
expectations of their likelihood of taking career breaks in order to have
children. This may explain why women's progression only accelerated when
they had moved beyond this career stage. Almost half of the women interviewed believed that discriminatory recruitment processes were ultimately
likely to lead to them having to seek positions in other sub-sectors of the
industry which did not present such resistance to their employment. These
included working for major clients of consultancy businesses which were
seen as offering greater workplace flexibility and career potential than
contracting work.

I have decided that I will never be given the chance of getting my own
project and that I would never do well as a QS in this company. I have
been offered a role in a small cost consultant. ... I don't really want to
leave the company, but I have had to face the reality that women are
not going to get on in the contracting side of the industry.

(Female QS – 32 yrs)

Operational roles

Site-based roles were acknowledged to offer greater scope for gaining
responsibility and rapid promotion in the early career stages by both
women and men. Even women in their later career stages tended to
advocate maintaining a close proximity to technical work and to their
chosen professional role. From these positions they derived a greater level
of intrinsic satisfaction and could demonstrate their ability to male
colleagues who were sceptical of their professional competence and commitment. However, despite women's preferences, in many cases the organizations had prescribed gendered roles by allocating female staff to
office-based support positions and men to operational site-based roles.
Almost a third of the women interviewed had been allocated to office-based
support positions, as opposed to front-line operational management positions on site. These women found it difficult to convince their managers of
their career conviction and professional competence and hence, found it
more difficult to develop their careers than their predominantly site-based
male colleagues.

Structural and cultural organizational processes

De-layered structures and short reporting lines were prevalent at both organizational and project levels within each of the companies investigated. They had positive effects in terms of creating lateral developmental opportunities, autonomy and empowerment amongst site-based managers. However, they also promoted informality in career structures and organizational processes. Notably, under such approaches project managers retained control of the staff development of those under their charge and senior office-based managers also retained a great deal of control over their site-based teams.

There were several elements of this middle management control and power that directly affected women's careers. Key among the structural aspects which produced and maintained women's under-achievement were performance management systems, used for monitoring performance, assessing training needs and allocating staff development opportunities. Male managers maintained control and power through their understanding and control of such organizational systems and procedures. Unfortunately, some unscrupulous male managers used them as a vehicle to undermine women's careers. Examples were given of women being given lower appraisal scores or being subjected to unfair assessments of their training needs. This is an example of how a structural policy measure can actually maintain women's subordination within companies.

> I think that managers assess people in terms of how they see a successful managerial style, which in reality will mean who gets closest to their own style. It is a concern for women I think, because the male managers carrying out the assessments may overlook the different skills that women bring to the job, and look for carbon copies of themselves.
> (Female QS – 30 yrs)

As well as having to confront such structural barriers, women were also excluded and undermined through cultural aspects of the workplace environment and particularly through discriminatory behaviour perpetrated by their male colleagues. Notably, the male middle managers, who retained direct control of HRM issues at a project level, demanded that women complied with ingrained, traditional, male-oriented work practices. For example, they insisted upon demanding work schedules, long working hours and employees gaining experience within international divisions. These doctrines rendered the combination of a fulfilling family life and a successful career very problematic, particularly for women. Some women also faced overt resentment in the form of sexual harassment and discrimination, and through being excluded from male-focused out of work social activities which were acknowledged to enhance career development. Most of the female informants had experienced poor treatment and generally felt isolated and unable to turn to senior colleagues for support. Thus, the resulting workplace presented an alien and problematic environment for non-traditional entrants, and one which was resistant to change due to the mutually supporting structural and cultural aspects which were found to underpin it.

> I complained so much at the mundane work that I was given that in the end that they started to give me tasks, but then it got ridiculous, I had work coming out of my ears. In the end I just couldn't cope any more, I thought that I was going to have a nervous breakdown. ... I tried to speak to some of my colleagues about it but they weren't really interested ... if your manager has it in for you there's not too much you can do but grin and bear it.
>
> (Female assistant construction manager – 26 yrs)

Work–life balance as a career determinant

Site-based roles are demanding and were seen as impinging on life-cycle activities and family responsibilities. Most women therefore perceived that they had to make a choice between a career *or* a family-oriented lifestyle. In contrast, men attempted to combine their work and family lives by relying on their female partners to undertake the major share of domestic responsibilities. Many of the women placed in specialist or major project divisions found it difficult, if not impossible, to combine work with their family lives. This was because the businesses had projects throughout the country, often a great geographical distance from the women's home. Thus, whereas the majority of men had supportive non-working partners who took on the responsibility for childcare and other domestic duties, women tended to have retained this responsibility or have to forego having children for the sake of their career. Several women talked of their marriages and relationships breaking down because of the inherent strains of working within the industry.

> I used to pick my child up on Tuesdays, drop her off at a childminders for three days and then drive back on Fridays. In the end it drove me and my husband apart because he was unemployed at the time and I was working away with the baby, and the child minder's fees used up nearly half my salary. By the time I got home I was tired and irritable and I lost out on a lot really. Now I get home to an empty flat and I have to make food for me and the baby and do the house work. ... you are expected to travel, you are expected to do the hours, you are expected to manage and you don't get any financial help from the company ... I would say that in this industry you have to make a choice between having a family and having a career.
>
> (Female sub-agent – 32 yrs)

Many women coped with the geographically dispersed nature of construction projects by gaining allocation to regional divisions where they could remain close to their homes. However, this often precluded them from working on large and/or prestigious projects which tended to be managed by the national divisions. Unsurprisingly, women's career motivation tended to decline in relation to their experience, as the salience of family issues became more significant, and as they realized the difficulties inherent in combining their work and domestic lives.

Individual career strategies

Organizational loyalty was found to be more likely to facilitate rapid vertical career progression than frequent inter-organizational mobility. Thus, men's career strategies focused on proactive intra-organizational career development within their existing organizations. Their informed preference for developing their careers in a familiar cultural and structural environment promoted a greater level of understanding and control of their careers and, ultimately, improved vertical advancement. In contrast, women's career development tended to be characterized by inter-organizational mobility within a multi-organization framework. Women expended considerable effort in coping with barriers to their continued presence in the contracting environment, or more commonly to seeking employment with other companies. This gave them less time and energy to spend in proactively developing their careers *within* their employing organization. Consequently, women's significantly greater inter-organizational mobility patterns had disadvantaged their progression.

> I would have liked to have developed my career a bit further in this company as there is plenty of scope within the work I am doing now, but I think it is acknowledged that to get on as a woman you have to move on when you can't overcome the barriers present themselves ... it is true to say that I have worked for a lot of firms considering the short time I have been in the industry and I know that this probably hasn't done me much good in terms gaining promotion, but its easier to move on than to confront discrimination.
>
> (Female QS – 28 yrs)

Future expectations, opportunities and threats

It was particularly noticeable that women tended to be in favour of and fully embrace the structural organizational and industry-wide change which they saw as progressively taking effect on the industry since the economic recession of the early 1990s. This had taken the form of structural re-organization and initiatives to improve work efficiency through a reduction in conflict and adversarial relationships. In contrast, men opposed such changes and openly stated that they sought to maintain existing hierarchies and work practices. Women's preference for structural change stemmed from a belief that it may lead to improvements in the cultural environment of their organizations in the long-term. However, they also feared that men's resistance to change could lead to a continued lack of recognition of the benefits that diversified workplaces could provide. Many also suggested that male entrants to the industry, however accepting of change at the outset of their careers, were likely to be indoctrinated into maintaining current attitudes and practices by those with influence and control over their development in the early years of their careers. As such, there was a general scepticism over the industry's likelihood of developing a more equitable workplace environment.

> I don't think women can succeed in the industry because there are too many traditional stigmas, they are seen as a threat to their working style. Maybe as this generation of managers comes through we will be accepted, but I think that it will take a few years yet.
>
> (Female QS – 32 yrs)

Discussion of the findings

The UK construction industry has made a conscious effort to increase the number of women entering it in response to impending skills shortages, and, some would argue, a recognition of the advantages that work-force diversification has been shown to provide in other sectors. However, the findings outlined within this chapter demonstrate that women are under-achieving relative to men within some large construction firms. If these findings are reflected within other construction organizations, then this could threaten the current upward trend in the number of women entering the sector.

Two recurring and interrelated themes emerge from women's career accounts which may go some way to explaining their under-achievement. The first concerns the problems of achieving work/life balance working in the sector (discussed in more detail in Chapter 9), whilst the second relates to the discriminatory actions of male professionals within the workplace. Both sets of problems could be seen to relate to women's initial career choice, where they tended to enter construction without an in-depth knowl-edge of or preparation for the informal aspects of the workplace culture and the inherent demands of the industry. Having entered the sector, the negative attitudes that women confront necessitated them having to focus their efforts on coping with the hostile work environment. Conversely, men were able to concentrate on using the informal nature of the industry to further their career opportunities. This leads to a self-reinforcing cycle;

men use their organizational power to improve their own careers whilst simultaneously using their positions to detrimentally affect women's careers. The only way women found to break this cycle was to change their employer, which was also found to impede their progression in the long-term given that the most successful people in the sample had developed their careers within a single organization.

The most marked disparity between women's and men's vertical advancement occurred in their early career stages. During the first ten years of their development, men progressed at markedly more rapid rates than their female colleagues. Women experienced particular difficulties at the transition from site-based, project-specific positions to senior head-office roles. Beyond this point, although women tended to progress in parity with men, the number that had actually remained in the industry for this length of time was proportionally very small. This suggests that obstacles encountered during women's early careers and in the transition to senior management need to be addressed if women are to remain in the industry long enough to reach senior levels. Moreover, the personal compromises over fulfilling family lives that women have to make in order to remain in the industry in the long-term also require further investigation. The reality of these sacrifices may present too fundamental a barrier for many women considering a construction career. Younger women may be unlikely to remain within the sector in the long-term. Such women could have acted as role models and mentors for potential women entrants in the future, so women's continued under-achievement may also contribute to their continued under-representation.

Significant barriers exist to implementing flexible working practices in the industry, as construction projects tend to be short-term, geographically diverse and subject to dynamic change. This means that staffing decisions have to be taken quickly and employees have to be adaptable in order to meet the changing needs of the organization and its individual projects. However, more consideration and acknowledgement of employee's personal lives could be made for both men and women. There is a fundamental need to change attitudes within the industry towards workforce diversity and the benefits that this could bring in terms of its future operation. Working towards such change should be prioritized along with structural change initiatives in the future.

Conclusions

The findings of our study reviewed in this chapter have potentially serious implications for an industry which is consciously trying to increase the number of women it employs in professional and managerial roles. Construction companies at present form arenas for discriminatory behaviour and the eventual exclusion of women. Problems for women stem from the prejudice of male managers who act as gatekeepers to successful careers and from work practices which effectively militate against their acceptance and progression. The root of many of these problems was the de-layering of organizational hierarchies and the devolution of responsibility for dealing with HRM related issues to line managers, some of whom have vested interests in maintaining current systems in order to protect their own positions.

The result of the mutually reinforcing structural and cultural environment that construction creates is that women effectively face a fundamental choice of whether to confront barriers to their careers (which risks further retribution and marginalization), conform to male life-cycle patterns (which strengthens existing structural and cultural barriers and does not guarantee equality of opportunity), leave their organization (which leads to poorer opportunities for succession management and slower vertical development), or leave the sector (which reinforces male stereotypes of women's lack of occupational commitment to the industry). In each case women's actions contributed to reproducing the attitudes and structures in need of change.

It is clear that the exclusionary and discriminatory aspects of construction have developed in response to the industry being almost exclusively male in the past. Consequently, men's opposition to change is likely to maintain existing hierarchies, work practices, and cultures for the foreseeable future, which in turn are likely to maintain a poor level of retention of women. This would result in construction companies continuing to experience skills shortages and other competitive disadvantages in the millennium through a continued lack of recognition of the benefits that diversity brings to modern businesses. Furthermore, it seems morally indefensible for the industry to market itself as an appropriate employer of women unless companies are prepared to make appropriate modifications to their employment practices and culture. Accordingly, steps must now be taken to create a more equitable work environment, and produce the cultural change necessary for diversification of the work force. This will require determined senior leaders from large companies to champion a movement for change within the sector, and to drive equal opportunities issues to the top of the strategic agenda.

Acknowledgement

This work was supported by a grant from the Economic and Social Research Council (ESRC). Any opinions, findings, conclusions, or recommendations expressed herein are those of the authors and do not necessarily reflect the views of the ESRC.

Discussion questions

1   How can large construction firms benefit from attracting, recruiting and retaining a workforce more representative of the wider labour market?
2   Compare and contrast the interplay of factors that determine the career progression of men and women in the construction.
3   How could large construction companies begin to address the structural and cultural barriers that this chapter has revealed to detrimentally affect equality of opportunity in the industry?

References

Agapiou, A. and Dainty, A.R.J. (2004). 'Client-led approaches to increasing participation of women, ethnic minorities and disabled people in the construction workforce: A framework for change', *Journal of Construction Procurement*, 9(2), 4–16.

Armstrong, M. (2001). *A Handbook of Human Resource Management Practice* (8th edn). Kogan Page, London.

Building (1995). 'QS Awarded £9000 Compensation for sexual discrimination by contractor'. *Building*, 20 January, 13.

Cargill, J. (1996). 'Rising sum'. *Building Magazine* (UK), 15 November, 42–47.

CITB (Construction Industry Training Board) (2002). *CITB Skills Foresight Report February 2002*. *CITB*.

CITB CS (Construction Industry Training Board ConstructionSkills) (2004). *Stats and Facts*, http://www.citb.co.uk/equal_ops/

Construction Industry Board (CIB) (1996). 'Tomorrow's team: Women and men in construction', *Report of the CIB's Working Group 8 (Equal Opportunities)*. Thomas Telford, London, UK.

Construction Industry Council (CIC) (1995). 'Construction Industry Council membership census'. CIC, London, UK.

Davey, C., Davidson, M., Gale, A., Hopley, A. and Rhys Jones, S. (1998). 'Building equality in construction: good practice guidelines for building contractors and housing associations'. University of Manchester Institute of Science and Technology/ Department of the Environment, Transport and the Regions (UK), Manchester, UK.

DTI (Department for Trade and Industry) (2003). *Construction Statistics Annual 2003 Edition*, The Stationary Office, London.

Equal Opportunities Commission (1996). 'Facts about women and men in Great Britain 1996'. *Equal Opportunities Commission*, Manchester, UK.

Evetts, J. (1992). 'Dimensions of career: avoiding reification in the analysis of change'. *Sociology*, 26(1), 1–21.

Evetts, J. (1994). 'Women and career in engineering: Continuity and change in the organisation'. *Work, Employment and Society*, 8(1), 101–112.

Evetts, J. (1996). *Gender & Career in Science & Engineering*. Taylor & Francis, London, UK.

Gale, A.W. (1994). 'Women in Construction: an investigation into some aspects of image and knowledge as determinants of the under representation of women in construction management in the British construction industry'. PhD thesis, University of Bath, Avon, UK.

Greenhaus, J.H. and Callanan, G.A. (1994). *Career Management* (2nd edn). Dryden Press, Fort Worth, PA.

Khazanet, V.L. (1996). 'Women in civil engineering and science: it's time for recognition & promotion'. *Journal of Professional Issues in Engring Education and Practice*, ASCE, 122(2), 65–68.

Kirk-Walker, S. and Isaiah, R. (1996). 'Undergraduate student survey: A report of the survey of first year students in construction industry degree courses, entrants to academic year 1995–1996'. *Institute of Advanced Architectural Studies*, York, UK.

Knutt, E. (1997). 'Thirtysomethings hit thirtysomething'. *Building* (UK), 25 April, 44–50.

Kvande, E. and Rasmussen, B. (1994). 'Men in male-dominated organisations and their encounter with women intruders'. *Scandinavian Journal of Management*, 10(2), 163–173.

Labour Market Trends (2001). *Labour Market Participation of Ethnic Groups*, January 2001, pp. 29–41.

Latham, M. (1994). *Constructing the Team*. HMSO, London, UK.

Rethinking Construction (2000). A Commitment to People 'Our Biggest Asset', Report of the Rethinking Construction working group on Respect for People, Strategic Forum, London.

Ross, R. and Schneider, R. (1992). *From Equality to Diversity: A Business Case for Equal Opportunities*. Pitman Publishing, London, UK.

Sonnenfeld, J. and Kotter, J.P. (1982). 'The maturation of career theory'. *Human Relations*, 35(1), 19–46.

Srivastava, A. (1996). 'Widening access: women in construction higher education'. PhD thesis, Leeds Metropolitan University, Leeds, UK.

Wilkinson, S.J. (1993). 'Entry to employment: choices made by qualified women civil engineers leaving higher education'. PhD thesis, Oxford Brooks University, Oxford, UK.

Yates, J.K. (1992). 'Women and minorities in construction in the 1990s'. *Cost Engineering*, 34(6), 9–12.

Young, B.A. (1988). 'Career development in construction management'. PhD thesis, University of Manchester Institute of Science and Technology, Manchester, UK.

Chapter 6

Women in civil engineering

Suzanne Wilkinson

Introduction

This chapter discusses, against the background of their under-representation, the experiences of women civil engineers from entering higher education through to leaving education and then analyses the employment opportunities and challenges they face in the construction industry. Much of the discussion in this chapter is centred on women in western countries. On the basis of an analysis of equality of opportunity, it then draws some conclusions about why something should be done about women's minority status, who should do it, and what should be done.

On entering higher education, women on civil engineering courses throughout western countries find that they are in a minority of usually 10–20 per cent (Gale, 1995; Wilkinson, 1996; Devgun and King, 1999; Issacs, 2001). The low percentage of women on civil engineering courses presents some unique challenges for women, such as coping with isolation, low confidence and reactions of male colleagues and staff, as they navigate through the higher education process. Near completion of their courses of study, women civil engineers are faced with employment decisions that may have different consequences compared to their male counterparts. These include whether to continue as an engineer, and if so, to consider which branch of the construction industry and engineering would most suit their needs. Many factors come into play at this stage. Women's preferred factors, such as type of employment, location and benefits can be different from the ones their male colleagues may choose. Once in the workforce, women civil engineers face some new challenges, by virtue of the fact that they are women. Documented evidence (Wilkinson, 1996; Issacs, 2001) of the attrition rates of women civil engineers cause concern and point to problems than women may face, such as discrimination, harassment and a misunderstanding of their needs. In terms of the impact on the construction industry, the educational experiences, the employment decisions and the treatment of women can have significant effects on the nature of the construction industry.

Women in civil engineering – their place in the construction industry

Civil engineers are responsible for the planning, design, construction and maintenance of many types of construction projects. These can vary from small projects such as designing and constructing rural roads or small traffic systems to the design of large-scale infrastructure, such as sewage systems, dams, bridges, airports and harbours. Women who choose to enter the construction industry as civil engineers will usually be university graduates or will have a civil engineering certificate or a diploma from a tertiary institution. When women enter civil engineering they may work for one of the many engineering consultancies or contractors, specialise in an area of civil engineering such as geotechnical engineering or structural engineering, work for government organisations, smaller companies, or as individual consultants. Choices for employment are based on many factors, such as the opportunities presented by the company and the type of work. Once in the workforce, women civil engineers may choose to gain professional recognition through the Institution of Civil Engineers (UK) or one of the many other professional engineering organisations such as The Institution of Professional Engineers, New Zealand, most of which are affiliated to the Institution of Civil Engineers.

Causes of minority status

The minority status of women in construction related disciplines including civil engineering in many western countries have been well documented (Gale, 1994; Court and Moralee, 1995; Dainty et al., 2000a). Women make up around half the population, but constitute only a small minority (about 10 per cent) in civil engineering (Wilkinson, 1996; Dainty et al., 2000a). Why are women such a small minority in civil engineering? The explanation can be found in some mix of the following: early socialisation within the family; schooling; higher education; civil engineering itself; the division of domestic labour within the family; and the influence of wider social attitudes. It is of course difficult to work out the separate effects that each has. For example, given that even before they enter the industry, women are a small minority in civil engineering courses one might be tempted to conclude that the explanation cannot be found in the industry. But this would not follow: it might be well-founded expectations of life within the industry that feed back into women's decisions to choose university courses besides civil engineering.

The educational experiences of women civil engineers

In most Western countries when women enter a civil engineering course at university they are in a minority of usually 10–20 per cent (Wilkinson,

1994; Murray *et al.*, 1999). This minority proportion continues, and often declines, as women proceed through the course and enter the construction workforce. The experiences that women have on civil engineering courses can affect their choices in their future civil engineering careers (Fitzpatrick and Silverman, 1989). For instance, if women do not enjoy being civil engineering students, feel isolated and insecure on the courses then this can lead women to consider other careers during or shortly after qualifying (Fitzpatrick and Silverman, 1989). Researcher findings on the educational experiences of women when they are on courses where they are a minority suggest that women face a number of similar problems (Carter and Kirkup, 1990; Tizzard and Chiosso, 1990; Chinn, 1999) and that these problems can change their view of the industry and career they have chosen.

The results from studies have also suggested various problems that women as civil engineering students face. These can be broadly categorised into the following areas: women's confidence, the lecturing staff, their male student colleagues, their minority status (Carter and Kirkup, 1990; Tizzard and Chiosso, 1990; Chinn, 1999). It is hypothesised that if any of these four areas are problems for women on civil engineering courses, then women would be more likely to leave the courses. However, if these areas are not problems, women are more likely to feel positive about a career as a civil engineer and continue into the construction industry.

Confidence

When women enter civil engineering programmes, they do so with a certain understanding about their intended career. If they find that they are not as sure about civil engineering as a career as they first thought, or are led to believe that they do not possess the necessary background, this can manifest itself in a lack of confidence in their own abilities. Many researchers have mentioned this as a problem for women on courses where they are a minority, including civil engineering (Carter and Kirkup, 1990; Tizzard and Chiosso, 1990; Bakos, 1992; Chinn, 1999; Murray *et al.*, 1999). Initial disadvantage, with respect to performance and confidence, has been particularly highlighted, together with a sense that, as women, they do not fit in and that they lacked the necessary practical knowledge, or an application of theoretical knowledge to practical situations (Thomas, 1988; Chinn, 1999; Murray *et al.*, 1999). Summing this up, Bakos (1992: 20), from his study of women civil engineering students wrote that some women 'felt at a disadvantage during early college examinations (in status and strength of material) since many problems were written with the assumption that all the students knew what a torque wrench was or what a cam shaft did'. If in the initial stages of a civil engineering programme, women experience anxiety about their abilities they may choose to leave the programme. Women at this stage of their careers can feel a lack of confidence, especially with regard to their lack of prior engineering experience (Carter and Kirkup, 1990; Wilkinson, 1994). Educational institutions need to be proactive in addressing the needs of women through specific initiatives such as networks, newsgroups, funded meetings and mentoring. Feelings of confidence in ability is an important factor for success for women intending to enter the construction industry. If educational institutions can engender a sense of success and confidence in ability, and many do by their various initiatives directed at women students, then women are more likely to enter the profession feeling that they can become successful practicing engineers.

Lecturing staff

Educational institutions provide a buffer between school and the industry. For civil engineering students they can reflect the practices of the profession, and, if found wanting, students will seek careers elsewhere. One problem that presents itself in the literature (Thomas, 1988; Gale, 1991; Chinn, 1999) is the attitudes that women have to the lecturing staff of educational institutions on civil engineering programmes and the impact that these attitudes can have on future career choices. If their attitudes towards staff are poor, women are more likely to leave the courses or not enter the industry, believing that staff reflect current industry norms (Gale, 1991). The experience of women in these minority classes and their relationship with their (usually male) lecturers has been well documented (Thomas, 1988; Chinn, 1999). Thomas, for example, suggested that women did not find the lecturing staff very approachable and that some women felt that they were picked on or harassed (Thomas, 1988). Another later study found that women generally experience a 'chilly campus climate', receiving less support and encouragement than male students (Murray et al., 1999).

Researchers (Tizard and Chiosso, 1990; Greed, 1991; Usher and Ward, 1991; Wilkinson, 1994; Devgun and King, 1999) have also reported that some construction and technology lecturers use sexist and demeaning comments and/or have a hostile attitude towards women. In particular, for women civil engineers, Usher and Ward (1991: 127), wrote that, 'Anecdotal evidence suggests that sexist language, gender-biased comments and examples are commonplace in the classroom and that sexual harassment is not unknown.' That the use of sexist language was fairly widespread on civil engineering courses in the United Kingdom was confirmed by a national survey of graduating women civil engineers (Wilkinson, 1994) and continues to be problem in engineering schools (Devgun and King, 1999). Certainly the attitude and behaviour of lecturing staff is important to the future careers of women civil engineers. Women who experience unwanted attention and are confronted with inappropriate behaviour and language are more likely to choose another career. In the United States, this has been investigated by Fitzpatrick and Silverman (1989) who suggested that professors' influence could be important in encouraging women to remain in a field of study and could be critical to the level of career aspiration. Clearly higher educational institutions need to make sure that they are not inadvertently causing women to leave their engineering studies and depriving the construction industry of valuable staff through the behaviour of their lecturing staff. Information, training and direction on appropriate behaviour for lecturing staff should be provided together with procedures for the prevention of harassment and discrimination procedures for students and staff.

Male colleagues and minority status

Women who find it difficult to be in a minority are unlikely to find a career in civil engineering rewarding since currently women civil engineers are, as women, likely to spend much of their working lives in a minority. However, some of the strategies that higher education can use to make women on minority courses feel less isolated can also be applied to women who enter the construction industry, such as encouraging networks with other women, either in industry or on other courses where women are in a minority (Chesler and Chesler, 2002). Where there are more women on courses, those women are likely to develop networks and coping strategies which will be useful in the construction industry post-graduation (Chesler and Chesler, 2002). Many researchers have explained the effects of minority status on women on courses such as civil engineering in countries like the United Kingdom and United States. For example, Thomas (1988) described a co-operative strategy between women on a physics course who provided help and support during the course, and noted that other women devised similar strategies for dealing with being in a minority, by sticking together and forming a 'collective identity'. Carter and Kirkup (1990) also remarked on this, explaining that some women students provided support networks. Srivastava (1992) found that the perceptions of their courses by women students were more positive with the existence of role models such as women lecturers and women students. Moreover, Tizard and Chiosso (1990) suggested that loneliness was the biggest problem faced by women studying on male-dominated courses, and that initially women tended to make contact with other women on the course. Bakos (1992: 21) reported on the experiences of such women and suggested that 'A few women wished that they had had more female classmates or co-workers in order to establish friendships.' Not surprisingly Court and Moralee (1995) found that a substantial minority of women in the United Kingdom would like to see more women students and lecturers on their courses and that the environment could be more supportive of them.

In addition to the perceived need of women students to have more women role models and women students on the courses, womens' attitudes towards male colleagues have been documented and, in some cases, found to be problematic. The literature suggests that the women in their engineering classes in the United Kingdom and the United States are not taken seriously by their male peers and that, in some cases, a certain antipathy existed between women and men on the course and that female engineers described their male classmates as 'dull', 'immature' and 'unfriendly' (Newton, 1987; Chinn, 1999). Nevertheless, it should be noted that these feeling towards their male colleagues may be a defence mechanism against being in a minority where women seek to justify any feelings of isolation and loneliness that they may be encountering (Wilkinson, 1994).

Leaving education – women civil engineers' preferences for an employer

Leaving higher education presents the graduating civil engineering student with a number of choices. The first choice is whether to enter the construction industry as a civil engineer. The fact that women make significantly more job applications per job offer than men (men made on average six applications, whilst women made on average eight) in the United Kingdom has been found by Wilkinson (1994), so this may well have an impact on the decision to stay or leave. Negative responses and unfriendly interview experiences may put women civil engineers off entering the industry. In addition, each year about 15 per cent of women civil engineering graduates choose other employment (Wilkinson, 1996), with common alternative choices being teaching, banking and general management. This is often after having had some work experience within the construction industry. If a woman civil engineer completing a degree decides to enter the construction industry, then she will choose between several types of employment options. In most western countries, these broad categories are between an engineering consultancy, where she will spend time predominantly undertaking design and (probably) construction supervision; contracting, where she will spend a large part of her time supervising construction work on site and tendering for construction projects; local authorities and government agencies where she will be involved in regional construction projects, both design and managing construction; and utilities (e.g. water, gas electricity) where she would again be involved in usually regional design and managing construction. There is undoubtedly a difference between the choices women make and the choices men make at this stage (Wilkinson, 1994). Women and men choose engineering consultancies in approximately equal numbers (about 45 per cent of the graduating year) (Wilkinson, 1994). Moreover, they are more likely than men to choose local authorities and utilities (17 per cent women, 13 per cent men) and less likely to choose contractors (23 per cent women, 34 per cent men) as places to work when completing their studies (Wilkinson, 1994). That there are some significant differences between male and female graduates in factor choice have been found to exist (Wilkinson, 1994; Bennett *et al.*, 1999) and these differences could be impacting on the choice of employer.

Factors important for employment

In the United Kingdom, some researchers (Wilkinson, 1996; Bennett *et al.*, 1999) have analysed the factors important for male and female civil engineering and construction students when choosing employment and found that male and female engineers find different factors important when choosing their first employment. The impact of this on the industry needs to be considered, as when recruiting engineers companies need to offer a range of incentives to attract and retain graduates of different genders. Bennett (1996) found that the top five factors important to engineering students, male and female, as motivators for employment were opportunity to do interesting work, personal satisfaction, good pay, respect and chance for promotion. In the United States, a Coopers Union survey (1989) reported that the most important considerations for women accepting their first job were opportunity to do interesting work and professional advancement. They went on to say that for a large number of women, ethical considerations were important factors in job decision (Coopers Union, 1989). Other research, such as a UK survey on engineering graduates suggested that when selecting employers, females place slightly more emphasis on training and the quality of the company, while males place more emphasis on salary and promotion (IVL, 1989).

In the United Kingdom, Bennett *et al.* (1999) found that there were few significant differences between male and female construction students on employment choice. Where significant differences existed, they tended to be related directly to women, more than men, such as raising the profile of successful women, flexible working/childcare/career breaks and a commitment of industry to equal opportunities. When UK civil engineering students preferences were examined, of the top five variables for employment choice for male and female civil engineering students, the first three were ranked the same: opportunity to do interesting work, opportunity to do varied work and the organisations training programmes for graduates. The fourth most important ranking for males was salary and the fifth opportunity for quick advancement whereas the fourth most important ranking for women was childcare policies and programmes, followed by opportunity for quick advancement (Wilkinson, 1996). Other significant differences were found to exist between male and female civil engineers. Ethical considerations about the type of work, location of organisation to family and close friends and involvement in new developments were all significantly more important to women. The usefulness of these differences can be found in the importance to the recruitment policies of employers. An awareness of the differences could enable employers who would like to employ more women civil engineers to target women civil engineering students more effectively.

The results may also give an indication as to the reasons why less women prefer to work for contractors compared to men. These could include the fact that the work may not be attractive, and the programmes and policies may not be flexible enough to accommodate women, specifically those relating to childcare (Francis and Lingard, 2002) and contracting may be perceived as less ethical than other industry sectors. It is also possible that women find location of the work close to friends and family reduces their choice of contracting as a viable employment option. What is clear is that women civil engineers are joining the industry with certain impressions about their future role (Wilkinson, 1994). Some of the common and widely held impressions are that they believe consultants treat women better than contractors; that local authorities have more favourable policies than consultancies; that employers will not provide adequate support if they have children; that they are unlikely to have the same promotional opportunities as men; and that sexual harassment is a problem (Wilkinson, 1994). That women civil engineers enter the construction industry at all is surprising, given these apparent widely held beliefs.

The workforce – the challenges and opportunities for women civil engineers

Research into women's working lives and careers in construction (including women civil engineers) in the United Kingdom has found that there are significant problems to advancement and that the opportunities that they are given and the challenges they face differ for different sectors of the industry (Dainty et al., 2000a). For instance, contractors may choose not to hire women, and discriminate during recruitment (Duncan et al. 2002). Many companies have equal opportunities policies designed to ensure that women are not discriminated against in either the selection for jobs or for promotion. There exists the possibility that these are not meant sincerely, as some women in engineering companies, interviewed by McRae et al. (1991) believed, or that these well-intentioned policies are not put into practice at more junior levels. McRae et al. (1991: 62) found 'reports of bias and prejudice among personnel with recruitment responsibilities [and] that well-formulated equal opportunities policies were being sabotaged, effectively if not deliberately, by individuals in positions of power.' This was also found to be the case by Duncan et al. (2002: 254) who found that for contractors 'the recruitment and selection process...serve to perpetuate the existing staff profile and culture (largely white, male, macho etc.).' That discrimination is an explanatory factor in the under-representation of women in engineering has been repeatedly demonstrated (Harding, 1999;

Dainty *et al.*, 2002b; Duncan *et al.*, 2002; Whittock, 2002). There is little doubt that in some cases this will cause women to leave the industry, or choose alternative careers before entering, or simply not be hired in the first instance.

Sexism, childcare and the nature of the work

Another reason for the choice that women make to avoid contracting has to be found in the nature of the work. Contracting still requires long hours (despite various countries attempts to shorten the hours, such as the UK's Working Time Directive (Duncan *et al.*, 2002)), it is often in remote locations and, even for site project managers (work often undertaken by civil engineers), the work is heavy and dirty. Even if women initially choose site-based contracting, it is unlikely to be compatible with having children. Mid to late pregnancy is extremely difficult if a woman civil engineer is managing a construction site, especially complex sites requiring climbing ladders, crawling through spaces and lifting. Supposing a woman continues with a site-based contracting career after childbirth, the likelihood is that she will be required to regularly relocate and will have to work long hours and possibly in remote sites. This can be incompatible with being a parent. For instance, children, especially pre-school children, require the opposite of what a site-based contracting life can provide. Children need attention which is incompatible with parents working longer than normal hours; closeness to facilities (especially medical and childcare/schooling facilities) which is incompatible with remote sites; and stability, which is incompatible with regular moves. Coupled with this, when people become parents, there is usually a re-evaluation of priorities away from work and towards the family. Maintaining an interest in a highly stressed environment (of which contracting is one) becomes incompatible with the needs of the child (Squirchuk and Bourke, 2000). Even if a woman with a child is relocated to the head office, it is unlikely that a return to site-based project managing of construction projects is an option in the near future, given the needs of the child and the disproportionate amount of child-rearing women do compared to men (Francis and Lingard, 2002). To what extent this then has an impact on women's future careers is difficult to determine. However, it could be that women civil engineers choose other sectors of the industry in preference to contracting because they are in a better position to be able to cater for women with family responsibilities. This is not to say that women should not choose to work for contractors and have a family, but a rethink of the structure and nature of contracting needs to take place if women with family responsibilities are to be more equally represented. Francis and Lingard (2002) and Duncan *et al.* (2002) have made some suggestions of how the construction industry, including contracting, can alter their working environment to be more family orientated, which include increased childcare facilities, flexible or part-time work. For women in civil engineering, working for consultants, for local and government authorities or for utilities' in civil engineering, there is no obvious reason for company policies and practices not to be sufficiently flexible to cater for women with children, in line with other industries.

Analytically, if not in practice, it is possible to separate two main types of causes of the under-representation of women in civil engineering. The first is the culture of the industry reported by researchers as white, male, macho and competitive (Gale, 1994, 1995; Gilbert and Walker, 2001; Duncan et al., 2002). As the industry is relatively new to employing women in higher numbers in professional and trade roles there still exist 'dinosaurs' who believe that women have no role in the industry (Wilkinson, 1992; Dainty, 2002) and make suggestive remarks and comments (Dainty, 2002; Duncan et al., 2002) as well as pay women less than men (Harding, 1999). Problems related to women's career advancement due to these discriminatory attitudes have been suggested by Dainty et al. (2000a) and are further increasing inequalities in the industry. Employers and employees should, as a minimum, be aware of the equal opportunities legislation and employers should have policies in place to address these issues should they arise. Companies should put place training and mentoring schemes that are directed at all employees but that address specific issues, such as combining career and family, as part of the training. Companies should also be informed that women can access particular professional support organisations and groups such as the National Association of Women in Construction in Australia and New Zealand or the Institution of Civil Engineers women engineering group, Women's Engineering Society or the Association of Women in Science and Engineering in the United Kingdom.

The second cause of under-representation is the difficulty of combining family life with employment in civil engineering, a problem that particularly affects mothers, given their disproportionate share of child rearing. Why is this difficulty not overcome for construction companies? In essence, as research indicates (McRea et al., 1991; Francis and Lingard, 2002), it is because the measures to assist women in combining childcare with paid employment that are favoured by industry are not regarded as sufficient by women, whereas the solutions favoured by women are regarded as too difficult or expensive by industry. McRae et al.'s (1991) research into employers policies and practices for women in engineering showed that there was little evidence of any trend towards flexible employment, such as job sharing and home working and that managers did not consider these to be operationally possible. One of the most widely canvasses solutions to the problem of attracting more women into any industry is the provision of child care (McRea et al., 1991; Francis and Lingard, 2002). McRae et al. (1991) found that there was a division of opinion on childcare. Their research found that managers in engineering and science companies in the United Kingdom were generally hostile to the provision of on-site childcare, citing fluctuating demand and expense as disadvantages, but on the other hand, the idea that there should be state-led childcare was considerably more popular (since they did not have to directly pay for it). Francis and Lingard (2002) presented some possible solutions for construction companies, such as increased childcare, flexible working practices and permanent part-time work, and stressed that these provisions should be tailored to individual needs. Another suggestion (McRae *et al.*, 1991) was career breaks as these are popular with managers in industry because they are cheap to run. McRae *et al.* (1991) believed that managers and employees are aware of the potential difficulties with career break schemes, chiefly losing touch with changes in work, so both are eager to retain links with each other during the breaks. They also found that part-time work was more popular than career breaks with women employees but less popular with managers, as managers stressed that the introduction of part-time work would necessitate considerable changes in the organisation of the working practices of the company. This is a move advocated by Francis and Lingard (2002), although as McRae *et al.* (1991) noted, for all their formal opposition to part-time work, employers were prepared to consider it, frequently on an *ad hoc* basis.

Re-evaluating equality of opportunity for women in civil engineering

It is often asserted that it is in the interests of the industry, specifically to promote equality and diversity; it is important to employ more women. (Khazanet, 1996; Bennett *et al.*, 1999; Fielden *et al.*, 2000). However, whether this is so depends on the extent of the shortage of engineers, the (cyclical) demand for them, and the cost to firms of taking whatever steps are needed to recruit and retain women. Moreover, since firms that do not take these steps may benefit from a general improvement in the image of the industry created by any that do, they have an incentive to free ride that is, not to take those steps themselves but benefit from the efforts of others. So, while in general it may be in the interests of the construction industry to redress the position of women, individual companies may not see it as their responsibility. Certainly, as previously discussed, there are significant drawbacks for women civil engineers based on the fact that they are a minority in the construction industry. However, many of these drawbacks apply to any minority, including those – say Asians in the New Zealand construction industry – who are not underrepresented relative to their proportion of the general population. The special case for women is that the minority status of women in civil engineering is against the principles of equality of opportunity (Duncan *et al.*, 2002).

If women do not have equality of opportunity, then they are being treated unjustly. This then raises two related questions: who is responsible for the injustice? and what should be done about it? The questions are related for two reasons because generally the costs of remedying the injustice should fall on the parties responsible for it, and so in deciding what should be done, one would look to the 'guilty' parties; and because it is the parties responsible that can do the most to improve equality of opportunity.

The research reviewed in this chapter has revealed the contributions of higher education and industry to women's minority status. It found that higher education can contribute to reducing women's minority status and has a role to play in increasing the numbers of women entering and staying in engineering. However, higher educational institutions are one of the contributors that have made progress to reduce inequality through various initiatives such as networks, mentoring and access schemes, and many have policies in place to mitigate the effects of women's minority status.

The major causes of inequality are the culture of industry (Gale, 1994; Duncan et al., 2002) and the nature of the work and its interaction with the women's predominant role in childrearing (McRae et al., 1991; Isaacs, 2001; Francis and Lingard, 2002). The industry is responsible for its culture and needs to change its sexist attitudes and macho image. Until this happens it is likely that inequalities and discrimination will continue leading to more women civil engineers leaving the industry.

What is less clear is the responsibility of the construction industry to make it easier for mothers to work within the industry. Women are put off sectors of the industry, especially contracting, by the difficulties of combining work with child rearing. In the case of some types of work, such as contracting, these difficulties can only be mitigated and not removed by the availability of childcare. For other work, such as those predominantly undertaken by consultants and government organisations, childcare, flexible working, and job sharing could substantially remove the problems. But these have costs. Who should bear them? Industry plays a major role in causing problems of inequality, but so does the still-existing expectation in families that it is women who look after children. It is not obvious how to decide who the responsible parties are, that is, the ones who should fairly bear the costs. It is also not obvious, even if industry is largely responsible, that it should pay the costs. This is because a legal or moral insistence that companies should provide or pay for childcare, career breaks, flexible working etc. raises the costs to firms of employing workers, and specifically parents. This may reduce the demand for workers, whether workers generally or parents specifically. Duncan et al. (2002) showed how companies avoid the issue of equal opportunities by only employing men.

This raises the question of the role of the state in providing childcare. While there needs to be more research on the effects on employment in construction, including female employment, and on attempts to recruit and retain more women, on the face of it there seem to be two reasons to expect the state to play a major role. The first is that it would be able to subsidise any costs out of general taxation without raising the costs to firms, in terms of, say, on-site childcare facilities for their staff. The second is that, to the extent that the problem is not with industry but in the family, the state is in a better position than the construction industry to remedy the problem. Nevertheless, in the interests of equality and diversity, the construction industry does need to accommodate changing society expectations.

Conclusions

This chapter discussed the experiences of women civil engineers from entering higher education through to leaving education and then analysed the employment opportunities and challenges they face in the construction industry. The research findings to date indicate that the construction industry still has a long way to go in improving its record of recruiting and retaining women. Until the industry provides the necessary facilities able to cater for women civil engineers at any stage of their working life, inequalities are likely to remain.

Discussion question

1 Identify any problems, and discuss potential solutions, for improving the status of minority groups in engineering education.
2 Imagine you are a woman civil engineer working in the construction industry for a major contractor and are about to have your first child, discuss what issues would you be facing and assess what you would want from your employer in order to assist you with continuing employment.
3 Devise a recruitment strategy specifically aimed at recruiting graduating women civil engineers.

References

Bakos, J. (1992) 'Women in Civil engineering – graduates perspectives', *Journal of Professional Issues in Engineering Education and Practice*, 188(1): 16–28.
Bennet, F.L. (1996) *The Management of Engineering*, John Wiley and Sons, New York.
Bennett, J.F., Davidson, M.J. and Gale, A.W. (1999) 'Women in construction: a comparative investigation into the expectations and experiences of female and male construction undergraduates and employees', *Women in Management Review*, 14(7): 273–291.
Carter, R. and Kirkup, G. (1990) *Women in Engineering: a Good Place to Be?* Macmillan Education Ltd, London.
Chesler, N. and Chesler, M. (2002) 'Gender-informed mentoring strategies for women engineering scholars: on establishing a caring community', *Journal of Engineering Education*, 91(1): 49–56.

Chinn, P. (1999) 'Multiple worlds/mismatched meanings: barriers to minority women engineers', *Journal of Research in Science Teaching*, 36(6): 621-636.

Cooper Union, (1989) *The Cooper Union 1989 National Survey of Undergraduate Women Engineering Students*, Cooper Union, New York.

Court, G. and Moralee, J. (1995) *Balancing the Building Team: Gender Issues in the Building Professions*, The Institute of Employment Studies, Brighton.

Dainty, A.R.J., Neale, R.H. and Bagilhole, B.M. (2000a) 'Comparison of men's and women's careers in U.K. construction industry', *ASCE Journal of Professional Issues in Engineering Education and Practice*, 126(3): 110–114.

Dainty, A.R.J., Bagilhole, B.M. and Neale, R.H. (2000b) 'A grounded theory of women's career under-achievement in large UK construction companies', *Construction Management and Economics*, 18(1): 239–249.

Dainty, A.R.J., Bagilhole, B.M. and Neale, R.H. (2002) 'Coping with construction culture: a longitudinal case study of a woman's experience of working on a British construction site', *Perspectives on Culture in Construction*, CIB report 275, CIB, Netherlands, 221–237.

Devgun, B. and King, P. (1999) 'Where have all the girls gone?', *New Zealand Engineering*, 54(7): 55–57.

Duncan, R., Neale, R. and Bagilhole, B. (2002) 'Equality of opportunity, family friendliness and UK construction industry culture', *Perspectives on Culture In Construction*, CIB report 275, CIB, Netherlands, 238–257.

Fielden, S.L., Davidson, M.J., Gale, A. and Davey, C.L. (2000) 'Women in construction: the untapped resource', *Construction Management and Economics*, 18(1): 113–120.

Fizpatrick, J. and Silverman, T. (1989) 'Women's selection of careers in engineering: do traditional-non-traditional differences still exist?', *Journal of Vocational Behaviour*, 34(3): 266–278.

Francis, V. and Lingard, H. (2002) 'The case for family-friendly practices in the Australian construction industry', *Australian Journal of Construction Economics and Building*, 2(1): 28–36.

Gale, A.W. (1991) 'What is good for women is good for men: theoretical foundations for action research aimed at increasing the proportion of women in construction', in Barrett, P. and Males, R. (eds) *Practice Management: New Perspectives for the Construction Professional*, Chapman Hall, London.

Gale, A.W. (1994) 'Women in non-traditional occupations: the construction industry', *Women in Management Review*, 9(2): 3–14.

Gale, A.W. (1995) 'Women in construction,' in Langford, D. Fellows, R., Hancock, M. and Gale, A.W. (eds) *Human Resource Management in Construction*, Longman Scientific and Technical, Essex.

Gilbert, G.L. and Walker, D.H.T. (2001) 'Motivation of Australian white collar construction employees: a gender issue?', *Engineering, Construction And Architectural Management*, 8(1): 59–65.

Greed, C. (1991) *Surveying Sisters: Women in a Traditional Male Profession*, Routledge, London.

Harding, S., Messer, D. and King, M. (1999) 'Women in a world of engineering', *New Zealand Engineering*, 54(7): 58–60.

Isaacs, B. (2001) 'Mystery of the missing women engineers: a solution', *ASCE Journal of Professional Issues in Engineering Education and Practice*, 127(2): 85–90.

IVL (Imperial Ventures Limited) (1989) *Survey of the Attitudes and Intentions of Graduating Engineers 1989*, Blackheath Publishing, London.

Khazanet, V.L. (1996) 'Women in engineering and science: it's time for recognition and promotion', *Journal of Professional Issues in Engineering Education and Practice*, 122(2): 65–67.

McRae, S., Devine, F. and Lakely, J. (1991) *Women in Engineering and Science: Employers Policies and Practices, Policy Studies Institute*, London.

Murray, S.L., Meinholdt, C. and Bergmann, L.S. (1999) 'Addressing gender issues in the engineering classroom', *Feminist Teacher*, 12(3): 169–183.

Newton, P. (1987) 'Who becomes and engineer? social psychological antecedents of a non-traditional career choice', in Spencer, A. and Podmore, D. (eds) *In A Man's World – Essays On Women In Male Dominated Professions*, Tavistock, London.

Squirchuk, R. and Bourke, J. (2000) 'From equal opportunity to family – friendly policies and beyond', in Hass, L., Hwang, P. and Russell, G. (eds) *Organisational Change And Gender Equity: International Perspectives On Fathers And Mothers At The Workplace*, Sage Publications, Beverley Hills, CA.

Srivastava, A. (1992) 'A case study of widening access to construction higher education', *Proceedings of the Eighth Annual Conference of the Association of Researchers in Construction Management (Arcom)*, Isle of Mann, UK.

Thomas, K. (1988) 'Gender and the arts science divide in higher education', *Studies in Higher Education*, Society for research into higher education, 13(2): 253–261.

Tizzard, J. and Chiosso, R. (1990) 'Women and girls on engineering and construction courses further education', *Vocational Aspect of Education*, 42(113): 101–107.

Usher, S. and Ward, K. (1991) 'Gender perspectives on engineering education', in *Proceedings of the National Conference on Women in Engineering*, New Brunswick, Canada, 125–136.

Weinreich-Haste and Newton, P. (1983) *A Profile of The Intending Woman Engineer*, EOC Research Bulletin No. 7, Equal Opportunities Commission, Manchester.

Whittock, M. (2002) 'Women's experiences of non-traditional employment: is gender equality in this area a possibility?', *Construction Management And Economics*, 20(5): 449–455.

Wilkinson, S. (1992) 'Career paths and childcare: employer's attitudes towards women in construction', *Proceedings from the Women in Construction Conference*, University of Northumbria, UK.

Wilkinson, S. (1994) *Entry to Employment: Choices Made by Qualified Women Civil Engineers Leaving Higher Education*, PhD thesis, Oxford Brookes University.

Wilkinson, S. (1996) 'The factors affecting the career choice of male and female civil engineering students in the UK', *Career Development International*, 1(5): 45–50.

Chapter 7

Women in surveying and planning

Creating diversity

Clara Greed

Introduction

This chapter looks at the state of play in two key contrasting areas of the built environment professions – planning and surveying – in respect of 'gender' issues, with particular emphasis on women and planning. First a brief background is provided on the growth of the numbers of women in planning and surveying. The contrasts between surveying and planning are highlighted, not least the fact that a far greater percentage of female membership is to be found within planning than in surveying. Yet, paradoxically, because surveying is such a larger profession, the actual number of women surveyors in the United Kingdom is double that of women planners as found in the main professional bodies, namely the RICS (Royal Institution of Chartered Surveyors) and the RTPI (Royal Town Planning Institute) respectively (Watts, 2003) (Table 7.1).

Women surveyors and planners (and for that matter the men too) are likely to be motivated by rather different objectives and aspirations in that 80 per cent of surveyors work in the private sector, whereas around 70 per cent of planners are to be found in the public sector (RICS, 2003). The total membership of the RTPI is about a quarter of that of the RICS (or for that matter the Institution of Civil Engineers) but planners have a disproportionate influence on the nature and design of the built environment with policy making and development control powers within every local authority in the country (Greed, 1999a).

Not only are women planners concerned with career problems as those in other built environment professions, but they are also critical of current planning policy. It has been demonstrated by research and human experience that women suffer disadvantage within a built environment that is developed by men, primarily for other men, as explained in published research (Greed, 1994; Little, 1994, 2002; Booth *et al.*, 1996; Darke *et al.*, 2000). A key factor in seeking to change the nature of the built environment is being able to reach a senior position in the planning profession in order to shape policy making and participate equally in high level decision-making.

Table 7.1 Membership of the Built Environment Professions

Professional body	Total membership	Women members
Royal Institution of Chartered Surveyors	107,817	11,173
Institution of Civil Engineers	78,641	3,678
Chartered Institute of Building	37,511	1,181
Royal Institute of British Architects	28,328	2,858
Institution of Structural Engineers	20,173	950
Royal Town Planning Institute	17,924	4,714
Total	290,394	24,554

Source: Watts, 2003.

Thus, the twin issues of 'women in planning' (as professionals) and 'women and planning' (as the planned as citizens of our towns and cities) are inextricably linked.

This chapter focuses on potential means of change in light of this situation, both in respect of 'who' is doing the planning policy making and 'what' is being planned. Drawing on recent developments in both the European Union and United Kingdom, this chapter presents means of generating change to improve the situation for women as fellow professionals and as citizens and users of the built environment by means of gender mainstreaming (Booth, 2000, Greed (eds), 2003a,b). A brief background is provided on the gradual growth of female participation in the surveying and town planning professions in the United Kingdom and Europe followed by an overview of the situation today. Both quantitative data and qualitative issues are provided to illustrate the problems. Furthermore, a summary of the 'urban problem' of the lack of integration of women's needs into the city of man is explained. Clearly change is long overdue. The main body of the chapter discusses the new 'Gender Mainstreaming' requirements and methods, as currently coming into force within EU member states, as a result of the Amsterdam Treaty of 1997 (which came into force from 1999) and UK governmental initiatives. The material presented is based on work recently undertaken at the University of the West of England by the author and research team for the Royal Town Planning Institute to produce a 'toolkit' and explanatory report to enable local planning departments, responsible for spatial planning in cities and countries within the United Kingdom, to mainstream gender issues into planning policy and practice (Reeves and Greed, 2003). Whilst the purpose of the original toolkit was to address town planning issues in particular, in this chapter the account is written in a way that seeks to show the applicability of principles and methods to the needs of the wider built environment professions, most of whom are also trying to mainstream gender and other equal opportunities (EO) issues into the professional arena. At the end of the chapter a list of key web site references is given to help readers find out more about gender mainstreaming.

The position of women in planning and surveying

Although there were exceptional and rare individuals who participated in the built environment professions in the historical past, in the United Kingdom it was not until the 1919 Sex Disqualification (Removal) Act of 1919 that women were first allowed into the professions. The numbers of women entering were very low until the 1960s, when with the growth of higher education, EO policy and the second wave of feminism more women began to enter surveying (Greed, 1991), planning (Greed, 1994), architecture (De Graft-Johnson *et al.*, 2003) and to a lesser extent civil engineering (Watts, 2003). Growth tended to occur in particular specialisms rather than across the board. The construction and built environment professions may be seen as a spectrum running from the hard site-based engineering and building areas at one end and the softer housing, social policy and urban policy areas at the other, with the middle ground being the more office-based area of commercial estate agency, valuation and general practice surveying (Thomas, 1999). Traditionally in the United Kingdom women were more likely to enter either the soft 'planning' end as members of the RTPI or as RICS members specialising in Planning and Development, or they were to be found in the smarter office-based commercial sector of general practice. Generally, women were more likely to go into the public sector, especially local government, which was well funded and powerful in the post-war years right up until the late 1970s. With the opening up of erstwhile male-dominated management and business specialisms from the 1980s, women increasingly took the option of working in private sector practices and consultancies too as the phenomenon of the 'business woman' became established under Thatcherism. However, the image of 'the woman engineer' proved far less attractive, and thus far less women were tempted into to the more technical and site-related end of surveying and the wider built environment professions (Greed, 1999b; Turrell, 2003).

Thus, by the 1980s women in the United Kingdom had established a presence in the planning and surveying professions. Nevertheless it proved much more difficult for them to rise within the organisation as they experienced the problems of the glass ceiling both in the professions and academia (Morley, 1994). Ethnic minority professionals were also likely to experience invisible glass walls around coveted specialisms too (Ismail, 1998). As more women and other minorities entered these professions certain niches and levels seemed to be allocated to them whereas high status areas remained reserved for high flying men (Williams, 1992; Evetts, 1996).

By 2000 there was, overall, a greater sprinkling of women throughout many professional firms within the built environment professions, and a few exceptions to the rule became partners in surveying firms and chief planning officers in local government. However, whether this should be seen as a triumph for feminism or women in general is another matter as the author has long argued that, 'more [women] does not necessarily mean better' (Greed, 1988). Whilst some women may be promoted as the result of individual merit, progressive thinking and EO policy and support the careers of women employees, others may not. Such are the powers of professional socialisation into the dominant values of the planning fraternity that one should not assume that a woman planner will necessarily hold more enlightened views than her male counterparts (Greed, 1994a: 164). Indeed it was found from recent research that increased numbers of women and ethnic minority planners – without enforcement of supportive equalities policies – can in fact lead to 'sexism by proxy' and 'racism by proxy' in which minority individuals themselves perpetrate discriminatory practices and planning policies often because of pressure from senior management, bullying from majority individuals in positions of power, and simply fear about their jobs (Onuoha and Greed, 2003). Nevertheless whatever the reasons for 'more women' or the personal perspective of the women themselves, an increase in female numbers certainly results in their firms and professional practices having to confront and effectively manage the demands and requirements of women's 'different' and more complex employment patterns and dual roles (Kandola and Fullerton, 1994; Kirton and Greene, 2003).

But, one must ask, in giving this potted history, and comparing 'then and now', are we comparing like with like? The built environment professions themselves have changed and evolved and the situation 30 years ago is not necessarily comparable with their 'meaning' or 'status' today. As for the surveyors, the RICS has undergone a series of restructuring and reorganisation and the property market itself has evolved with new specialisms and focal points of power emerging, and arguably the RICS does not have the monopoly over jobs in property. Nowadays a whole range of legal, banking and development organisations have emerged, many at the international level, who deal with property and construction matters but whose employees are not necessarily 'chartered professionals' in the traditional British sense.

There has been a noticeable drop in the numbers of young women and men entering surveying as other more attractive careers have developed, for example, in computers, media, business studies, law and environmentalism (but curiously not 'built' environment careers, indeed the contrast in the speed of growth between say law and architecture is stark) (De Graft-Johnson et al., 2003). As for the RTPI, the planning profession, and indeed local government, never quite recovered from the Thatcher era and although hopes of a return to a powerful, well-funded public sector were briefly raised by the advent of New Labour any optimism soon proved unfounded. Jobs in local planning departments were increasingly poorly paid and professional staff shortages were commonplace particularly in London. Increasingly local government vacancies have been taken by women who receive neither the salary or the status of their male predecessors and who are often working on a part-time, short-term contract or agency basis and who do not think of their work as a safe municipal job for life (Reeves, 2002; Greed *et al.*, 2003a).

The situation has not been helped by the fact that the New Labour government has, arguably, tended to prioritise the more glamorous policy issues of urban regeneration, rather than concentrate upon the more mundane but highly important matter of day-to-day planning for ordinary towns and cities (Brownill, 2000). Although 'gender' issues have remained high on the agenda of many local authority equalities officers, especially in relation to the fields of employment, health and education policy, there has not been a commensurate emphasis upon women's issues in local planning departments. The situation has not been helped by central government failing to refer specifically to gender issues in its national planning policy guidance documents, in spite of frequently referring in generic terms to community issues, social exclusion and the needs of minorities (CRESR, 2003). However, greater opportunities are offered by the more progressive large urban local authorities, whilst devolution in Scotland (Fitzgerald, 1999), Wales and the Greater London Council (GLA) has created new opportunities for women to exert greater influence on policy making (GLA, 2003a,b). Overall there is a greater emphasis on equal opportunities requirements in local government as a whole (EOLG, 2001) and the means of applying these principles (GLA, 2003c,d).

The RTPI has also battled on, and nowadays is positively seeking to support gender equality objectives as promoted by the EO Women's panel nationally and like other professional bodies has a Code of Professional conduct that includes gender and other issues. Whether this is because of genuine concern about EO or because the existing membership is ageing and not growing as anticipated, because of reduced student intakes and lack of interest among young people for town planning and other construction-related careers, is another matter for discussion. The Royal Town Planning Institute's 'New Vision for Planning' (RTPI, 2001) has led with an inclusive approach to spatial planning and integration of gender awareness into the planning process which takes into account the needs of other groups such as ethnic minorities and people with disabilities. Cynics might say 'too much too late' as it would seem much of the general public, and especially women, are already disillusioned with 'planning' having suffered from its effects for many years, but at least they are trying now, not least because of the huge surge in generic and generic equality initiatives that have been coming in at both national and European level which all have implications for the operation of the town planning system (Booth, 1999, 2001).

The urban problem for women

Research and human experience has shown that women and men have different needs in respect of the design and nature of the built environment (Greed, 1994; Agrest *et al.*, 1996; Darke *et al.*, 2000). Women are still the main people responsible for childcare, shopping, homemaking and other essential caring activities, as well as going out to work nowadays. Whilst women use public transport twice as much as men, those women with cars tend to use them for more family-related journeys. Forty per cent of car journeys undertaken by women are for escort purposes such as taking children to school, spouses to the station and other family members to the shops, doctors and to leisure activities (compared with 11 per cent in the case of men). In contrast a typical male journey to work is likely to be uninterrupted, a mono-purpose work journey in the mornings, whilst in the evenings if stop-offs are made they are more likely to be for the purpose of going to the pub, to the video store or to bring a take-away meal back with them (Hamilton, 2000). The division of work and home, that is between employment and residential areas, has been exaggerated by land use zoning, policies of dispersal and decentralisation. This makes it difficult for parents to combine work and home duties. The problem increasingly affects those men who try to take on a greater role in family care too. The situation is made worse by the fact that transport policy has tended to favour the needs of those commuters who make simple mono-purpose journeys, to and from work. Women and other carers are far more likely to trip-chain journeys in order to fulfil their many roles and duties along the way. Rather than travelling simply from home to work, their journey may comprise home to childminder, to school to work to food shopping to school again to childminder and back home again (Bichard *et al.*, 2004). Therefore, land use transportation policy designed to meet the needs of male workers may prove impractical for many women whose travel patterns and activities are complex and diverse. Clearly change is needed, and gender mainstreaming is seen as a means of ensuring that women's requirements, as well as men's needs, are considered at all stages and in all aspects of policy making. Not only will this result in better cities for women but arguably also for men, as 'gender' is not 'just' a women's issue but includes everyone (Greed, 1999b).

Gender mainstreaming

Gender mainstreaming may be defined as the process by which an acceptance of the importance of gender awareness is incorporated into all aspects of policy making, process and practice within an organisation (Dobbie and Purcell, 2002). The consideration of gender issues becomes a core feature of the planning process rather than an add-on. This approach challenges existing institutional approaches in both the public and private sector. It has implications for the composition and organisation of the planning department's staffing, as well as influencing policy content. For example, in respect of town planning policy, mainstreaming involves taking into account fully the fact that men and women have different roles in society and therefore they have different expectations and requirements from the planning system. Real mainstreaming can be achieved by incorporating a gender equality perspective into all stages and levels of policy and plan making. All this cannot be done in isolation from other priorities and day-to-day aspects of plan making. Rather, it is a matter of looking at existing and proposed policy areas through a gender lens and incorporating gender considerations into the planning process to achieve better policy outcomes for all. It should be seen as a strengthening factor which adds to the build up of critical mass to achieve greater equality and thus a more sustainable and accessible built environment. For example, gender considerations, such as lower car ownership among women, might be cited as a reason for promoting higher density, pedestrian friendly, mixed-use brownfield site developments that reduce the need for greater urban expansion, as is explained further in this chapter. In this process the need for planning policy-making bodies to be more reflective in composition of the population being planned soon becomes a key consideration. Therefore, all the issues (as discussed in this section of the book) of under-representation of women in the more senior levels of the built environment professions, the reasons for this, and the means of change, all come into play again in the process of achieving gender mainstreaming.

Within the European Union, the Amsterdam Treaty 1997 makes 'the elimination of inequalities and the promotion of equality between women and men' a central principle in all public policy making and activities at local authority level within the member states of the European Union. This Treaty and its related EU regulations and directives having been gradually introduced and implemented and the requirements are beginning to bite in the individual member states of the European Union – and that means in the United Kingdom too (Booth, 2001). Already European and single regeneration funding regimes require evidence of collection of gender specific data and evaluation of the differential gender impact of policy upon different members of the community. Any construction firms and contractors involved in inner city areas that are the subject of regeneration budgets, Objective 1 and 2 regional grant regimes, or for that matter Millennium and Lottery funding, will know that it is now essential to show 'contract compliance' that 'gender-proofing' has taken place in respect of its organisational policy, contracting practices and staffing composition. The Amsterdam Treaty when fully introduced effectively extends these requirements to all local authorities and all development projects (although the legal details are still being fought over furiously).

Public authorities already have a statutory duty to promote equality of opportunity within the workplace and in the delivery of services and this includes the work of LPAs (local planning authorities, i.e. all town planning departments). This duty is set out in the Race Relations Amendment Act 2001 and the DDA (Disability Discrimination Act, 1995) *inter alia* (ACAS, 2002) and in a range of other equality legislation (EOC, 1997, 2001). For example, the Stephen Lawrence Inquiry Recommendations for Implementation, namely PAET: Policy Appraisal for Equal Treatment are being widely used for this purpose and incorporate an impact assessment approach. Significantly they highlight the importance of multiple discrimination situations and the need to look at the relationship between race and gender in particular (MBI, 2001). PAET is based upon evaluating policy at three stages:

• Identify the impact of a policy on different groups
• Validate the legal requirements on minority individuals
• Amend the proposals accordingly.

Admittedly, there is more legislation covering race and disability – than gender – in the United Kingdom but, as stated earlier, this is changing with the introduction of the requirements of the Amsterdam Treaty. This no doubt reflects Continental European practices where in many countries, in contrast to the United Kingdom, gender tends to receive more statutory backing than race or disability (for better or worse) but the ultimate intention is to make it better for everyone. Authorities in Wales, Scotland, Northern Ireland and the Republic of Ireland are also covered by legislation which includes a duty to promote equality (ECNI, 2001; Reeves, 2001, 2002; NAW, 2001). This is particularly evident in the constitution of the new devolved government structures emerging within the British Isles.

Non-statutory equality standards and indicators now bind local authorities because they are used as measurements and benchmarks in current Best Value and National Audit inspections in local government. These include the Equality Standard for Local Government in England (EOLG, 2001) which requires LPAs to contribute towards their local authority's programme for achieving an assessed level of competence and similar requirements apply in Wales, Scotland and Northern Ireland. The Race Relations Amendment Act 2001 requires a similar process of formative appraisal.

The Government's emphasis upon 'Modernising Planning' (DTLR, 2001, 2002) stresses a more holistic approach to planning that is reflective of the composition of society. Mainstreaming equalities in plan-making will improve the quality and performance of plan policy and thus meet the planning needs of all members of society more appropriately. There have been a series of central government initiatives such as the Cabinet Office

Women's unit that sets out the requirements (Cabinet Office, 2002). A whole range of other policy guidance documents and programmes stress the importance of 'community involvement' including Local Agenda 21 (sustainability); the production of Community Strategies for local areas and the representation of community members in urban regeneration programmes. But whether 'women' are seen to be part of 'the community' (one of the spongiest words in government policy) is open to question as research has shown that many of the 'community representatives' on boards and *ad hoc* committees set up under New Labour's mass of urban initiatives, have been predominantly male, white and middle class in composition (Bennett *et al.*, 2000; Brownill, 2001; WDS, 2001).

As a result of the current emphasis upon gender mainstreaming the RTPI has sought to produce a toolkit to provide guidance to local authorities seeking to mainstream gender into planning policy and practice with tools and ideas to mainstream gender issues into statutory plan-making and development control. The research was undertaken by the author and a research team (Greed *et al.*, 2003a), and the toolkit was developed in association with the chair of the RTPI steering group responsible for gender and planning initiatives (Reeves and Greed, 2003). But why just gender? It may be argued there are already a whole range of generic EO standards and procedures around especially in local government but in the private sector too nowadays. Nevertheless the research team found that these did not tackle gender specifically. Indeed *gender* may get lost in amongst hundreds of other equalities issues within the wider process of *generic* mainstreaming (Greed (eds), 2003b). The whole exercise may be reduced to a complicated tick-box procedure that never gets to grips with any of the issues in any depth. In contrast, the research team sought to look at the gendered implications of specific planning policies and practices in considerable detail and develop practical examples in the toolkit (Reeves and Greed, 2003).

On the other hand it is true that gender does not exist in isolation. But gender is a cross-cutting and over-arching characteristic that overlays race, class and disability and that this complexity needs to be taken on board. This is because the lives of women and men are affected and influenced in different ways by the implementation of many planning policies. Thus the tools and techniques in place to deal with race and disability issues cannot necessarily be adapted to deal with gender issues and are no substitute for gender mainstreaming. Planning policy has a differential effect on men and women. Therefore it is important to use a gender lens when looking at all policy issues, including those relating to environment, transport, race and sustainability and not to see these as separate from gender. For example, figures show that there are significant differences between the way that men and women travel, live and work. Fifty two per cent of the UK population are women; and 45 per cent of employed people were women, the majority of public transport users are women (GLA, 2003b; ONS, 2003).

Three times as many men as women cycle to work, whilst twice as many women as men are responsible for escorting the children to school. About 75 per cent of white and black Caribbean women of working age are economically active, compared to 30 per cent of Pakistani women and 19 per cent of Bangladeshi women. Over 44 per cent of working women worked part-time, while only 8 per cent of men work part-time. Thus their experience of life and the built environment is sure to vary as is their potential role in the workforce (GSUG, 2003).

If gender mainstreaming were in operation in a local planning department this would result in a different set of priorities being embedded in planning policy. For instance, women are primarily the ones with caring responsibilities although nowadays the majority of women work too (Darke *et al.*, 2000). Since a far greater proportion of women depend upon public transport or walking to get between their various destinations, women are more likely to benefit from local planning policies that promote less dispersed land use patterns, with a closer relationship between employment and residential areas. All this has implications both at the city-wide strategic level of planning and direct effects on the local planning level. Cities based on short distances, public transport, mixed uses and a good distribution of local centres and facilities are favoured by many women. But at the detailed level, public participation exercises have also shown that women are (more) concerned about a wider range of street level issues, encountered by pedestrians, including personal safety, provision of better social amenities such as public toilets, crèches and public seating, plus a higher standard of integrated public transport provision. All these issues need to be mainstreamed into plan-making and urban design proposals. It is important to incorporate a gender equality perspective into all stages and levels of policy making. Although sustainability policies are to be welcomed, there is a need that they should be seen through a gender lens, so that all policies concerned with car control, re-zoning, public transport priorities and pedestrian priorities take into account women's essential journeys too. This would reduce the number of journeys and make places more convenient and better for everyone.

Strategies for implementing gender mainstreaming: the RTPI toolkit

Gender mainstreaming requires integrating relevant issues into each stage of the policy-making process. In order to produce the appropriate toolkit to meet the specific needs of town planners, the research team investigated management literature that might link to planning methodology requirements (Greed (ed.), 2003a). A range of case studies of local planning authorities where some form of gender mainstreaming was already underway was investigated. In due course a draft toolkit was developed and this was tested out with six representative local planning authorities. In addition a wide range of private and public sector planning practitioners were consulted in the process of developing the toolkit. As a result we structured the RTPI toolkit around the following five key activities:

- Preparation of the organisational framework
- Research and analysis
- Policy development
- Consultation and participation
- Monitoring and evaluation of outcomes.

It should be noted that the toolkit eventually was structured around a series of questions in respect of each stage, but also in relation to overarching issues such as staff composition, resources, timescales and overall management of the planning processes and practices. The toolkit included key questions that were intended to enable local planning departments identify the stages and components of their work that required particular attention in seeking to undertake gender mainstreaming (see Appendix).

Organisational framework

It was important to ensure that there was evidence of inclusion of women and other minorities at all stages and levels of the planning process, among the planners as well as the planned. Many mainstreaming schemes focus on to the institutional and organisational framework, and investigate 'who' is making the decisions and whose interests are represented at the highest policy-making levels. Particular attention should be given to ensuring that both groups are entirely and equally participating in gender mainstreaming at all stages. The intention to undertake mainstreaming should be fully discussed with members, appropriate training undertaken and full explanation given, particularly at the early stages of change. To achieve successful mainstreaming, attention needs to be given to programme preparation, training and awareness raising among planners. CPD and formal training, as well as sharing and comparing approaches and progress with other LPAs, are all valuable means of facilitating this process and generating culture change (Greed, 1999b). It is most important, throughout the process, to be aware of the huge effect that the institutional framework can have on the whole process. However, there is a danger in such equality exercises that the institutional and personnel matters become such a major issue that the purpose of planning the improvement, control and design of the built environment is sidelined. But it is extremely important that the particular perspective and needs of women professionals in construction, planning, surveying and architecture are acknowledged as otherwise, the whole exercise can become yet another vague, generic 'tickbox' exercise that completely misses the mark.

Research and analysis

Every plan is based on a survey and analysis of the issues facing the local area and of its needs for the future. If a plan is to meet the needs of the population it serves, then there has to be analysis of what is happening to that population and an identification of the issues that are important for particular groups. The sophistication and methodology for this analysis will vary depending on the time, resources and data available and the location of the planning situation. So the research process depends on the availability of disaggregated statistics for men and women within the planning area under consideration. This first stage requires the collection and analysis of such data on matters such as employment, travel patterns, recreation, shopping and home ownership, as well as on ethnicity, disability and age, as these are all cross-cutting factors. In fact, producing gender-specific data is the key to the exercise but one of the most difficult tasks. The analysis of gender-disaggregated data will highlight areas of difference that might affect future planning policy. Gender sensitive data analysis will also raise questions about previous policy approaches. Furthermore, data showing large discrepancies in the use of bicycles and cycle paths by men and women may suggest a flaw in the authority's cycling strategy and point to the need for further research. For example, issues of personal safety, poor lighting, isolated routeways may deter women from cycling to work, whilst the need to carry shopping, accompany children, or lack of workplace changing facilities may all be a deterrent.

Policy development

It should not be assumed that only women will benefit from gender mainstreaming. Men are often responsible for childcare in areas where there is high male unemployment and women work shifts. They are therefore more likely to experience similar problems in accessing the built environment when accompanied by small children. The development of policy encompasses a range of activities from the setting of objectives to processes of policy improvement known as auditing or appraisal. In evaluating a particular policy it is important to ask:

- Who does the issue affect?
- Does this issue affect women and men differently?
- Is there evidence of any existing inequality already in this area?
- Has the LPA received complaints that are specifically related to gender issues?

- Are previous complaints to the department particularly gendered in content?
- Where and what are the positive and negative impacts of the policy in question?
- Will the resulting policy make it worse/better for women/men after implementation?

Some LPAs, such as Plymouth, are using a matrix-based approach in which policy topics are put down the left hand column and a range of gender-sensitive checks are put along the top and then each policy is looked at and if necessary altered accordingly (Plymouth, 2001). Many policy areas offer the opportunity for linkages to be made with ethnicity, equality and disability policies too, alongside gender mainstreaming activities. This requires a greater depth of data collection at first, but with time a local planning department undertaking gender mainstreaming would, of necessity, require an inclusive data base and build up expertise in looking for factors that indicate gender-sensitive policy.

In reviewing each policy, it is also recommended that they are evaluated in terms of yielding

- direct benefits (policies that benefit women specifically)
- indirect benefits (disabled access benefits women with pushchairs too)
- neutral impacts (no worse no better)
- missed opportunities (a chance to improve something for women missed e.g. public toilets).

Publicity and consultation

Consultation and participation is central to the process of plan preparation and is governed by certain statutory requirements. Consultation also receives attention in the Equality Standard for local government and in the guidance on the preparation of community strategies (EOLG, 2001). Councils should ensure that consultation techniques do not discriminate against particular groups. It should be recognised that some groups may find some consultation techniques and methods difficult to understand, for example, because English is not their first language. Gender mainstreaming should be integrated into all stages of plan-making. It is concerned with planning process as well as planning policy. Attention should be given to the way in which participation is undertaken especially as to where and when it takes place. This is one of the most important stages of all in achieving social inclusion and equality in the planning process. But consultation procedures, alone, are not sufficient to ensure that policy is gender mainstreamed. They must be acted upon so that people can get tangible feedback and ideally a visible outcome for their effort, as otherwise they may feel their views were not taken seriously.

Monitoring and evaluation

The final vital stage in the process is that of carrying out monitoring and evaluation – that is, looking at the effects of the process and the resultant change. To judge whether there has been progress or not it is vital to have 'before and after' statistics so as to reveal 'the difference' that gender mainstreaming makes. One measure might be whether there is now more inclusion of women in decision-making committees and in the executive levels of the organisation. This would in turn affect the priorities and design of urban regeneration and renewal policies which are recognised as areas of under-representation of women's voices. Other evidence might be time-use surveys that show a better balance between family and work–life hours. At the building site level, simple quantitative evidence like an improved ratio of female to male toilets might be evidence of culture change, whilst qualitatively a more flexible approach to childcare and to time allowed off site might also suggest that progress is being made. Of course quantitative data as to the numbers of women and men at each level in the organisation, and within specific specialisms, are very telling as is the percentage of women as against men sent on conferences, training programmes and CPD (Continuing Professional Development events).

Incorporating a gendered perspective into the planning process

A gendered perspective should suffuse all aspects of the organisation's activities, such as policy formulation and development. One should not assume a topic is not gendered, as the most unpromising, boring technical topics may yield surprising results. It is very important that gendered issues should be integrated within the resultant planning documents, and that gender should not appear as a mere add-on or 'tick box' issue. Rather there should be a clear explanation in each policy topic section of 'how', 'where' and 'why' gender has been taken into account in the policy in question. Good intentions and lip service are not good policy-making tools. Instead there should be clear objectives, timescales, budgets and evaluation criteria set.

Staff and managers at all levels should be trained and sent on CPD courses to develop awareness and understanding of how to do all this. Thus, the LPA will soon be integrating equalities into the plan-making process, as part of their role in modern local government. It should not be left to a few keen women to shoulder the whole burden of changing the organisation, whilst senior male managers totally ignore the topic. Women should be given time, resources, encouragement and freedom to work within the organisation to develop mainstreaming implementation. They should not be expected to attend EO events in their own time or take it out of their leave (as we found was the case in some ostensibly progressive LPAs). There is nothing worse than some senior male manager being given extra money and responsibility for EO whilst women who really know about the subject are not even allowed time away from work to attend necessary gender mainstreaming meetings (as has happened in some organisations).

Conclusions

The purpose of this chapter was to give the reader an account of ongoing initiatives that might bring us higher up the mountain towards achieving EO in practice. A summarised selection of the principles and checklists that were incorporated into the RTPI Toolkit are set out in the following boxed appendix section in the form of indicative headings. Readers may consult the full toolkit and the supporting research document online at http://www.rtpi.org.uk. A key purpose of the RTPI Toolkit was to encourage planners to stop and think about the gender implications of their actions at each stage of the planning process, thus building in a repetitive and iterative process of gender-proofing into their working lives. Of course the construction industry has already produced a number of excellent toolkits of its own, including one developed by Helen Stone and team, entitled, *A Commitment to People, 'Our Biggest Asset': A Report and Associated Toolkits* (CIC, 2000 and see web page list).

Toolkits seem to be the way ahead in seeking to implement gender mainstreaming in organisations where a predominantly male (and sometimes hostile) staff need to know what to do, without necessarily understanding or believing in the reasons behind the exercise: although ideally with appropriate training and staff appraisal they may become sympathetic. However, in order to avoid the process becoming purely mechanistic it is important to seek to create change in the hearts, as well as the minds, of the built environment professions. In other words there is a need to change the professional subculture, so that a proactive and willing spirit prevails in respect of diversity issues, and a reactive, and often resentful mentality towards gender, race and disability issues becomes a thing of the past (Edwards, 2004). Thus for the future, it is vital to continue to seek to integrate gender and other diversity considerations into professional education so that the next generation takes equalities issues as part and parcel of the agenda of being a built environment professional. It is also important, as discussed in other chapters, to seek to ensure that the wider construction industry is itself composed of a greater range of types of people, so that it reflects the diversity of the nation itself, not least in respect of race and gender.

Discussion questions

1 To what extent do you consider that 'more women means better?' It may be argued that in professions where women compose less than 20 per cent of the membership one should concentrate on 'changing the men' rather than on 'getting more women'. Discuss.

2 Have you come across, or experienced, examples of gender-mainstreaming activity, either in respect of personnel matters, education, or urban policy making and design considerations? Identify the factors involved that contribute towards success (or failure).

3 Is the United Kingdom particularly 'bad' at equalities and diversity issues? Investigate the situation in one other EU country and any one other country? To what extent do you consider that national characteristics or cultural factors affect attitudes towards gender issues?

Appendix

Illustrative toolkit

The toolkit is based around a series of questions related to each stage of the planning process and the main components of planning practice and spatial policy making. In summary it can all be boiled down to asking the following questions, the answers to which might form the basis for focused action and change that will result in gender considerations being built into the planning department in question.

Mini version of toolkit

In order to mainstream gender into the planning process ask the following questions:

- Who are the decision-makers?
- Who forms the policy team?
- Which sorts of people are perceived to be the planned?
- How are statistics gathered and who do they include?
- What are the key values, priorities and objectives of the plan?
- Who is consulted and who is involved in participation?
- How are planning proposals evaluated? By whom?
- How is the policy implemented, monitored and managed?
- Is gender mainstreaming being integrated into all areas?

The following 'lists' comprise an aide memoire to undertaking gender mainstreaming within planning. It should be stressed that it is a continuous process, over-arching and cross-cutting all other planning department activities, processes and components. Thus some of the questions seek to identify weak areas, encourage and strengthen the process, and raise consciousness of the issues involved, thus creating proactive gender mainstreaming embedded within the planning process. Thus these lists are quite different from toolkits that take a more reactive approach of 'testing' the end product.

The following factors need to be considered to facilitate the process:

- Preliminaries: What supportive initiatives, resources and possibilities exist already?
- Planners: Who is doing the planning? Who are perceived to be the planned?
- Populations: How are statistics gathered and who do they include?
- Priorities: What are the key values and objectives of the plan?
- Participation: Who is consulted and who is involved in participation?
- Piloting: How is it evaluated? Are discussion groups used? How are they selected?
- Programme: How is the policy implemented, monitored and managed?
- Performance: Who benefits? who loses out? What side effects are there?
- Proofing: Is gender incorporated into each policy? Does it make sense?

The following questions need to be asked:

- Is the planning department's composition reflective of 'the planned'?
- What is the percentage of male/female decision-makers?
- What has been done to increase the representation of women?
- What links are there with other departments with gender mainstreaming experience?
- Can you draw on Human Resource Management and personnel guidance?
- What has been done in other similar professions or organisations?
- Can you use existing Best Value and Audit sources in the same local authority?
- Is staff training mandatory particularly in respect of consultation techniques?
- Are there hidden groups that became visible because of mainstreaming (e.g. off-site female prefabrication assembly workers in construction)?
- Is there evidence of existing inequality getting worse as a result of implementation?
- Are complaints to the LPA increasing and particularly gendered in content?

- Where and what are the positive and negative impacts of this issue?
- Is implementation of policies making the situation worse/better for women/men?
- Which are the worst and best aspects of the organisation, before and after?
- What makes them good/bad and do women and men agree on these reasons?

In undertaking both these tasks the following questions should be asked for each policy topic:

- What is the gender make-up of the people affected by the policy?
- Will the policy affect women and men differently?
- Has previous work uncovered gender inequalities or barriers arising?
- What do women's and men's organisations have to say about the objectives and proposals?
- Do the recommended policies benefit or disadvantage women or men to a greater extent?
- Who ends up with the most resources, advantages and benefits from this policy?

In dealing with individual planning policies planners consider whether each policy manifests (adapted from National Assembly of Wales, NAW, 2001):

- gender innocence (no awareness of gender implications of a policy)
- gender awareness (gender issues have been highlighted through participation)
- gender understanding (survey, analysis, policy, data, CPD all contribute to this)
- gender competence (the policy meets all gender auditing requirements)
- gender excellence (positive feedback and a real change to daily life is manifest).

In seeking to develop robust planning policy should:

- collect gender specific statistics and indicators
- rethink priorities and identify fundamental gaps
- develop equality 'know-how'
- develop appropriate tools supported by practical training
- develop new partnership structures with the community
- monitor progress in achieving objectives
- engage people at all levels in the process to take on the role of change agents
- keep the momentum going year after year.

In the feedback to the piloting of the Toolkit, local planning departments told us they did not want a prescriptive, 'one size fits all' type of toolkit as all planning departments are different, dealing with a range of sorts of geographical and urban situations, and all at different stages in their policy-making cycles. Therefore they preferred such questions, which in the final Toolkit were built up in relation to examples which showed the different that mainstreaming gender might make.

Note

1 All the statistics in this paper are taken from ONS, 2003, and confirmed by updates from http://www.womenandequalityunit.gov.uk and see GSUG (2003) and the gender pages of the National Statistics Online service at http://www.statistics.gov.uk/CCI/nugget.asp?ID=441).

References

ACAS (2002) *Equality Direct Initiative*, London: ACAS and see http://www.equalitydirect.org.uk

Agrest, D., Conway, P. and Kanes-Weisman, L. (eds) (1996) *The Sex of Architecture*, New York: Harry Abrams Incorporated.

Bennett, C., Booth, C. and Yeandle, S. (2000) *Gender and Regeneration Project: Developing Tools For Regeneration Partnerships*, Sheffield: Centre for Regional Economic and Social Research, Sheffield Hallam University.

Bichard, J., Hanson, J. and Greed, C. (2004) *Access to the Built Environment: Barriers, Chains and Missing Links: Initial Review*, London: University College London.

Booth, C. (2001) 'Managing Diversity in the Planning Process', *Journal of the Planning Inspectorate*, Autumn 2001, Issue 23, Autumn 2001, pp. 9–13.

Booth, C., Darke, J. and Yeandle, S. (1996) *Changing Places: Women's Lives in the City*, London: Paul Chapman.

Brownill, S. (2001) 'Regen(d)eration: women and urban policy in Britain' in J. Darke, S. Ledwith, and R. Woods, (eds) *Women and the City: Visibility and Voice in Urban Space*, Oxford: Palgrave (Macmillan), pp. 114–130.

Cabinet Office (2002) 'Women and men in the UK: facts and figures', The Women's Unit, London.2004 figures available at http://www.womenandequalityunit.gov.uk

CIC (2000) *A Commitment to People, 'Our Biggest Asset': A Report and Associated Toolkits'*, Movement for Innovation Working Group on Respect for People in association with the EO Taskforce of the Construction Industry. London: CIC and DETR. See http://www.rethinkingconstruction.org and http://www. m4i.org.uk

Clarke, L., Michielsens, E., Pederson, E.F., Susman, B. and Wall, C. (2004) *Women in Construction*, Brussels: Centre for Construction Labour Research.

CRESR (2003) *Planning for Diversity: Research into Policies and Procedures*, Research Report produced for ODPM by CRESR Research Centre (Centre for Regional, Economic and Social Research), Sheffield: Sheffield Hallam University.

Darke, J., Ledwith, S. and Woods, R. (2000) *Women and the City: Visibility and Voice in Urban Space*, Oxford: Palgrave, pp. 114–130.

De Graft-Johnson, A., Manley, S. and Greed, C. (2003) *Why do Women Leave Architecture*, RIBA commissioned study, London: Royal Institution of British Architects. http://www.riba.org.uk

Dobbie, S. and Purcell, K. (2002) *Gender Mainstreaming and Policy Practice*, Bristol: ESRU (Employment Studies Research Unit) University of West of England and at http://www.uwe.ac.uk/bbs/esru

DTLR (2001) *Modernising Planning*, London: The Stationery Office (TSO).

DTLR (2002) *Planning Green Paper: Delivering a Fundamental Change*, London: TSO.

ECNI (Equality Commission for Northern Ireland) (2001) *Practical Guidance on Equality Impact Assessment*, Section 75 of NI Act 1988 Belfast. Available to download (pdf format) at http://www.equalityni.org/home.html

Edwards, J. (2004) *Mainstreaming Equality in Wales: The Case of the National Assembly Building*, Policy and Politics, Vol. 32, No. 1, 33–38.

EOC (1997) *Mainstreaming Gender Equality in Local Government*, London: EOC (Equal Opportunities Commission) available at: http://www.eoc.org.uk

EOC (2001) *Good Practice Guidance: How to Set Targets for Gender Equality*, London: EOC (Equal Opportunities Commission).

EOLG (Employers Organisation for Local Government) (2001) *The Equality Standard for Local Government in England*, EOLG, London, lg-employers. gov.uk/equal-pol.html

Evetts, J. (1996) *Gender and Career in Science and Engineering*, London: Taylor & Francis.

Fitzgerald, R. (1999) *Toolkit for Mainstreaming Equal Opportunities in the European Union for Structural Fund Programmes in Scotland: Plan Preparation and Implementation*, Glasgow: European Policies Research Centre, University of Strathclyde.

GLA (Greater London Authority) (2003a) *Gender Equality Scheme*, London: Greater London Authority and at http://www.london.gov.uk

GLA (2003b) *Local Transport Plans for London Boroughs*, London: Greater London Authority and at http://www.london.gov.uk

GLA (2003c) *Equality Impact Assessments (EQIAs): How to do them*, London: Greater London Authority and at http://www.london.gov.uk

GLA (2003d) *Into the Mainstream: Equalities with the Greater London Authority, Informative Publicity Leaflet for the General Public*, London: Greater London Authority.

Greed, C. (1991) *Surveying Sisters: Women in a Traditional Male Profession*, London: Routledge.

Greed, C. (1988) ' "Is more better?": with reference to the position of women chartered surveyors in Britain', *Women's Studies International Forum*, Pergamon Press, New York, Vol. 11, No. 3: 187–197.

Greed, C. (1994) *Women and Planning: Creating Gendered Realities*, London: Routledge.

Greed, C. (1999a) *The Changing Composition of the Construction Professions*, (Bristol, University of the West of England), Faculty of the Built Environment, Occasional Paper No. 5.

Greed, C. (1999b) 'Changing cultures' in Greed, C. (ed.) *Social Town Planning*, London: Routledge.

Greed, C. (ed.) (2003a) Report on *Gender Auditing and Mainstreaming: Incorporating Case Studies and Pilots*, Research Report edited by Clara Greed, with research contributions by Linda Davies, Caroline Brown and Stephanie Duhr (London, Royal Town Planning Institute) and at http://www.rtpi.org.uk

Greed, C. (ed.) (2003b) *The Rocky Path from Women and Planning to Gender Mainstreaming*, Clara Greed (ed.) with Linda Davies, Caroline Brown and Stephanie Dühr, Faculty Occasional Paper, Faculty of the Built Environment, Bristol, University of the West of England, Occasional Paper No. 14, May 2003, Bristol: University of the West of England.

GSUG (2003) (Gender Statistics Users' Group) *Newsletter*, London, ONS (Office of National Statistics) and at National Statistics Online at http://www.statistics.gov.uk/CCI/nugget.asp?ID=441

Hamilton, K. (2000) *Public Transport Audit London*, Commissioned study for DETR Mobility Unit undertaken by University of East London (London, DTLR) and at http://www.uel.ac.uk/womenandtransport

Ismail, A. (1998) *An Investigation of the Reasons for the Low Representation of Black and Ethnic Minority Professionals in Contracting*, unpublished special research project report, available at University of the West of England, Bristol.

Kandola, R. and Fullerton, J. (1994) *Managing the Mosaic; Diversity in Action*, London: Institute of Personnel Development.

Kirton, G. and Greene, A. (2003) *The Dynamics of Managing Diversity: A Critical Approach*, Oxford: Butterworth Heinemann.

Little, J. (1994) *Gender Planning and the Policy Process*, London: Elsevier.

Little, J. (2002) *Gender and Rural Policy: Identity, Sexuality and Power in the Countryside*, Harlow: Pearsons.

MBI (2001) *Delivering Racial Equality in Diverse Organisations: Post-Lawrence Agenda London*, London: Millennium Britons Institute, pp. 16–28.

Morley, L. (1994) 'Glass ceiling or iron cage: women in UK Academia', *Gender, Work and Organisation*, Vol. 1, No. 4: 194–204, October.

NAW (2001) *Promoting Equality of Opportunity*, Cardiff: National Assembly of Wales.

NDP (National Development Plan) (2000) *Mainstreaming Equality between Women and Men in Ireland*, Gender Equality Unit of the Department of Justice, Equality and Law Reform, Republic of Ireland.

ONS (Office of National Statistics) (2003) *Social Trends*, London: TSO.

Onuoha, C. and Greed, C. (2003) *Racial Discrimination in Local Planning Authority Development Control Procedures in London Boroughs*, Faculty of the Built Environment, Occasional Paper No. 15, Bristol: University of the West of England.

Plymouth (2001) *Gender Mainstreaming for Policy Makers*, Plymouth City Council in association with University of Plymouth, Department of Architecture (led by Mhaire Mc.Kie).

Reeves, D. (2002) 'Mainstreaming gender equality: an examination of the gender sensitivity of strategic planning in Great Britain', *Town Planning Review*, Vol. 73, No. 2: 197–214.

Reeves, D. and Greed, C. (2003) *Gender Equality and Plan Making: The Gender Mainstreaming Toolkit*, with contributors, Linda Davies, Caroline Brown and

Stephanie Dühr (London, RTPI). Final web version edited by C. Sheridan and D. Reeves for RTPI and at http://www.rtpi.org.uk

RICS (2003) *Raising the Ratio*, Investigation of composition of the surveying profession (Led by Louise Ellman and Sarah Sayce) London: University of Kingston on Thames) and see http://www.rics.org.uk

RTPI (2001) *A New Vision for Planning: Delivering Sustainable Communities, Settlements and Places*, London: RTPI.

Thomas, H. (1999) 'Social town planning and the planning profession' in Greed, C. (ed.) *Social Town Planning*, London: Routledge.

Turrell, P. (ed.) (2003) *Getting Women into Built Environment Careers*, Sheffield: Sheffield Hallam University in association with JIVE (Joint Intervention Partners) and funded by the ESF (European Social Fund), Brussels.

Watts, J. (2003) *Women in Civil Engineering: Continuity and Change*, unpublished PhD thesis, London, Middlesex University.

WDS (2001) *Removing the Goal Posts: Perspectives on Women in Regeneration*, London: Women's Design Service (project led by Sarah Clemens).

Williams, C.L. (1992) 'The glass escalator: hidden advantages for men in "female" professions', *Social Problems*, Vol. 39, No. 3: 253–268.

Women's Unit (2000) *Gender Mainstreaming for Policy Makers*, available on-line at http://www.womens-unit.gov.uk/genderimpact assesment/mainstrm.html

Web pages

Cabinet Office, Women and Equality Unit: http://www.cabinet-office.gov.uk/womens-unit/

CIC (Construction Industry Council) Diversity Toolkit: http://www. cic.org.uk and see http://www.rethinkingconstruction.org and http:// www.m4i.org.uk

DIALOG (Diversity in Action for Local Government) The new generic Equality Standard for Local Government, see e-mail: Dialog@lg-employers.gov.uk and http://www.lg-employers.gov.uk/dialog/equality.html

DTI construction statistics including some gender statistics: http://www.dti.gov.uk/construction/stats/stats2001/pdf/constat2001.pdf

EOLG (Employers Organisation for Local Government) see: http://www. lg-employers.gov.uk/equal-pol.html

Equality Commission for Northern Ireland: http://www.equalityni.org/ home.html

Equal Opportunities Commission: http://www.eoc.org.uk

Gender and urban planning in Europe http://www.generourban.com and generourban-admin@Listas.net managed by University of Madrid, School of Architecture.

Greater London Authority: http://www.london.gov.uk

Irish Government: http://www.irlgov.ie/justice/equality/gender/NDP/ Gender1.html

Northern Ireland: http://www.equalityni.org/home.html

Oxfam gender and development programme: http://www.oxfam.org.uk

Royal Institute of British Architects Report with recommendations at http://www.riba.org.uk based on research on 'Why are women leaving architecture?'

Royal Institution of Chartered Surveyors: http://www.rics.org.uk

Royal Town Planning Institute: http://www.rtpi.org.uk
Scottish Executive: http://www.scotland.gov.uk/government/devolution/meo-00.asp
UK Equality Direct Initiative: http://www.equalitydirect.org.uk
Welsh Assembly Equality Unit: http://www.assembly.wales.gov.uk
Women and Transport Toolkit: http://www.uel.ac.uk/womenandtransport
Womens Design Service: http://www.wds.org.uk
Womens National Commission: http://www.thewnc.org.uk
Women's Unit: http://www.womenandequalityunit.gov.uk

Chapter 8

Women in manual trades

*Linda Clarke, Elisabeth Michielsens and
Christine Wall*

Introduction

Although women occupy a prominent position in the British labour market, making up 44 per cent of the labour force, they are subject to both horizontal and vertical segregation. The most severely male-segregated sector is the construction industry (90 per cent male) according to *Census 2001* (National Statistics, 2003a). This chapter examines the reasons for gender segregation in this sector and the impact of policies to combat social and economic structural discrimination. It draws on a programme of research undertaken over the last decade, Women in Construction in Europe, including surveys of tradeswomen employed by local authority building departments or DLOs (Direct Labour Organisations).

The construction sector, public and private, employs nearly 1.6 million people and accounts for at least 7.2 per cent and up to 12 per cent of all employment (EC, 2002; DTI, 2003). It is highly segregated at all levels, although in certain professions there is some change through 'the qualification lever', as greater numbers of women study construction subjects in universities: 14.5 per cent of all UK civil engineering students and 27 per cent of architecture, building and planning students are female (HESA, 2002). For the past decade women have made up about 6 per cent of the total membership of construction professions (Court and Moralee, 1995; Greed, 2000), but at skilled operative level, gender segregation remains typical of a 'sex-typed' manual occupation, with women in 2002 making up less than 1 per cent of the manual workforce, a proportion that has even declined in the past decade (CITB, 2003). As in the professions, there are occupational variations in the trade, with consistently higher numbers in painting (CIB, 1996; CITB, 2002).

In this chapter the relative significance of different factors affecting gender segregation in the manual trades are explored. It begins with a review of theoretical interpretations of gender-segregated labour markets, followed by an account of the empirical context of our research and the methodology used, and an historical overview of the role of the state in the formation of gender-segregated labour markets. The main body of the chapter consists of an analysis of our research findings, covering the areas of equal opportunities (EO) policies, training and qualifications, and work and employment conditions. We conclude with recommendations for the removal of the structural obstacles preventing women from working in manual occupations alongside men.

Approaches to gender segregation

The position of women in the labour market has been variously explained through theories based on the premise of rational, economic decision making and individual choice such as neoclassical or preference theories (e.g. Anker, 1998; Hakim, 1998; Blackburn et al., 2002), more integrated approaches such as that of Reskin and Roos (1990), focusing on the dynamic of change and emphasising the structural determinants of segregation rather than the characteristics of female workers; labour market segmentation (e.g. Doeringer and Piore, 1971; Kenrick, 1981; Rubery and Fagan, 1995); and more complex approaches that include social, institutional and political factors such as the role of the unions and productive systems (e.g. Crompton and Sanderson, 1990; Crompton and Harris, 1998; Rubery et al., 2003). Segregation has also been shown to be inscribed within the structure of the wage relations, particularly through the persistence of the 'family' wage and the male breadwinner (e.g. Black et al., 1999; Bruegel, 2000; Picchio, 2000). One consistent factor to emerge is the importance of the relation between unpaid work in the home and the position of women in the labour market with respect to the kind of work undertaken and employment conditions, including hours of work (Crompton, 1997; Blackwell, 2001).

In our studies of women in construction in the United Kingdom our concern has been both to examine the relative importance of different factors in determining segregation and to understand how the situation can be changed. The research, whilst designed to examine gender exclusion and segregation specifically in construction, has resonance for other male-dominated sectors, for example software engineering and printing (Beck et al., 2003). Why has occupational segregation, though declining overall, remained persistent in these traditional areas? According to Reskin and Roos (1990), women make inroads in an occupation principally as a result of demand factors, for example shortage of qualified male workers, declining attractiveness of the sector to male workers or reorganisation of the sector towards more 'feminine' occupations. However, in the construction industry endemic skill shortages have not been a catalyst for employers to make a greater use of the female workforce, implying that demand explanations are insufficient (Corcoran-Nantes and Roberts, 1995; Clarke and Wall, 1998; Clarke et al., 2004). Indeed, as skill shortages have worsened, the proportion of women in the manual trades has even declined, from 1.7 per cent in 1992 to under 1 per cent in 2002 (CITB, 2003).

Explanations based on masculinity and organisational sexuality offer valuable insights into men's resistance to female entry into male-dominated occupations (Cockburn, 1991; Pringle and Winning, 1998). There is no avoiding the fact that society views construction as epitomising a particular version of 'masculinity'. The ways in which occupations become gendered and the social construction of gender have, however, been well investigated elsewhere (e.g. Cockburn, 1985; Connell, 1995). Our intention here is rather to show how training and recruitment, as well as labour relations, the wage system and employment conditions in construction represent important forces of exclusion. They act as factors of structural discrimination, pre- and post-employment, embedded in the historical development of industrial relations in the sector. Through confronting these structural factors at local level, inroads – however small – have been made.

The research context

Our empirical studies have focused on UK local authority repair and maintenance departments because of their prominent position, compared with the private sector, in terms of training, implementing EO policies, employing the workforce directly, accountability and availability of information. Three London boroughs investigated in detail were involved in a European Commission-funded project co-ordinated by Women and Manual Trades and involving partners in Spain and Denmark. The aim was to identify the further training needs of the tradeswomen employed and to give them an opportunity to meet women elsewhere and learn about each other's good practice. In-depth interviews, of up to two hours, were made with 46 of the 61 tradeswomen employed by the 3 building service departments or DLOs. These interviews – covering their training and qualifications, working environment, employment conditions and aspirations towards further and higher qualifications in the industry – provided an invaluable source of socio-economic data as well as a unique record of their individual experiences. The findings were used as a basis for the design of in-house training courses and meetings for the tradeswomen. Re-interviewing of largely the same group of women 18 months later, provided the opportunity to analyse changes in their position and the effectiveness of the extra training provided (Wall and Clarke, 1996; Michielsens et al., 1997, 2000; Clarke et al., 1999).

To situate these in-depth survey findings, a national survey of DLOs was conducted with the co-operation of ADLO (Association of Direct Labour Organisations), providing the first detailed national overview of their workforces, training policies, EO policies and employment conditions. A total of 93 questionnaires were returned, giving a 43 per cent response rate. The DLOs' sampled varied greatly in workforce size, from 6 to over 2,000, and 40 per cent of them employed female operatives, in most cases just 1 or 2 women; only 7 per cent had two or more women working, including one exceptional case where a total of 7 per cent of tradespersons were women (42 in absolute numbers). Half of those DLOs not employing women were part of District Councils, whilst 81 per cent of Metropolitan District Councils and 75 per cent of London Boroughs did employ women. Although the DLOs have been drastically reduced, they still accounted at the time of the survey for 25 per cent of directly employed operatives and 13 per cent of the repair and maintenance output of the construction industry, though this has subsequently been reduced to 6 per cent (DTI, 2003). The survey focused on the tradeswomen and men employed and covered: their occupations, qualifications, ethnicity and gender; training and career development; EO policies and special support; and employment conditions.

The workforce employed in the construction trades in our national DLO sample, a total of 26,652, was almost entirely male; only 0.9 per cent, or 231 were women. Statistics on women in construction are sparse and this poor figure is proof of the extremely male-dominated nature of construction work. Indeed, the overall proportion of women in construction in the DLOs is the same as that in general. Broken down according to job status, however, the relative position of women within the DLOs differs significantly. Women operatives constitute 0.6 per cent of all operatives. Women trainees/apprentices are a much higher proportion of all trainees, at 4.5 per cent. And 1.2 per cent of supervisors are female, a higher proportion than for operatives, indicating perhaps that women are often well suited to supervisory positions because of their training and higher levels of qualification. Segregation is, therefore, less at trainee than at operative or supervisory levels and less than in the private sector, where only 1.1 per cent of the surveyed apprentices in 2001 (aged 17–24) were female (CITB, 2001). If we look at the sum total of female construction trainees, including those training full time in Further Education (FE) colleges, the disparity between the proportion of trainees and of operatives who are female becomes even starker: 3 per cent of trainees are female compared with only 1 per cent of operatives, suggesting that women trainees are confronted with very significant barriers to entry and to obtaining work experience in the industry (CITB, 2003).

The proportion of trainee entrants also rises considerably for females over 19 years old, to 8 per cent. The implication is that adult entry is particularly appropriate for women, who as 16-year olds may be unaware of opportunities or may find the minority position too daunting. This was clear from our interviews, especially among the older women (over 30 years):

> It's something I wanted to do over a long while. I tried everywhere but no-one knew anything. I spent 2 or 3 years looking for a traineeship in painting and decorating. I wish I'd done it when I was younger.
>
> (trainee painter, age 39)

> There was no guidance at school – it was either office work or nursing.
>
> (gas fitter, age 38)

I wanted to be a motor mechanic, but my Dad laughed – I wasn't allowed to stay on at school; I had to leave and work in a bank. This is what I wanted – I would have jumped at this given the chance.

(supervisor and former plumber, age 39)

The state and gendered labour markets

Historically, it could be argued that the mechanisms maintaining the extreme gender segregation in what remain 'craft' trades pre-date industrial capitalism and follow a course dating from the feudal development of the apprenticeship system and waged labour. Women have perhaps been successfully excluded from construction for so long precisely because wage labour originated so early, the craft tradition remains strong, and this is one of the few sectors with a statutory training board maintaining apprenticeships (Clarke, 1999). In the twentieth century women were only actively encouraged to learn construction skills during the two world wars, substantiating research theories of the reserve army of labour (Siltanen, 1994). During the First World War, through government policy, skilled workers were substituted by women, who were publicly acclaimed for successfully and rapidly acquiring male skills and being more productive workers. At the end of the war, these women were quickly ousted from their jobs in an expeditious reversal of policies (Briar, 1997). Government strategy for the recruitment of women during the Second World War was careful to maintain a gendered hierarchical division of labour. Men were employed in key positions such as foremen and women either placed in semi-skilled jobs or, where they did gain access to skilled occupations, warned that their skills would not be needed for post-war reconstruction work.

The collusion of the trade unions with both the employers and the state in the exclusion of women from skilled work, particularly in the second post-war period in an attempt to provide 'full (male) employment' was a major factor in the reassertion of occupational segregation in construction (Boston, 1987). With the reversion to traditional training, wage and industrial relations systems, training and employment for women were severely restricted, being confined to industries deemed suitable by the government, which, despite an acute scarcity of building labour, did not include construction (Phelps Brown, 1968; Wilson, 1980).

Only with the implementation of the Sex Discrimination Act in 1976 were women given access to state-run training schemes in construction trades under the Training Opportunities Programme (TOPs), aimed at unemployed adults. Women were not actively encouraged onto these schemes, but nevertheless, quite inadvertently, the state had provided a means to acquire craft skills independently of the patronage of an employer. The timing coincided with the height of the second wave of British feminism and, although the actual numbers of women succeeding in training in non-traditional areas were extremely small, they provided valuable role models (Payne, 1991). Subsequent programmes superseding TOPs – Employment Training and Training for Work – involved employers in training provision, a factor that, together with a restrictive funding arrangement routed through the Training and Enterprise Councils (TECs), reinforced patterns of inequality for women in non-traditional occupations (Felstead, 1995).

In general, national equality legislation has been negligible in changing the nature of occupational segregation in construction, partly due to the notion of equality implied. This has been based on providing opportunities for women in what remains an unequal labour market (Michielsens *et al.*, 2000). Government policy has tended to be advisory, market-led and individualistic rather than enforcing and collective, with the result that the implementation of EO initiatives has been extremely backward, especially in the private sector. Since the United Kingdom's adoption of the Maastricht Social Charter (1998) and the Amsterdam Treaty (1999) with positive action at its core, minimal collective equality and employment policies have been enacted by the Labour government, but have made little significant difference to EOs in construction in the United Kingdom.

Equal opportunities policies at local level

If state intervention has been limited and relatively ineffective at national level apart from during the wars, the local level shows what can be achieved – albeit only in pockets – if active EO policies are pursued. EO policies were pioneered by the urban metropolitan councils in the early 1980s and by the late 1980s the effects were obvious in Inner London DLOs: in Haringey alone over 100 tradeswomen were employed and a total of 266 in just 7 boroughs.

The urban legacy of EO policies is reflected in our DLO survey. The majority of urban authorities had measures in place to encourage women to train or seek employment. Many, though considering themselves 'Equal Opportunity' employers and perhaps even having a women's unit, were very passive in their approach and targeted women in much the same way as they would men. Such authorities were also found to be the least successful in recruiting women into the construction workforce. In contrast, those succeeding took more active measures such as:

- targeting recruitment (including through positive images and a range of outlets for advertising) and having women in the recruitment team. For example, as described by one DLO manager:

  When we recruit trainees, an advertisement plus a radio advert is produced stating that we are keen to train new people – with applications particularly welcome from women and ethnic minorities.

- providing 'Taster' days, work experience for trainees, information on entry, guidelines on harassment and working alone, flexibility in working arrangements and support for pregnant women and childcare;
- establishing links between the women's unit and the DLO and with women-only training workshops;
- running training programmes aimed at women;
- facilitating women-only meetings with elected women representatives. One DLO, for example, gave pre-employment training and close support through a craftswomen's group. Another described how:

> Proactive monitoring of employees' daily experience in a male-dominated environment is conducted by female personnel officers outside the management chain of command.

All such support takes into account the fact that women are entering a male-dominated workforce and undertaking work traditionally seen as a man's. It suggests that, where EO policies are systematically and very proactively implemented and positive action measures taken, then success is much greater. However, the fact that results remain relatively poor indicates that the typical instruments of EO policy leave untouched key societal determinants of gender segregation.

Though the implementation of EO policies in the 1980s had a positive effect on the training and employment of women in the DLOs, by the early 1990s numbers had plummeted. This can be attributed to other factors, in particular to the demise of new local authority housebuilding and the introduction of compulsory competitive tendering, which meant drastic reductions in the DLO workforces (often on a last in-first out principle, disproportionately affecting the women) and the eradication of new-build training schemes – the best environment for work experience – and of adult traineeships. By 1994 over two-thirds of the 266 women employed by 7 councils in 1989 had disappeared. This well illustrates the devastating effects, particularly for women, of government deregulation (Escott and Whitfield, 1995).

Training and qualifications

Women's entry into construction is especially sensitive to any changes or weakness in the training regime. Our findings show a clear link between the level of training and the proportion of women employed: the average proportion of trainees in the DLO workforce was 6.5 per cent but, for those DLOs employing women, it was slightly higher at 7.8 per cent and for the top 7 in terms of female employment it was very significantly higher at 11.7 per cent. Male trainees and operatives dominate overwhelmingly in each of the trades, with the proportion of females in most cases under 7 per cent for trainees and under 1 per cent for operatives. Only with painting and decorating are higher proportions to be observed, at 24 per cent for trainees, constituting over half of all female new entrants in construction, and

1.6 per cent for operatives. This higher share of women in the decorative trades is also found in the private sector, with women making up 0.8 per cent of employees, followed by wood trades at 0.6 per cent (CITB, 2002). Throughout much of Europe, painting and decorating emerges as the most popular trade for women, employing as many as 27 per cent of women operatives and 40 per cent of trainees in Denmark and 6 per cent of operatives in Germany (Beck *et al.*, 2003). As a trade painting is confronted with similar dangers as the traditional trades, including dangerous chemicals in use, working at heights and the hard, physical work involved (CLR News, 1997; Clarke *et al.*, 1999). One explanation for its accessibility to women is a weaker craft tradition than carpentry and other traditional trades and often lower earnings, as well as a greater familiarity in the domestic environment.

A key difference in the characteristics of the male and female DLO workforces relates to the level of qualifications. In comparison with their male colleagues, women were on average found to be more qualified: while most of the tradesmen, about 60 per cent, had a qualification to at least craft level (NVQ2 or City and Guilds Craft) and 37 per cent were qualified to advanced level or above, *all* tradeswomen were qualified to craft level and most (65 per cent) also to advanced level (NVQ3 or City and Guilds Advanced) or above. It was found that 53 per cent of local authorities employed men with no trade qualifications at all, confirming the supposition that, whereas for men informal means of entry still remain important, aided by male social networks and the possibility of informally picking up construction skills, for a woman being able to prove her ability through her qualifications is the only way to get a job. Women are largely excluded from the informal network that acts as a 'gate-keeper' giving men inside information about job openings, promotion and other opportunities (Reskin and Padavic, 1988).

Just as initial training, usually to an advanced level, is essential for women to prove their ability, so also are they reliant on further training for advancing their careers. Thus it is no accident that those authorities with the greater variety and number of further training courses – ranging from trades-related multi-skilling, management, trades training, health and safety, information technology and EO courses – are also those employing women. Similarly, those authorities employing women scored highest in terms of providing incentives to career development – including career guidance, appraisals, college day release, fee payment and flexibility in choosing courses.

Case study

The success of Leicester City Council's building department

Thirty per cent of apprentices appointed by Leicester City Council's building or DLO between 1985 and 2002 have been women. Today, 1 in 12 of the

480-strong workforce are women, employed in all trades, including as carpenters, electricians, plasterers, painters and decorators, bricklayers, heating and ventilating engineers, gasfitters and metalworkers. Private firms in the area complain that they cannot attract women or even apprentices generally into the industry. So what is the secret of the Leicester DLO's success?

One reason for success is training, with many women coming through the apprenticeship programme, which has over 50 apprentices at any one time and is vastly oversubscribed. Many apprentices are adults and receive a good trainee allowance over their three-year apprenticeship programme. This consists of college-based training, training in the DLO's own centre and work experience up to NVQ Level 3. It proves that where good training is in place there is no lack of applicants.

Another reason for the DLO's success is its careful recruitment and selection procedures. The DLO goes out of its way to attract women, working closely with schools, giving girls work experience and encouragement and running one-week 'taster' courses annually, attended in 2001 by 64 women and involving one day in different trades. Advertising shows female and ethnic minority construction workers from different trades and is targeted to particular audiences, such as women's centres. All new recruits into the skilled workforce have to have served an apprenticeship and worked in their trade and the interviewing panel always involves one women and one person from an ethnic minority. And a final reason for success is the employment provisions in place. These include stable working hours, good maternity leave and a wide range of support groups, such as a tradeswomen's group and a gay, lesbian and bisexual group. There is ample opportunity for further training and promotion and a retrain and re-entry scheme. All tradeswomen are in the trade union and two are shop stewards. (Based on an interview with Leicester DLO by Linda Clarke and Barbara Susman (Housing Forum, 2004 Housing Skills: Approaches to The Current Challenges, Constructing Excellence).)

Employment and working conditions

The first and major obstacle for women, once having been trained, is to obtain employment. Many never achieve this; others may disappear into the alternative economy that flourishes in Britain, particularly for small maintenance work for homeowners, thereby escaping official statistics and perhaps explaining the large gap between the number of women trained in construction and those actually employed.

In response to a question on what tradeswomen in the three London DLOs liked best about working in construction, typical replies were:

I get immediate satisfaction and gratification from fixing something that's broken.

It's rewarding – the flats look better after they've been decorated.

They frequently described their jobs positively in comparison with more traditional women's occupations, for example:

> I like the work that we do. I don't like to be sitting down all the time....I couldn't sit in an office all day and pick up the phone.

They were well aware of the restrictions under which women work 'indoors' and cited the freedom and autonomy of skilled construction work as an extremely positive aspect of their working conditions:

> The freedom and the variety of jobs.

> Freedom – no manager breathing down my neck.

> Out and about – not stuck in one place.

> I plan my own work, I'm independent.

'Earning a man's wage' was also an important incentive for women to enter construction. Nevertheless, there has been a significant widening in earnings differentials nationally between white-collar and manual workers within construction and between construction and other industries. In spite of the relative decline in male manual construction earnings, there remains a large difference between female and male skilled trades (all sectors) workers' earnings: respectively £287.3 and £420.5 per week in 2003 (National Statistics, 2003b). For construction, average hourly pay rates in 2003 were £11.17 per hour compared with £9.61 for women (National Statistics, 2004). Significantly, our research indicated there was usually parity in the wages of the men and women working in the DLOs.

For tradeswomen, the public sector offers more favourable employment conditions than the private sector. Whereas for men the opportunity for self-employment within the private sector is a valid option, it is far less so for women. Not only does the private labour market rely on male networks for the supply of labour, even within the alternative economy, but the private sector's record on EO employment is very poor. The attraction for women of work in male-dominated occupations in the public sector is also found in other research, for example Hansen's study in Scandinavia, which concluded that 'especially attractive for women is the male-dominated part of the public sector, where women may obtain large earnings and there is no punishment for caring responsibilities' (Hansen, 1997).

The majority of tradeswomen interviewed in the three London boroughs lived in or near the borough where they were employed. Women, once in employment, were found to be a particularly stable group in the workforce: 37 per cent had already been working in construction for over ten years and 63 per cent anticipated remaining in construction-related occupations for another ten years. Another incentive for women seeking employment in the public sector is the regulated hours compared with private sector construction firms. Nationally, most DLOs (73 per cent of the sample) have a working week of 39 hours; 7 per cent had a longer week and these were unlikely to employ women. Indeed, the women working in the DLO cited the fact that the job was local, with regulated hours and a fixed wage (as opposed to piecework or bonus systems), as its main attractions, reducing uneven competition between workers and advantageous for their work/family arrangements. The implication is that stable, localised and time-regulated employment is more conducive to women's inclusion than the extreme flexibility of employment, lack of continuity and long working hours found in much of construction, especially in the private sector.

The importance of good employment and working conditions to women also suggests a potentially significant role for trade unions. There is much evidence that unions are an influential factor in the position of both men and women in the labour market (Lane, 1993; Anker, 1998). A very high proportion, 81 per cent, of tradeswomen interviewed in the three London boroughs were union members; non-union members were mostly all trainees on an adult training scheme. In one borough, the women's representatives were also union representatives and played an important role in workplace policies on health and safety issues.

Case study

Jacky, carpenter

I've been a carpenter working in London for the last 18 years, from the age of 25 to 43. I think my interest in doing something manual or physical started young. I was aware that boys were allowed to do things that girls weren't and that this was unfair, and after the Sex Discrimination Act of 1975 three friends and I took the opportunity to do metalwork at school. I've always liked physical activities and I played a lot of sports at school.

When I left school I went to Liverpool Polytechnic, where I studied social sciences. Then I moved to London. I didn't want to be a social worker or a teacher; I was 22 and it was a relief to get out of full time education. In the 1980s, I found it an exciting time to be in London and I worked for a while in the first women's bookshop in the United Kingdom and at the same time I was going to a carpentry evening class. After some advice from a worker in a Job Centre I went on a ten-week introductory course, 'Women into manual skills', at a Skill Centre. I was very committed and got up at the crack of dawn to get across London to be there at 8 a.m. After the introductory course I did a six-month TOPs course in carpentry and joinery at the Skill Centre. I absolutely loved it; there was a great group of people, both men and women, and I'm still in touch with some of them.

My first job as a carpenter was in Battersea in a small workshop with a man making sash windows. I didn't turn out to be the person he was looking for; I was new to the job and he wanted someone who would have enough experience to run his business in his absence, which at that stage I didn't have.

My next job was with a small builder and I worked there for one and a half years. They mainly did renovations and work for housing associations. I was the only woman and, although they were curious at first about why I was there, I soon made friends. I learned a lot, particularly from a foreman, George, a perfectionist, who was very positive and encouraging about my work. He made sure I knew how to do the work. For instance, when we had a contract to hang new front doors in every flat in a tower block, he said that he'd seen smaller people than me do this work and that there was a knack to lifting heavy doors without straining, and he showed me how. During this time I continued my training at Hackney College, where I got my City and Guilds qualifications and then my advanced City and Guilds and then left to work for Hackney Council as an improver.

Since I started working there 16 years ago, I've seen enormous changes in the type of work we undertake and in the conditions of work. My first job at Hackney was building a new council housing estate. In those days Hackney had a big training programme and I learned a lot from some older men who were happy to share their knowledge, even if it meant the work went more slowly. Today we do only maintenance work and there has been a massive reduction in the workforce from 1,000 directly employed people to 260 operatives and from 50 women to 7, which includes 2 women adult trainees who have just joined.

An important aspect of my work at Hackney has been my involvement in the trade union. In my first job in Battersea I knew I wanted to join the trade union. I became a TGWU (Transport and General Workers' Union) women's representative and, after a couple of years, a shop steward and have been one ever since. I have spent a couple of years in the local Hackney building union office working full time as the deputy convenor. For the past four years I've become more involved with the national union; I'm on the TGWU regional women's committee and on the building and construction trade group.

I now work in the Rapid Response team and the disabled adaptation team. I enjoy the satisfaction of making people's homes more accessible. We are given our job tickets in the morning which specify the jobs and the time allowed for them. Our pay is determined by the so-called 'Fair Pay' scheme, which was brought in about four years ago, whereby basic wages were reduced and a new bonus scheme introduced. Some people have made this work for them, but I'm not one of them. I'm earning less than I was 10 years ago and I now consider myself fairly low paid in relative terms. The amount you earn does vary and some months you can earn a lot of money, but it depends on the types of jobs you get given. An output-based scheme like this one doesn't work when you are doing small maintenance jobs, as I am, in different venues. If I was fixing handrails all day on a site I could make money. The 'Fair Pay' scheme meant loads of people left, including some of the few remaining women. Quite a lot of the women have gone on to other jobs, such as supervisory work within the department, or have moved into technical jobs such as surveying and clerk of works. One former colleague is teaching bricklaying at Hackney College.

I've chosen to stay. I enjoy working in people's homes and I still get pleasure from the work. Yesterday afternoon one of our jobs was hanging a new gate with an adult trainee; it was a warm November afternoon, we were outside and it was magic. I really like doing the trade union work as well, as fundamentally I want to change conditions for all workers. My ambitions are not about getting ahead. As a woman I feel very confident, particularly now that I'm older, about being a carpenter and in my skills. What I'd like to say to any woman who wants to work in the building industry is to remember that you can do whatever you want and that we are capable of learning anything we choose, we can have everything. (Based on an interview by Barbara Susman.)

Societal context

The majority of women interviewed were aware that they were breaking new ground and many emphasised the important role of families and friendship networks in providing continuous support.

> Both [parents] are over the moon and want me to stick at it. My Dad is really good – he works in the borough.
>
> (apprentice)

> At first I didn't want to tell anyone – now everyone knows and asks me for a job. I would recommend it to other young women – I talk about work with my girl friends – some of them are in offices or not going anywhere in shops.
>
> (painter)

Attributes commonly associated with heavily sex-typed occupations, such as manual dexterity and physical strength, are closely linked to societal notions of 'masculinity'. However, when tradeswomen in the three London boroughs were asked if they considered their size any impediment, an overwhelming 74 per cent replied in the negative, only 9 per cent considered it was an impediment and 17 per cent were of the opinion that it was 'sometimes'. One remarked: 'I am the perfect size for my job'. Similar responses might be anticipated from male colleagues.

A number of studies of women in non-traditional occupations have focused on workplace relationships, in particular harassment and the dynamics of exclusionary group male behaviour (e.g. DEET, 1993; Jones, 1995). Our research showed the majority (96 per cent) to be 'reasonably happy' in the relationships with the men they worked with. Fifty-two per cent reported they had in the past been subject to harassment (a third of which related to the behaviour of tenants, not co-workers); the majority had complained and been satisfied with the outcome. They had reached an equilibrium in their workplace relations and their immediate concerns related to career advancement and worries about redundancies. This echoes the findings of Padavic and Reskin (1990) that relationships with male co-workers are less a barrier to women's interest in 'blue-collar' jobs than is usually presumed.

Despite the general shift in the labour market away from craft and manual occupations, the arguments for women manual workers gaining construction skills now are as strong as they were in the 1970s, especially within the public sector where women make up the majority of all manual workers. As one woman expressed it:

> I'll stay as long as the council has a job for me. It would be hard to keep these wages on the outside. It's about confidence as well – you'd have to face all those blokes again and prove yourself. I'm lucky here with all the policies on equal opportunities.

Conclusions

To change the position of women requires policy regulation. However, the very lack of regulation and abdication to employers' prerogative in Britain helps to perpetuate and reinforce the male-segregated nature of construction. Women's entry depends on the accidental good will of individual employers rather than on any real and concerted social change. Important aspects of the limited success in women entering skilled construction occupations are the financial reward in comparison with female-dominated manual jobs, stable working time and short travelling time. In such strongly sex-typed occupations it is, however, necessary to provide additional leverage to integrate women, particularly proactive recruitment and special support. None of these aspects usually applies in the private sector, so the public sector DLOs are especially accommodating to women; in this sense they represent a special case.

Another critical factor identified in our studies is training. It is no accident that DLOs with a good training record are also those with higher proportions of women. Not only are women totally dependent on training for entry and progression, but authorities with good training are inevitably those prepared to invest more in their workforce. A question for further research is how far the dependence of women on proving their ability through qualifications implies a different and more explicit notion of skill and recognition of this in the wage structure. Rather than implicit skills associated with the crafts, passed on through 'learning on the job', the skills of the tradeswoman are visible and defined, as required with formal training. In the same way recruitment of tradeswomen relies on regulated and transparent procedures, as opposed to the intangible practices attached to informal networks and apprenticeships.

Our research has concentrated on the socio-economic factors that enable women to be successful as skilled tradespeople. The findings indicate that horizontal occupational segregation is not constant and immovable. Tradeswomen were found where effective overall EO policies in both recruitment and employment were combined with good employment conditions, training schemes, career development and support mechanisms. We conclude, therefore, that the excessive male domination of the construction trades can be changed through a focus on these structural factors; just as high numbers of women have entered in the past, so they can again.

Discussion questions

1   Does apprenticeship act as a barrier to women's entry into the skilled construction trades and, if so, can this be overcome?
2   What constitutes good practice in the inclusion of women in construction at manual level?
3   How would you explain the desperately low proportion of women in construction at operative level?
4   Why do young women not consider a career in the construction industry in favour of other industries?

Acknowledgements

We are grateful to the organisation Women and the Manual Trades (WAMT), as coordinator of the European NOW project on women in construction, for supporting these surveys, as well as to all those within and associated with WAMT for their invaluable discussion and enthusiastic commitment. We also acknowledge the support of ADLO (Association of Direct Labour Organisations) in carrying out the national survey of DLOs.

Note

1 SOC2000 category – Skilled construction and building trades.

References

Anker, R. (1998) *Gender and Jobs – Sex Segregation of Occupations in the World*, Geneva: ILO.

Beck, V., Clarke, L. and Michielsens, E. (eds) (2003) *Overcoming Marginalisation: Structural Obstacles and Openings to Integration in Strongly Segregated Sectors*, Final Report submitted to European Commission, London: University of Westminster.

Black, B., Trainor, M. and Spencer, J.E. (1999) 'Wage protection systems, segregation and gender pay inequalities: West Germany, the Netherlands and Great Britain', *Cambridge Journal of Economics*, 23: 4, 449–464.

Blackburn, R.M., Browne, J., Brooks, B. and Jarman, J. (2002) 'Explaining gender segregation', *British Journal of Sociology*, 53: 4, 513–536.

Blackwell, Louisa (2001) 'Occupational sex segregation and part-time work in Modern Britain', *Gender, Work and Organization*, 8: 2, 146–163.

Boston, S. (1987) *Women Workers and the Trade Unions*, London: Lawrence and Wishart.

Briar, C. (1997) *Working for Women? Gendered Work and Welfare Policies in Twentieth Century Britain*, London: UCL Press Ltd.

Bruegel, I. (2000) 'The restructuring of the family wage system, wage relations and gender relations in Britain', in Clarke, L., de Gijsel, P. and Janssen, J. (eds) *The Dynamics of Wage Relations in the New Europe*, New York: Kluwer 214–227.

CIB (Construction Industry Board) (1996) *Tomorrow's Team: Women and Men in Construction*, Working Group 8, London.

CITB (Construction Industry Training Board) (2001) *Construction Apprentices Survey 2000 National Report*, London.

CITB (Construction Industry Training Board) (2002) *Survey of Employment by Occupation in the Construction Industry 2001*, London.

CITB (Construction Industry Training Board) (2003) *CITB Skills Foresight Report*, London.

Clarke, L. (1999) 'The changing structure and significance of apprenticeship', in Ainley, P. and Rainbird, H. (eds) *Apprenticeship: Towards a New Paradigm of Learning*, London: Kogan Page, 25–40.

Clarke, L. and Wall, C. (1998) *A Blueprint for Change: Construction Skills Training in Britain*, Bristol: The Policy Press.

Clarke, L., Pedersen, E. and Wall, C. (1999) 'Balancing acts: women in construction: a study of women painters in Denmark and Britain', *Nordic Journal of Women's Studies (NORA)*, 2–3 November, 7: 138–150.

Clarke, L., Pedersen, E., Michielsens, E., Susman, B. and Wall, C. (2004) *Women in Construction*, Brussels: CLR Studies/Reed International.

CLR – European Institute for Construction Labour Research (1997) *CLR News 3*, on Women in Construction, Brussels.

Cockburn, C. (1985) *Machinery of Dominance: Women, Men and the Technical Know-How*, London: Pluto Press.

Cockburn, C. (1991) *In the Way of Women – Men's Resistance to Sex Equality in Organisations*, London: Macmillan Press Ltd.

Connell, R.W. (1995) *Masculinities*, London: Policy Press.

Corcoran-Nantes, Y. and Roberts, K. (1995) 'We've got one of those: the peripheral status of women in male dominated industries', *Gender, Work and Organization*, 2: 1, 21–33.

Court, G. and Moralee, J. (1995) *Balancing the Building Team: Gender Issues in the Building Professions*, Brighton: Institute for Employment Studies, University of Sussex.

Crompton, R. (1997) *Women and Work in Modern Britain*, Oxford: Oxford University Press.

Crompton, R. and Harris, F. (1998) 'Explaining women's employment patterns: "orientation to work revisited" ', *British Journal of Sociology*, 49: 1, 118–136.

Crompton, R. and Sanderson, K. (1990) *Gendered Jobs and Social Change*, London: Unwin Hyman.

DEET (Department of Employment, Education and Training) (1993) *Women in Building – The Missing 51%*, Canberra: Australian Government Publishing Service.

Doeringer, P. and Piore, M. (1971) *Internal Labor Markets and Manpower Analysis*, Lexington, Massachusetts, MA: D.C. Heath & Co.

DTI (Department of Trade and Industry) (2003) *Construction Statistics Annual*, London: National Statistics.

Escott, K. and Whitfield, D. (1995) *The Gender Impact of CCT in Local Government*, Manchester: Equal Opportunities Commission.

European Commission (EC) (2002) *European Social Statistics: Labour Force Survey Results 2002*, Eurostats.

Felstead, A. (1995) 'The gender implications of creating a training market: alleviating or reinforcing inequality or access', in Humphries, J. and Rubery, J. (eds) *The Economics of Equal Opportunities*, Manchester: Equal Opportunities Commission, 177–201.

Greed, C. (2000) 'Women in the construction professions: achieving critical mass', *Gender, Work and Organization*, 7: 3, 181–196.

Hakim, C. (1998) 'Developing a sociology for the twenty-first century: preference theory', *British Journal of Sociology*, 49: 1, 137–143.

Hansen, M. (1997) 'The Scandinavian welfare state model: the impact of the public sector on segregation and gender equality', *Work, Employment and Society*, 11: 1, 83–99.

HESA (Higher Education Statistics Agency) (2002) *Student Tables by Subject of Study 2001/2*, London: National Statistics.

Jones, A. (1995) *Hammering it Home: A Report on Women's Experiences in Manual Craft Trades in the Construction Industry in Sheffield*, Sheffield: Sheffield City Council.

Kenrick, J. (1981) 'Politics and the construction of women as second class workers', in Wilkinson, F. (ed.) *The Dynamics of Labour Market Segmentation*, London: Academic Press, 167–191.

Lane, C. (1993) 'Gender and the labour market in Europe – Britain, Germany and France compared', *Sociological Review*, 4: 2 (May), 274–301.

Michielsens, E., Wall, C. and Clarke, L. (1997) *A Fair Day's Work – Women in the Direct Labour Organisations*, London: London Women and Manual Trades and Manchester: Association of Direct Labour Organisations.

Michielsens, E., Clarke, L. and Wall, C. (2000) 'Diverse equality in Europe', in Noon, M. and Ogbonna, E. (eds) *Equality and Diversity in Employment*, Basingstoke: Palgrave, 118–135.

National Statistics (2003a) *Census 2001*, London.

National Statistics (2003b) *New Earnings Survey 2003: Analysis by Occupation*, London.

National Statistics (2004) *Labour Market Trends*, London.

Padavic, I. and Reskin, B. (1990) 'Men's behaviour and women's interest in blue collar jobs', *Social Problems*, 37: 4, 613–628.

Payne, J. (1991) *Women, Training and the Skills Shortage: The Case for Public Investment*, London: Policy Studies Institute.

Phelps Brown, S.H. (1968) *Report of the Committee of Inquiry under Professor E.H. Phelps Brown into Certain Matters concerning Labour in Building and Civil Engineering*, London: HMSO.

Picchio, A. (2000) 'Wages as a result of socially embedded relationships between production and reproduction', in Clarke, L., de Gijsel, P. and Janssen, J. (eds) *The Dynamics of Wage Relations in the New Europe*, New York: Kluwer, 195–213.

Pringle, R. and Winning, A. (1998) 'Building strategies: equal opportunity in the construction industry', *Gender, Work and Organization*, 5: 4, 220–229.

Reskin, B. and Padavic, I. (1988) 'Supervisors as gatekeepers: male supervisors' response to women's integration in plant jobs', *Social Problems*, 35: 5, 536–550.

Reskin, B. and Roos, P. (1990) *Job Queues, Gender Queues – Explaining Women's Inroads into Male Occupations*, Philadelphia, PA: Temple University Press.

Rubery, J. and Fagan, C. (1995) 'Gender segregation in societal context', *Work, Employment and Society*, 9: 2, 213–240.

Rubery, J., Humphries, J., Fagan, C., Grimshaw, D. and Smith, M. (2003) 'Equal opportunities as a productive factor', in Burchell, B., Deakin, S., Michie, J. and Rubery, J. (eds) *Systems of Production: Markets, organisations and performance*, London: Routledge, 236–262.

Siltanen, J. (1994) *Locating Gender – Occupational Segregation, Wages and Domestic Responsibilities*, London: UCL Press.

Wall, C. and Clarke, L. (1996) *Staying Power: Women in Direct Labour Building Teams*, London: London Women and Manual Trades.

Wilson, E. (1980) *Only Half Way to Paradise: Women in Post-war Britain*, London: Tavistock.

Chapter 9

Promoting diversity in construction by supporting employees' work–life balance

Helen Lingard and Valerie Francis

Introduction

The construction industry's poor performance in the areas of equal opportunity and diversity is well documented, as evidenced by the research findings documented throughout this book. Women and ethnic minorities are seriously under-represented among the industry's workforce (Gale, 1994; Fielden *et al.*, 2001; Australian Bureau of Statistics, 2003). The lack of diversity is likely to be due, in part, to the construction industry's image as a white, 'masculine' work environment, characterized by chaotic and dangerous work practices and high levels of inter-personal conflict. However, there is also evidence that exclusive and discriminatory employment practices prevail. For example, those women who do enter the industry experience barriers to career success (Dainty *et al.*, 1999). Loosemore and Chau (2002) report that racial discrimination is rife.

In this chapter, based predominantly on Australian research studies, we suggest that the inflexible employment practices adopted by construction firms do not meet the needs of today's workforce and that these practices act against equality of opportunity and stifle diversity. We identify some dramatic changes to the demographic profile of the workforce, to family structures and to social expectations that have occurred in most developed countries since the 1950s. We suggest that traditional management practices, which were based upon the notion of a homogeneous workforce, comprising of males of European ancestry married to full-time home-makers, who would take on the sole responsibility for child care and domestic duties, are no longer relevant. Instead, we argue that employment practices must be flexible to accommodate the needs of employees in a wide variety of 'non-traditional' family structures and with differing values about work and non-work life.

We argue that the provision of a work environment that is more supportive of employees' work–life balance would improve the construction industry's image and attract employees from currently under-represented groups. However, we also suggest that the inflexible employment arrangements adopted by construction organizations are discriminatory, serving to exclude or disadvantage some groups of employees. We propose that the introduction of flexible work arrangements designed to help employees to better balance their work and non-work lives is an essential step towards eliminating this discrimination.

In order to achieve this end, we identify organizational work–life balance policies that construction organizations could implement and discuss how employees' needs are likely to differ according to their life-stage. However, we emphasize that the introduction of 'paper' policies is insufficient because benefits may not be accepted as part of the construction workplace culture. Moreover, we argue that the use of such policies must be seen as legitimate and available to all employees, irrespective of sex, age or race. Furthermore, we emphasize that employees who utilize work–life policies should not be subject to informal or institutional sanction. We conclude that work–life balance policies will only promote diversity in construction, if they are accompanied by cultural change within construction organizations.

Social changes

Changing attitudes about women's roles

The roles and expectations of women and men have changed significantly over the past 60 years. More women than ever before are in the workforce reflecting rising educational levels, changing societal attitudes and declining birth rates. The Australian Bureau of Statistics (ABS) report that women's overall participation in the paid work force in Australia has risen from 43.7 per cent in 1978 to 55.5 per cent in 2004 (ABS, 2004a). While women are still highly under-represented in non-traditional industries, such as construction and engineering, their participation rates in Australia have risen. For example, the proportion of female engineers rose from 0.5 per cent in 1985 to 6.6 per cent in 2000 and women now represent about 15 per cent of the undergraduate engineering cohort (Yates et al., 2001). However this growth has substantially flattened since the mid-1990s to an annual growth rate of 0.5 per cent (Greenwood, 2002). Currently women comprise 7.9 per cent of all managers and professionals in the Australian construction industry (ABS, 2003).

Whilst we have seen wide acceptance that childless couples both work, attitudinal and institutional barriers to women's employment resurface upon child-bearing (Bourke, 2000). Professional women tend to have fewer children than non-professional women (1.6 compared to 1.8 for all mothers) and also bear children at an older age (30.6 years compared to 27.6 years for all mothers) (various sources cited in Bourke, 2000). It is possible that professional couples are forgoing or delaying child-bearing in order to maintain their careers. As a consequence, the number of dual income couples overall has increased. In 57.5 per cent of Australian two-parent families (with children under the age of 15) both parents were in paid employment in 2003 (ABS, 2004b). Perhaps unsurprisingly, a report prepared by the ABS found, that in dual income couples, 70 per cent of all mothers and 56 per cent of all fathers reported that they always/often felt rushed or pressed for time. Only 25.2 per cent of couples without children reportedly experienced this feeling with the same frequency (ABS, 1999a). Half of all fathers and one-third of all mothers in Australia regularly work overtime (Glezer and Wolcott, 2000).

Although family responsibilities influence an individual's participation in the workforce, many young, married women, who may be considered in the child-bearing and early child-rearing age group are also now in employment. For example, employment among Australian women in the 25–34 years of age group has risen from 30 per cent in 1960 to 66 per cent and is predicted to rise to 79 per cent by 2011 (ABS, 1994, 1998).

Changing attitudes about men's roles

However, it is not just women whose work–life balance issues must be managed. The increasing number of dual income couples mean that men and women now share, to some degree, parenting and family responsibilities. Although very few studies have looked at the effect of paternal employment demands on children, it would appear that work hours and job stress have at least an indirect effect by increasing the parenting burden on the mother and decreasing the perception of fathers in a nurturing role. However, concurrently there has been a substantial shift in the expectation of fathers' involvement in parenting. A recent Australian study showed that fathers now spend more time with and are closer to their children than they were 15 years ago. However, 68 per cent of fathers said they did not spend enough time with their children and 53 per cent felt that job and family interfered with each other. Interestingly, 57 per cent of fathers identified work-related barriers, such as expectations of longer hours and inflexibility, as being the critical factor preventing them from being the kind of father they would like to be (unpublished report by Russell cited in Russell and Bowman, 2000).

With the increasing acceptance of gender equity among current and future generations, family is being seen more as a joint responsibility, both from a financial and a nurturing perspective. A survey of Australian men under 35 years of age, with young children and partners in the work force, reported that they were feeling more stress and are keen to change the corporate world to enable them to better balance work–life issues (Russell and Bowman, 2000). In yet another unpublished survey by Russell, 63 per cent of young men said they would refuse a job or promotion that had a negative impact on their family or their partner's career or they would refuse to transfer for the same reason (cited in Russell and Bowman, 2000). These findings are consistent with those of Loughlin and Barling (2001), who report that the new generation of younger workers, both male and female, is not motivated by the same rewards as their parents' generation. Instead they place greater value on 'non-standard' work models that enable them to enjoy a more satisfactory work–life balance.

Increase in the number of aged dependents

The recent change from institutional aged care to home and community-based care means that responsibility for caring for elderly relatives now rests more heavily with family members. With Australia's ageing population and increasing life expectancy, the number of workers with filial responsibilities is likely to rise. In 1998, older persons (aged 65 years or more) comprised 12 per cent of the Australian population and are projected to form 24 per cent of the total population by 2051 (ABS, 1998; ABS, 1999b). In fact it was predicted that, between 1996 and 2041, the aged dependency ratio will double from 18.1 to 34.8, thus, for every 100 workers there will be 34 aged dependents (Gorey et al., 1992).

In 1994, 70 per cent of all providers of personal care and home help for the aged, terminally ill or disabled persons were also in the Australian work force (ABS, 1994). Furthermore, together with the trend towards delayed child-bearing and the increasing employment rate among older women, the availability of grandparents to provide informal child care has reduced. In addition, an increasing number of couples face the additional responsibility of dependent children and filial care simultaneously or sequentially – the so-called sandwich generation.

The changing workforce has also forced changes to work practices to accommodate those with family responsibilities, most notably by taking time off work. The ABS reported that, during 1998, 58 per cent of people with dependent care responsibilities took time off to meet family responsibilities. The average duration of this absence was 9.4 days in a 12 month period (Glezer and Wolcott, 2000). Furthermore, it appears that both men and women take time off work on an almost equal basis for this purpose (ABS, 1999c).

Employment of ethnic minorities

In the United Kingdom at present, ethnic minorities are severely under-represented in construction. (See Chapter 10 for more details.) The situation in Australia is somewhat different in that a relatively high proportion of Australia's immigrants enter the construction industry, often to perform unskilled, menial or manual tasks (Loosemore and Chau, 2002). However, Loosemore and Chau (2002) report the alarming result that 40 per cent of Asian construction workers interviewed felt they had suffered discrimination in that some aspect of their employment rights had been detrimentally affected because of their race, while 61 per cent indicated they had experienced racial intimidation at work. Racial discrimination not only poses a threat to cultural diversity but also has a negative impact upon the family life of its victims (Murry et al., 2001). Thus, racial discrimination in the workplace is likely to be one source of negative work–life spillover among employees who are discriminated against. Perhaps worse, Loosemore and Chau (2002) report that Australian construction managers appeared ill-informed and unable to address blatant racial discrimination and harassment when it was reported in their workplaces, suggesting they are not sensitive or sympathetic to the needs of employees from ethnic minority groups.

Cultural differences exist in the national holidays, rituals and special events celebrated by families. For example, Chinese New Year is likely to be of special significance to a Chinese employee. Different ethnic groups also differ in terms of their values and attitudes towards cohabitation and non-marital child-bearing (Berrington, 1994; Manning and Landale, 1996) and responsibility for elder care (Liu and Kendig, 2000). These differences are likely to have a significant influence on their work–life balance needs. Traditional management practices fail to recognize the needs of an ethnically diverse workforce, for example, in forcing recognition of Judeo-Christian festivals to the exclusion of those celebrated by other religions. However, in the context of the construction industry's assimilationist culture, as described by Loosemore and Chau (2000), it is likely that management acknowledgement of the work–life balance needs of ethnic groups would generate further 'ghetto-isation' and racial prejudice.

The need for non-traditional management approaches

Changes in workforce characteristics require a shift in management approach to re-examine the values, roles and stereotypes and to meet the increasing expectation that a balance between work and non-work life be achieved. In order to achieve greater equity and diversity, there is a need to challenge career structures and work practices that favour full-time workers with minimal family responsibilities. The artificial boundary between employees' work and non-work life domains must be broken down. This boundary should be regarded as permeable because what is positive or negative in one domain, is known to affect the other, a phenomenon known as 'spillover' (Grzywacz et al., 2002). The remainder of this chapter describes construction industry employees' experience of negative work–life spillover and examines ways in which construction organizations could better meet the work–life balance expectations of their employees.

The work–life interface

The links between employees' experiences at work and in other life domains are well researched. For example, organizational variables, such as schedule flexibility and supervisor support have been found to directly influence employees' sense of work–family conflict (Thomas and Ganster, 1995). Employees' perceptions of 'overload' are also important predictors of negative work–life experiences. Role overload is the perception of having too many things to do in a particular role (i.e. employee or parent) and not enough time to perform them (Caplan, 1975). Overload arising as a result of combining paid work and parenting has been found to be related to work–family conflict (Parasuraman et al., 1996; Frone et al., 1997) and less positive family relationships (Crouter et al., 2001). However, role overload can be 'buffered' by spousal support (Beutell and Greenhaus, 1983), indicating the benefits of a supportive home environment. Frone and his colleagues also found supervisor support to be indirectly related to work–family conflict through its effect on work distress and work overload (1997).

Organizational benefits

A growing body of evidence suggests that organizations have much to gain from assisting employees' to achieve a better balance between their work and non-work lives. For example, Grover and Crooker (1995) found that employees in companies with family-supportive benefits expressed higher levels of affective commitment and lower intention to leave, regardless of whether the employee individually used the policy. They explain this positive influence on terms of the symbolic value of work–life balance policies in demonstrating corporate concern for employees' well-being. Flexibility of work location has also been reported to reduce turnover intention among employees with dependent children (Rothausen, 1994).

It seems that work–life benefits are valued by male as well as female employees, perhaps reflecting the social changes identified earlier in this chapter (Butruille, 1990; Pleck, 1993). A study by Burke (2000) found that male managers who report working in an organization that allows them to achieve a satisfactory work–family balance also report experiencing less job stress, a greater joy in work, a lower intention to quit, enhanced career and life satisfaction, fewer psychosomatic symptoms and more positive emotional and physical well-being. Well-integrated employees can also be more effective in their jobs through combining skills developed in domestic life, such as addressing emotions etc, with those traditionally valued by work places, such as linear thinking (Bailyn et al., 2000). Thus, supporting employees' work–life balance could not only bring about a more diverse and satisfied workforce but may also facilitate employee development leading to a more innovative, 'well rounded' workforce.

The legal imperative

Laws now exist at the federal level and in most Australian States and territories which impose requirements on employers in respect of the family responsibilities of employees. Such laws fall under the categories of industrial relations laws, anti-discrimination laws and affirmative action laws and were stimulated by domestic and international trends (Squirchuk and Bourke, 2000). They present a compelling reason for organizations to address the work–life balance concerns of employees.

In Australia, the Federal *Equal Opportunity for Women in the Workplace Act 1999* (previously the *Affirmative Action (Equal Employment Opportunity for Women) Act 1986*) places a positive obligation on organizations to address discrimination against women by both eliminating discrimination and taking special measures to promote women's equal opportunity in employment. These special measures are often in the form of positive discrimination designed to directly redress the disadvantage that women have experienced in the past and are 'exempt' under anti-discrimination laws. The Federal *Act* operates in relation to employers of more than 100 employees and requires employers to lodge an annual report recording their performance. Some State laws also impose affirmative action obligations on State Government departments and agencies. In some States, for example New South Wales, the State-based legislation extends beyond women and addresses the needs of Aboriginal and Torres Strait Islander peoples, people from non-English speaking backgrounds and people with disabilities. Construction organizations that remain unconvinced of the business case for supporting employees' work–life balance should take heed of legal precedents that clearly link the failure to provide this support to unlawful discriminatory practices. For example, in *Hickie vs Hunt and Hunt Solicitors*, an Australian court found that the termination of a legal partner's contract on the basis of her part-time work status amounted to indirect sex discrimination (Bourke, 2000). Part-time work options can help female employees to balance the joint demands of child rearing and work. Drobnic *et al.* (1999) suggest that the availablility of part-time work enables women to continue working thus minimizing depreciation of their human capital that would otherwise occur as a result of lengthy career interruptions.

Work–life imbalance in construction

A recent study in the Australian construction industry, investigated employees' perceptions that work negatively impacts upon aspects of their personal lives – a phenomenon known as negative work–life spillover. The data were collected from within a single construction contracting organization. A questionnaire was distributed to a total of 600 employees including, senior managers, middle managers, secretarial and administrative employees, employees in technical support roles, foremen and site supervisors. 281 completed questionnaires were received, representing a response rate of 47 per cent.

Negative work–life spillover was measured using Small and Reilly's spillover scale (1990). This scale measures respondents' subjective sense that work–life interferes with personal life in a negative way. The scale measures work spillover into four aspects of respondents' personal lives:

- relationship with spouse/partner (my marriage/relationship suffers because of my work);
- relationship with children (because I am often irritable after work, I am not as good a parent as I would like to be);
- home management (my job makes it difficult to get household chores done); and
- leisure activities (the amount of time I spend working interferes with how much free time I have).

Each question is rated on a scale ranging from 1 (strongly disagree) to 5 (strongly agree), thus a high score represents a high level of negative spillover from work to life.

Comparisons were made to determine whether negative spillover from work to other life domains was experienced differently by male and female employees. Comparisons were also made between employees reporting involvement in different categories of family structure. For the purpose of these comparisons the sample was divided into four family structure categories, previously used by Burke (1997). These are

- single, no children;
- single, children;
- partnered, children; and
- partnered, no children.

Table 9.1 shows the distribution of male and female respondents in each group.

Table 9.2 presents the results of comparisons of the average spillover scores for employees in the different demographic groups. In almost all family structures and life domains the average spillover score was higher for men than it was for women. The only instance in which women reported higher negative work–life interference than men was in the domain of home management among respondents who were partnered with children, probably reflecting the fact that, irrespective of employment status, women still

Table 9.1 Respondents by gender and family structure

Status	Women		Men	
	n	%	n	%
1 Single, no children	9	21.4	19	8.5
2 Single, children	0	0.0	4	1.8
3 Partnered, children	13	31.0	134	59.8
4 Partnered, no children	20	47.6	67	29.9
	42		224	

Table 9.2 Negative work–life spillover by gender and family structure

Work–life interference measure	Single, no children	Single, with children	Partnered, with children	Partnered, with no children
1 *Parent–child relationship*				
Men	N/A	14.0	17.2	N/A
Women	N/A	N/A	15.7	N/A
2 *Leisure activities*				
Men	18.5	15.0	16.3	16.8
Women	14.1	N/A	15.1	13.7
3 *Home management*				
Men	16.7	15.7	16.0	16.1
Women	14.2	N/A	17.1	14.8
4 *Spouse/partner relationship*				
Men	N/A	N/A	17.2	17.1
Women	N/A	N/A	13.6	13.6

perform more domestic chores than their male partners (Demo and Acock, 1993; Presser, 1994).

A closer examination of the role of female respondents in the organization was conducted. The results revealed that women are seriously underrepresented, particularly in site-based jobs. Of the respondents who worked predominantly on-site in direct construction activity, 98.3 per cent were male and 1.7 per cent were female. Of the respondents who worked mostly on-site but in the office, 86.8 per cent were male and 13.2 per cent were female. Of the respondents who worked in the head or a regional office, 70.1 per cent were male and 29.9 per cent were female.

In terms of job role description, most female respondents fulfilled a role with a lower average number of hours worked each week. 45.5 per cent of female respondents worked in a clerical/secretarial role and a further 25.0 per cent were in support services. The average hours worked each week for these two groups was 43.2 and 49.1 respectively. This is substantially lower than in positions dominated by men, such as site/project engineer (60.2 hours per week) and foreman/supervisor (62.1 hours per week). Thus it seems that, although men suffer from greater negative work–life interference than women, this may be a reflection of the fact that female employees perform less demanding roles and are seriously under-represented in the mainstream construction and engineering roles associated with greater time-based demands.

A statistical comparison of mean spillover scores was undertaken using a series of one-way analyses of variance (ANOVAs) to determine whether site-based employees reported higher negative work–life strain (spillover) than office-based employees. The results are shown in Table 9.3. The importance of work location in determining construction industry employees' work–life experiences was clearly demonstrated in this analysis.

In all categories of work–life strain (spillover), employees engaged in direct construction activity reported greater negative interference than those based in a site office and employees based in a site office reported greater interference than employees in a head or a regional office. These differences were all statistically significant.

Exploratory analysis of the conditions of work required of site and office-based employees revealed that site-based employees work longer, more irregular hours and enjoy less schedule flexibility than office-based employees. All of these time-based job demands were strongly correlated with all four dimensions of work–life strain (spillover).

The study confirms the findings of other researchers in revealing that women are significantly under-represented in site-based roles. A closer examination of the demographic characteristics of female employees in the sample suggests that women in site-based roles tend to be younger than female employees based in the head or regional offices. Sixty-one per cent of site-based female respondents were below the age of 35, whereas 65 per cent of female employees in the head or regional office were 35 years of age or over. An analysis of the job categories of female respondents who worked predominantly on-site revealed that, out of a modest total of 18, 8 respondents were in clerical or secretarial roles and 3 performed a support service role. This indicates women in site-based jobs are usually not engaged in work directly relating to the construction process.

Table 9.3 Work–life interference by life domain and work location

	On-site	Site-office	Head/regional office	p
1 Parent–child relationship	18.8	17.3	14.9	.000
2 Leisure activities	18.5	16.7	14.0	.000
3 Home management	16.9	16.9	14.1	.000
4 Spouse/partner relationship	18.5	17.5	14.1	.000

These results clearly demonstrate the seriousness of the gender imbalance in construction. However, a second important finding related to gender is that male construction employees also experience serious difficulties in achieving a work–life balance. These findings are probably related to broader social changes described earlier in this chapter. It is interesting to note that our sample included several lone fathers, a trend which is likely to further enhance men's interest in work–life balance initiatives.

Work–life balance initiatives

A growing interest in work–life balance issues has created pressure on organizations to respond by implementing work–life policies. In Australia, leading construction organizations are beginning to recognize a need to provide a work environment that is supportive of employees' personal lives. Once industry leaders implement such policies, institutional pressures will force their competitors to follow suit in an attempt to be perceived as 'good' employers (Goodstein, 1994). It is likely that unresponsive organizations will experience difficulties in attracting and retaining high calibre employees.

There are many ways companies can assist employees to achieve a better work–life balance. Some of the options are presented later. However, the needs of individual employees will differ and change over time. It is therefore important that companies examine the needs of their employees and ensure that policies address these needs. Consultation with employees through surveys, focus groups, newsletters, notices or workshops is also recommended.

Child care

Australian companies' provision of child care surpasses that available in most other OECD countries (Cass, 1993) but company sponsored child care is still rarely available to construction industry employees. While it may be difficult to provide on-site child care centres due to the limited space and temporary nature of construction work, there are other options for child care provision which construction companies may be able to provide. These include employer contributions towards employees' child care fees, provision of off-site single employer child care centres for company employees, joint venture child care centres or purchase or lease of places in existing centres (Napoli, 1994). Needs for the care of children outside school hours, during school holidays and when they are sick should also be considered.

Elder care

As noted earlier in this chapter, an ageing population will have a serious impact on the workplace. Elder care obligations may actually come to eclipse child care obligations in the number of employees affected. If elder care is a need, then support in the form of special family leave or an information and referral service may be helpful.

Flexible work practices

Flexible work arrangements are one of the most frequently used ways to assist employees with family responsibilities (Fernandez, 1986). Flexible work arrangements cover a range of practices including flexible work hours, job sharing or working from home or telecommuting. While there will always be the need to have supervisory personnel on-site, advances in the capabilities and use of information technology mean that working remotely is now feasible for many professionals.

Permanent part-time work

Part-time work can assist employees to maintain a balance between their work and personal life. Squirchuk and Bourke (2000) note that flexible work practices and part-time work options can greatly increase employee retention rate after maternity leave and reduced their overall staff turnover. From a company's point of view part-time work can provide flexibility to cater for peak periods (Napoli, 1994). This flexibility may help construction organizations to cope with the cyclical nature of construction demand. Permanent part-time work differs from casual work in that employees have a 'permanent' contract of employment with the company and retain benefits such as annual leave, sick leave, maternity and long service leave. However, it is important that part-time workers are valued, not marginalized, and that they enjoy access to identified career paths.

Parental leave

Parental leave allows employees with a new child, either natural or adopted, to care for their child at home on a full-time basis in the child's first year and still retain employment and accrue entitlements. In Australia the *Workplace Relations Act* of 1996, the primary legislative instrument at the federal level which regulates employee entitlements, provides employees with the opportunity to take 52 weeks of unpaid combined paternity and maternity leave, where an employee has had 12 months' continuous service with the same employer. However Australia and the United States are the only two OECD countries that do not provide paid maternity leave (Sex Discrimination Unit, 2002). The provision of at least 12 weeks paid maternity leave is typically only available in public sector jobs (Cass, 1993). Some private sector firms attempting to recruit and retain female employees are reported to offer between 6 and 12 weeks paid maternity leave and a lesser amount of paid paternity leave (ILO, 2001). Construction firms serious about attracting and retaining staff may consider the provision of paid parental leave or offering part-time work to their male and female employees.

Other initiatives

Companies that actively seek to support employees with family responsibilities do not limit themselves to meeting employees' immediate needs for child care and leave. Other initiatives intended to elicit commitment and loyalty from employees include, but are not limited to

- salary packaging of child care costs, school fees or elder care costs to provide a tax benefit to employees;
- health and dental insurance;
- family related phone calls to enable employees to check on children or elderly relatives;
- employee assistance programmes offering counselling for employees with personal or family difficulties; and
- corporate wellness programmes, including such benefits as gym membership, health and fitness assessments, smoking cessation programmes, dietary advice, stress reduction programmes etc.

Work–life policy preferences

In the Australian work–life study described earlier in this chapter, employees of a participating construction organization were asked to indicate the extent to which they valued each of a number of work–life balance benefits. Preferences were rated on a scale ranging from 1 (high priority) to 4 (not applicable). Thus, a low score indicates a strong preference for this benefit, while a high score indicates a low preference for this benefit. The preferences of participants who differed by family structure were compared. The results of these comparisons are presented in Table 9.4.

Significant differences were observed for child care facilities, information and referral service for dependants, child care cost assistance and special family leave. As might be expected, employees with dependent children placed a significantly higher priority on the work–life benefits associated with child care, than those without dependent children. These results indicate that employees' work–life balance issues, and, by extension, the work–life balance policies that they value will vary according to life-stage. In deciding what types of policies to implement, the needs of key constituent groups of employees should be considered (Bardoel et al., 1999).

It is important to recognize the legitimacy of the needs of employees without family responsibilities. In the Australian work–life study, negative work interference with leisure activities was particularly high among single

Table 9.4 Work–life balance preferences by family structure

Work–life initiative	Single, no children	Single, with children	Partnered, with children	Partnered, with no children	p
Child care facilities	3.98[a,b,c]	3.33[a]	3.00[b,d]	3.57[c,d]	.000
Information/referral service for dependants	3.48[e,f]	2.33[e]	2.82[f,g]	3.21[g]	.000
Child care assistance costs	3.75[h]	3.33	2.66[h,i]	3.46[i]	.000
More flexible work hours	1.86	2.33	1.85	1.87	.850
Part-time work option after childbirth/adoption	3.64	3.67	3.10	3.10	.057
Job sharing	2.69	3.00	2.73	2.76	.939
Special family leave	2.61[j,k]	1.67	1.82[j]	1.95[k]	.002
Permanent part-time work option	2.68	2.67	2.69	2.71	.998
Temporary part-time work in family crisis	2.15	2.67	1.76	1.94	.092
Extended parental leave after childbirth/adoption	3.26	2.67	2.76	2.81	.226
Flexibility in work location	2.57	2.33	2.36	2.30	.678
Employee assistance programme for family problems	2.04	2.00	1.80	1.93	.446

Notes
Groups with the same superscript are significantly different.
Scale used; 1 = high priority, 2 = medium priority, 3 = low priority, 4 = not applicable.

male employees (see Table 9.2). The impact of work on employees' abilities to pursue a healthy lifestyle emerged as an important theme in the qualitative comments provided by employees who returned the questionnaire. For example, one employee wrote 'Joining a Saturday sports club is impossible, as you would be letting all the other team members down on a regular basis and training during the week after work is unreliable as you don't know what time you would be finishing work.' This is an important issue because failure to address this is likely to lead to feelings of resentment on the part of the employee. Also, frequent and regular respite from work has been shown to be important in reducing the impact of job stress and preventing employee burnout (Westman and Eden, 1997).

Construction organizations should examine their employees' needs both at present and in the future to ensure that work–life issues are adequately addressed. In analysing workforce needs, it is important not to focus solely on the needs of any single group of employees. To do so, is likely to raise issues of equity. For example, Medjuck *et al.* (1998) suggest that many employees perceive a child care bias in work–life balance policies and, in the United States, a national organization called the 'Childfree Network' has been established to provide a support network for childless employees. This group calls for equal treatment of employees' regardless of parental status and even describes work–family policies as discriminatory (Kirkpatrick, 1997 (reported in Kirby and Krone, 2002)). In order to prevent a negative 'backlash' it is therefore very important that the work–life balance experiences of employees in different life-stages and family structures be considered and policy decisions clearly communicated to all employees.

The need for cultural change

The implementation of work–life balance policies that are matched to the demographic profile of the workforce is very important. However, even if construction organizations implement work–life balance policies, the mere implementation of these policies does not mean they will automatically create a work environment that meets the needs of a diverse workforce, especially if they are just 'added on' to the existing culture (Rapoport and Bailyn, 1996). Work–life balance policies are only effective to the extent that they are used. Unfortunately, policies available 'on paper' may remain de-coupled from the actual workings of an organization and thereby have no substantive effect on employees' experiences or behaviour. Haas and Hwang (1995), for example, comment on the lack of take-up of work–life balance initiatives, particularly by men. There remain significant cultural impediments to the use of work–life balance policies and Lewis (2001) argues that, only in the context of dramatic cultural change, will the negative effects of work on non-work experiences be overcome.

Many writers focus on the role of the work environment in explaining the low take-up of work practices designed to enable women to balance their work with family-care responsibilities. For example, Schwartz (1996) suggests that company culture and supervisors' attitudes are key determinants of women's use of part-time work options. Higgins *et al.* (2000) also explore the impact of part-time work on women's experiences in balancing work and family. They suggest that career women are less likely to benefit from part-time work than non-career women because part-time work is more likely to be stigmatized in professional roles. Consequently, professional women taking advantage of part-time options will suffer career penalties. Higgins *et al.* (2000) recommend that characteristics of the work environment be reviewed and part-time work be given credibility through formalized career management processes.

The development of family-friendly cultures is certainly important and is critical to enabling employees to balance their work and family lives (Galinsky and Stein, 1990). Thompson *et al.* (1999: 394) define work–family culture as the 'shared assumptions, beliefs, and values regarding the extent to which an organization supports the values and integration of employees' work and family lives.' They consider a negative work–family culture to have at least three components: organizational time demands or expectations that employees prioritize work over family; negative career consequences associated with utilizing work–family benefits; and lack of managerial support and sensitivity to employees' family responsibilities. They feel that these components are interdependent as each reflects an aspect of the overall supportiveness of the culture for integrating employees' work and family lives. Comments made by employees in the Australian work–life study, described earlier in this chapter, suggest that their work context is not family friendly. For example, a female employee commented: 'A major contributing factor to me not having children is the fact that in such an unstable industry I am not convinced that I would have a position to return to at the end of my maternity leave.' This comment suggests that job insecurity, a stressor also linked with diminished family functioning (Larson *et al.*, 1994), is also a concern.

The focus on families is also important from a policy perspective. Carnoy (1999) argues that high levels of work–family conflict pose a threat to social cohesion because families are the most basic and enduring social unit. Reflecting similar concern, in 1990, Australia ratified ILO Convention 156. This obliges Australia to aim to enable people engaged in work with family responsibilities to work without being subject to discrimination and without conflict between work and family. Schwartz notes that one immutable difference between men and women is that women experience pregnancy and childbirth. Even after childbirth, women maintain primary responsibility for child care and household duties. Though this latter difference between the sexes is not immutable, Gorman (1999) offers social, economic and political explanations for its persistence and Hochschild (1997) has referred to women's domestic work, on return from a day in paid employment as 'the second shift'. Given the persistence of this domestic division of labour, it is likely that women will be more likely to enter and remain in employment that is supportive of their family and home responsibilities. Therefore, the development of a family-friendly work culture should help to attract and retain female employees. However, focusing exclusively on women's experiences may reinforce attitudes about gender roles and the assumption that women should take primary responsibility for children and family matters.

Also, the family-bias in many policies fails to address the experiences and needs of employees without family responsibilities. Instead, the aim should be to create a positive work–*life* culture. The features of such a culture are similar to those originally conceived of as characteristics of a family-friendly culture. For example, Bailyn (1997) identifies three characteristics of family-friendly work cultures: flexible work scheduling, flexible work processes and an understanding by organizational leadership that family needs are important. This flexibility and managerial understanding is required to assist employees balance their work with all aspects of their non-work lives. This includes, but is not limited to, family matters.

Research suggests that a supportive work–life culture contributes significantly to the utilization of work–life initiatives and reduces employees' perceptions of work to non-work conflict, over and above the mere provision of such 'benefits' (Thompson et al., 1999). Work–life initiatives might be available in an organization but an employee's decision to use such initiatives might be influenced by the prevailing organizational culture, in which traditional methods of working may be strongly encouraged (Perlow, 1995; Blair-Loy and Wharton, 2002). Lewis (2001) suggests that traditional models of work construct the ideal worker as one who works long hours on a full-time basis and does not allow family issues to interfere with work. However, she argues that this assumption is no longer applicable to men's or women's experiences in the contemporary labour market. In the light of changing workforce demographics and employees' expectations, Lewis suggests that traditional models of work, embedded in organizational cultures are increasingly questionable. The creation of a work environment, in which employees' achievements are recognized, rather than the number of hours they spend at work was recommended by employees in the Australian work–life study.

It could be argued that the need for cultural change is particularly acute in the construction industry, in which the prevailing 'masculine' culture is likely to create dissent as to whether work–life balance policies are appropriate or desirable. Indeed, to some organizational members, the notion that organizations should assist employees with their personal lives is unacceptable because it acknowledges that 'non-work' issues inevitably have an impact in the workplace. For example, in the Australian work–life study, one manager commented 'Our industry is high pace, high risk and highly paid. If you can't stand the conditions, go and work in a soft job.' And another said, 'Before becoming a family, the [construction industry] participants well know the pace of the construction industry.' Such individuals are likely to be more comfortable with management practices based upon the traditional separation of the work and non-work spheres of an employees' life (Kanter, 1977).

The social context in which work–life policies are implemented is also of great importance to the take-up of these policies. There is strong evidence to suggest that support from one's immediate supervisor or manager is a strong predictor of employees' work–life experiences (Thomas and Ganster, 1995). In the Australian work–life study, when asked what their employer could do to make it easier to achieve a work–life balance, one employee wrote, 'A more understanding employer would be helpful. Perhaps one who listens to problems and gets things done... one who takes time out to ask "how are you?" or "how are things going?" ' Such supervisory support is likely to boost employees' morale. For example, mothers who work for flexible bosses are reported to be 7 times less likely to want to quit and 4 times more likely to say they love their jobs (Friedman and

Galinsky (1990), cited in Watkins, 1995). Schwartz (1996) also reports how supervisors affect the take-up of work–life balance initiatives. Supervisors have considerable discretion as to how to resolve employees' work–life dilemmas, a situation which is likely to be particularly true in the decentralized construction industry. Watkins (1995) studied supervisors'/managers' responses to a hypothetical work–life dilemma and found that, in almost every case, the supervisor/manager adopted a 'defensive reasoning' approach in which the validity of the employee's problems was denied in favour of production goals. In the production-oriented construction industry, supervisor sensitivity training may be needed to overcome such problems.

Kirby and Krone (2002) found that how co-workers structure work and family issues in their discourse plays a key role in the take-up of family-friendly employment practices, including parental leave and leave to provide care to sick dependents. They report perceived inequities and resentment among single or childless employees who feel that family-friendly benefits are 'abused' by others. In the organization studied by Kirby and Krone (2002), these feelings generated peer pressure not to use work–life benefits which served as a control mechanism over employees who wanted to use these benefits. Co-workers' ability to control fellow employees' behaviour, through the development of informal norms is reported to be high in self-managed teams (Barker, 1993). Peer pressure to conform to cultural norms of long and rigid work schedules may be particularly acute in construction project teams, whose members must work to meet tight, client-imposed deadlines or suffer monetary penalties. Similarly, Fielden and her colleagues (2000) suggest that in the construction industry a lack of compliance with norms can adversely affect employees. For instance, if the underlying cultural assumption is that an employee's presence at work indicates their commitment and contribution, employees who utilize work–life benefits that make them less visible at work (e.g. working from home), may be regarded as being lacking in commitment. This can have a negative impact upon relationships with co-workers and may result in institutional sanctions, such as career penalties or unfavourable performance evaluations.

Cultural change is not easy and will not be achieved over night, especially in construction, in which male-centred attitudes structure the workplace and a belief in the separation of employees' work and personal lives is deeply ingrained. Furthermore, these attitudes appear to be transmitted from one generation of managers to the next. Dainty et al. (2000) report that, however accepting of change they may be at the start of their career, male entrants to the construction industry inadvertently reinforce current attitudes and practices by emulating the behaviour of the managers who influenced their own career development (Dainty et al. 2000). Given the strength of the influences that perpetuate the status quo at middle management level, it is likely that the adoption of the non-traditional management approaches that are required to accommodate the needs of the workforce in the twenty-first century will have to be driven from the top down.

Rodgers (1992) provides an insight into organizational environments in which work–life policies have been successfully integrated into company culture. In such organizations, the implementation of such policies has been accompanied by clearly communicated support from senior management. It is also essential that all employees be provided with an explanation for the introduction of such policies. It may also help to communicate to all employees the organizational benefits expected to be achieved by the implementation of work–life policies, such as reduced turnover, tardiness and absenteeism and enhanced job satisfaction, organizational commitment and productivity. In order to minimize perceived inequity and legitimize the use of work–life balance policies among male employees, it is particularly important that gendered assumptions are not reinforced and that it is made clear that the policies are not just aimed at female employees with children. Prejudices may be broken down if both male and female employees utilizing work–life policies are profiled and presented in a positive light in company newsletters. Given the newness of work–life policies, particularly in the construction industry context, managers should be trained in how to respond with sensitivity to employees' work–life balance issues and in the specific skills they will need in implementing flexible work practices (Bruce and Reed 1994).

Conclusions

Dramatic social changes have occurred since the middle of the twentieth century. Most notably women's and men's workforce participation rates have converged resulting in more dual earner households. Traditional family structures have also become less prevalent and employees now participate in different family structures, including *de facto* couples, lone parents and blended families. Young employees have also adopted different attitudes to work to those of their parents, displaying less company loyalty and expecting greater flexibility.

These changes render irrelevant, traditional management theories and practices that presumed the male 'breadwinner' model to be the norm. In their place, more flexible models of work should be developed to reflect the needs of the new, more diverse workforce. These models must be adaptable in order to match the expectations of employees who differ in terms of age, life-stage, race, religion, family structure, sex and sexuality.

It is recommended that the implementation of flexible work practices be integrated with an affirmative action programme to attract, into construction, employees from groups that are currently under-represented. Work–life policies should also be integrated with anti-discrimination policies to ensure that no single group of employees should benefit at the expense of others. Mutual respect and tolerance of others' circumstances should be the goal. Once in place, work–life policies should serve to improve the image of the construction industry and increase the retention of employees, who may otherwise leave due to intolerable levels of work–life conflict.

However, the implementation of work–life policies cannot meet these needs unless more fundamental assumptions and attitudes are addressed. In short, organizational cultures that support employees' personal lives must be fostered before the construction industry will be in a position to enjoy the benefits associated with workforce diversity.

Discussion questions

1   Briefly describe how social and demographic trends have served to increase the importance of work–life balance. Why do construction organizations need to respond to these trends?
2   Why is it unlikely that human resource managers will find a 'one size fits all' solution to work–life balance in the construction industry?
3   Why is managing employees' work–life balance important in the promotion of a diverse workforce?
4   Why is the implementation of formal work–life benefits, on their own, unlikely to resolve the work–life balance difficulties experienced by construction industry employees?

References

Australian Bureau of Statistics (1994) *Focus on Families: Work and Family Responsibilities*, Cat No. 4422.0, Canberra. Australian Government Printing Service.
Australian Bureau of Statistics (1998) *Population Projection, Australia 1997 to 2051*, *Australia*, Cat No. 3222, Canberra. Australian Government Printing Service.
Australian Bureau of Statistics (1999a) *Australia Social Trends*, Cat No. 4102, Canberra. Australian Government Printing Service.
Australian Bureau of Statistics (1999b) *Older People, Australia: A Social Report*, Cat No. 4109, Canberra. Australian Government Printing Service.
Australian Bureau of Statistics (1999c) *Balancing Work and Caring Responsibilities*, Cat No. 4903.6, Canberra. Australian Government Printing Service.
Australian Bureau of Statistics (2003) *Labour Force*, Cat No. 6203, February, Canberra. Australian Government Printing Service.
Australian Bureau of Statistics (2004a) *Labour Force Australia, Spreadsheets*, Cat No. 6202.0.55.001, July, Canberra. Australian Government Printing Service.
Australian Bureau of Statistics (2004b) *Australian Social Trends*. Cat No. 4102, Canberra. Australian Government Printing Service.
Bailyn, L. (1997) 'The impact of corporate culture on work–family integration'. In Parasuraman S. and Greenhaus J.H. (eds) *Integrating Work and Family: Challenges and Choices for a Changing Worlds*. Westport, Quorum Books, pp. 209–219.

Bailyn, L., Rapoport, R. and Fletcher, J.K. (2000) 'Moving corporations in the United States towards gender equity; a cautionary tale'. In Haas, L., Hwang, P. and Russell, G. (eds) *Organizational Change and Gender Equity – International Perspectives on Parents at The Workplace*. Beverley Hills, CA. Sage Publications.

Bardoel, E.A., Moss, S.A., Smyrnios, K. and Tharenou, P. (1999) 'Employee characteristics associated with the provision of work–family policies and programs', *International Journal of Manpower*, 20, 563–576.

Barker, J.R. (1993) 'Tightening the iron cage: concertive control in self-managing teams', *Administrative Science Quarterly*, 38, 408–437.

Berrington, A. (1994) 'Marriage and family formation among white and ethnic minority population in Britain', *Ethnic and Racial Studies*, 17, 517–546.

Beutell, N.J. and Greenhaus, J.H. (1983) 'Interrole conflict among married women: the influence of husband and wife characteristics on conflict and coping behavior', *Journal of Vocational Behavior*, 21, 99–110.

Blair-Loy, M. and Wharton, A.S. (2002), 'Employees' use of work–family policies and the workplace social context', *Social Forces*, 80, 813–845.

Bourke, J. (2000), 'Corporate women, children, careers and workplace culture: the integration of flexible work practices into the legal and finance professions', *Studies in Organisational Analysis and Innovation Monograph*, Number 15, Industrial Relations Research Centre, University of New South Wales, Sydney.

Bruce, W. and Reed C. (1994) 'Preparing supervisors for the future workforce: the dual-income couple and the work–family dichotomy', *Public Administration Review*, 54, 36–43.

Burke, R. (1997) 'Are families damaging to careers?', *Women in Management Review*, 12, 320–324.

Burke, R. (2000) 'Do managerial men benefit from organizational values supporting work–personal life balance?', *Women in Management Review*, 15(2), 114.

Butruille, S.G. (1990) 'Corporate caretaking', *Training and Development Journal*, 44, 48–55.

Caplan, R.D. (1975) *Job Demands and Worker Health: Main Effects and Occupational Differences*. US Department of Health, Education, and Welfare, Cincinnati, OH and Washington, DC.

Carnoy, M. (1999) 'The family, flexible work and social cohesion at risk', *International Labour Review*, 138, 411–429.

Cass, B. (1993) 'The Work and Family Debate in Australia', Paper presented at *AFR/BCA Conference on Work and Family: The Corporate Challenge*, Melbourne, 1 December 1993.

Crouter, A.C., Bumpus, M.F., Head, M.R. and McHale, S.M. (2001) 'Implications of overwork and overload for the quality of men's family relationships', *Journal of Marriage and Family*, 63(2), 404–416.

Dainty, A.R.J., Neale, R.H. and Bagilhole, B.M. (1999) 'Women's careers in large construction companies: expectations unfulfilled?', *Career Development International*, 4(7), 353–357.

Dainty, A.R.J., Neale R.H. and Bagilhole, B.M. (2000) 'Comparison of men's and women's careers in U. K construction Industry', *Journal of Professional Issues in Engineering Education and Practice*, July 2000, ASCE.

Demo, D.H. and Acock, A.C. (1993) 'Family diversity and the division of domestic labor: how much have things really changed?', *Family Relations*, 42, 323–331.

Drobnic, S., Blossfield, H.P. and Rohwer, G. (1999) 'Dynamics of women's employment patterns over the family life course: a comparison of the United States and Germany', *Journal of Marriage and the Family*, 61, 133–146.

Fernandez, J.P. (1986) *Child care and Corporate Productivity: Resolving Family/Work Conflicts*. Lexington Books, Lexington, MA.

Fielden, S.L., Davidson, M.J., Gale, A.W. and Davey, C.L. (2000) 'Women in construction: the untapped resource', *Construction Management and Economics*, 18, 113–121.

Fielden, S.L., Davidson, M.J., Gale, A.W. and Davey, C.L. (2001) 'Women, equality and construction', *Journal of Management Development*, 20, 293–304.

Frone, M.R., Yardley, J.K. and Markel, K.S. (1997) 'Developing and testing an integrated model of the work–family interface', *Journal of Vocational Behavior*, 50, 145–167.

Gale, A.W. (1994) 'Women in non-traditional occupations: the construction industry', *Women in Management Review*, 9, 3–14.

Galinsky, E. and Stein, P.J. (1990) 'The impact of human resource policies on employees: balancing work/family life', *Journal of Family Issues*, 8, 368–383.

Glezer, H. and Wolcott, I. (2000) 'Conflicting commitments: working mothers and fathers in Australia', in Haas, L.L., Hwang, P. and Russell, G. (eds), *Organisational Change and Gender Equity: International Perspectives on Fathers and Mothers at the Workplace*, California, Sage Publications.

Goodstein, J.D. (1994) 'Institutional pressures and strategic responsiveness: employer involvement in work–family issues', *Academy of Management Journal*, 37, 350–382.

Gorey. K., Rice, R. and Brice, G. (1992) 'The prevalence of elder care responsibilities among the workforce population', *Research on Aging*, 14, 399–418.

Gorman, E.H. (1999) 'Bringing home the bacon: marital allocation of income-earning responsibility, job shifts and men's wages', *Journal of Marriage and the Family*, 61, 110–122.

Greenwood, P. (2002) 'Renewed focus on women in engineering', *Engineers Australia*, 74(6), 3.

Grover, S.L. and Crooker, K.J. (1995) 'Who appreciates family responsive human resource policies: the impact of family friendly policies on the organizational attachment of parents and non parents', *Personnel Psychology*, 48, 271–288.

Grzywacz, J.G., Almeida, D.M. and McDonald, D.A. (2002) 'Work–family spillover and daily reports of work and family stress in the adult labor force', *Family Relations*, 51, 28–36.

Haas, L. and Hwang, P. (1995) 'Company culture and men's usage of family leave benefits in Sweden', *Family Relations*, 44, 28–36.

Higgins, C., Duxbury, L. and Johnson, K.L. (2000) 'Part-time work for women: does it really help balance work and family?', *Human Resource Management*, 39, 17–32.

Hochschild, A.R. (1997) 'When work becomes home and home becomes work', *California Management Review*, 39, 79–97.

ILO (2001) 'E.Quality @ work: an information base on equal employment opportunities for women and men', www.ilo.org/public/english/employment/gems/eeo/main.htm

Kanter, R.M. (1977) *Men and Women of the Corporation*, Basic Books, New York.

Kirby, E.L. and Krone, K.J. (2002) 'The policy exists but you can't really use it: communication and the structuration of work–family policies', *Journal of Applied Communication Research*, 30, 50–77.

Larson, J.H., Wilson, S.M. and Beley, R. (1994) 'The impact of job insecurity on marital and family relationships', *Family Relations*, 43, 138–143.

Lewis, S. (2001) 'Restructuring workplace cultures: the ultimate work–family challenge?', *Women in Management Review*, 16, 21–29.

Liu, W. and Kendig, H. (2000) *Who should care for the Elderly*: An East-West Value Divide. Singapore University Press, London.

Loosemore, M. and Chau, D.W. (2002) 'Racial discrimination towards Asian operatives in the Australian construction industry', *Construction Management and Economics*, 20, 91–102.

Loughlin, C. and Barling, J. (2001) 'Young workers' work values, attitudes and behaviours', *Journal of Occupational and Organisational Psychology*, 74, 543–558.

Manning, W.D. and Landale, N.S. (1996) 'Racial and ethnic differences in the role of cohabitation in premarital childbearing', *Journal of Marriage and the Family*, 58, 63–77.

Medjuck, S., Keefe, J.M. and Fancey, P.J. (1998) 'Available but not accessible: an examination of the use of workplace policies for caregivers of elderly kin', *Journal of Family Issues*, 19, 274–299.

Murry, V.M., Brown, P.A., Brody, G.H., Cultrona, C.E. and Simons, R.L. (2001) 'Racial discrimination as a moderator of the links among stress, maternal psychological functioning and family relationships', *Journal of Marriage and the Family*, 63, 915–926.

Napoli J. (1994) *Work and Family Responsibilities: Adjusting the Balance*, CHH Australia, North Ryde.

Parasuraman, S., Purohit, Y.S., Godshalk, V.M. and Beutell, N.J. (1996) 'Work and family variables, entrepreneurial career success and psychological well-being', *Journal of Vocational Behavior*, 48, 275–300.

Perlow, L. (1995) 'Putting the work back into work/family', *Group and Organisation Management*, 20, 227–239.

Pleck, J.H. (1993) 'Are "family-supportive" employer policies relevant to men?', from Hood, J.C. (ed.), *Men, Work and Family*, Newbury Park, Sage.

Presser, H.B. (1994) 'Employment schedules among dual-earner spouses and the division of household labor by gender', *American Sociological Review*, 59, 348–364.

Rapoport, R. and Bailyn, L. (1996) *Rethinking Life and Work: Toward a Better Future*, New York, The Ford Foundation.

Rodgers, C.S. (1992) 'The flexible workplace: what have we learned?', *Human Resources Management*, 31, 83–199.

Rothausen, T.J. (1994) 'Job satisfaction and the parent worker: the role of flexibility and rewards', *Journal of Vocational Behavior*, 44, 317–336.

Russell, G. and Bowman, L. (2000) *Work and Family: Current Thinking, Research and Practice*, Department of Family and Community Services. Commonwealth of Australia.

Schwartz, D.B. (1996) 'The impact of work–family policies on women's career development: boon or bust?', *Women in Management Review*, 11, 5–19.

Sex Discrimination Unit (2002) 'Valuing parenthood – options for paid maternity leave: interim paper 2002', Sex Discrimination Unit, Human Rights and Equal Opportunity Commission. Sydney.

Small, S.A. and Riley, D. (1990) 'Toward a multidimensional assessment of work spillover into family life', *Journal of Marriage and the Family*, 52, 51–61.

Squirchuk, R. and Bourke, J. (2000) 'From equal opportunity to family–friendly policies and beyond', In Haas, L.L., Hwang, P. and Russell, G. (eds), *Organisational Change and Gender Equity: International Perspectives on Fathers and Mothers at the Workplace*, Beverley Hills, CA: Sage Publications.

Thomas, L.T. and Ganster, D.C. (1995) 'Impact of family-supportive variables on work–family conflict and strain: a control perspective', *Journal of Applied Psychology*, 80, 6–15.

Thompson, C.A., Beauvais, L.L. and Lyness, K.S. (1999) 'When work–family benefits are not enough: the influence of work–family culture on benefit utilization, organizational attachment and work–family conflict', *Journal of Vocational Behavior*, 54, 392–415.

Watkins, K.E. (1995) 'Changing managers' defensive reasoning about work/family conflicts', *Journal of Management Development*, 14, 77–88.

Westman, M. and Eden, D. (1997) 'Effects of respite from work on burnout: vacation relief and fade-out', *Journal of Applied Psychology*, 82, 516–527.

Yates, A., Agnew, J., Kryger, S. and Palmer, M. (2001) *The Engineering Profession: A Statistical Overview 2001*. Institution of Engineers, Canberra, Australia.

Part III

Race, disability and equality

Chapter 10

Race, ethnic minorities and the construction industry

Andy Steele and Dianne Sodhi

Introduction

It was estimated that the UK construction industry would need to recruit 360,000 construction workers by 2004 to replace those leaving the industry and to respond to the recent recovery in this sector (Building, 1999). As a result of this growth, the industry is expected to experience skills shortages in both traditional and new skill areas. Mackenzie *et al.* (2000) highlighted the factors that have contributed to this skills shortage, including: a fall in the number of young people available to enter the labour market, thus increasing competition to attract new employees (Ashworth and Harvey, 1993; Druker and White, 1996); the changing nature of construction markets and, in particular, a decrease in the need for construction workers with traditional skills (CITB, 1991); the introduction of new technologies impacting on skill requirements (Gruneberg, 1997); the cyclical nature of the industry and fluctuating employment patterns; the growth of self-employment and labour-only sub-contractors (Fellows *et al.*, 1995); the fragmentation of the industry (Rainbird, 1991) and the decline in construction training and associated resources (Agapiou *et al.*, 1995; Morton and Jagger, 1995; Thomas, 1996).

While the traditional response to such a skilled labour shortage has been to increase remuneration, Mackenzie *et al.* (2000) argue that this response strategy can only represent a short-term solution to this problem. The Construction Industry Training Board (CITB), conscious of this skills shortage, suggests that 'the industry must look to increase recruitment of ethnic minorities and women if it is to survive and grow' (Building, 2001). This is in recognition of the potentially large pool of new entrants to the industry. Ethnic minorities form 5.5 per cent of the total population of Britain. However, a higher percentage of the working population is of a Black and Minority Ethnic (BME) background (6.4 per cent) than in the total population figures. This difference reflects the younger profile of most of the BME groups compared with the white population. In the case of the 16–24-year old age range, one of the key target groups for entry into trade training in the construction industry, 7 per cent are BME. The figure increases to 9 per cent and 9.2 per cent in respect of the current 5–15 and 0–4 age cohorts (CITB and Royal Holloway University of London, 1999). At the same time, unemployment rates among ethnic minorities are higher than for the white population, for example, while young African men are better qualified than white men, they are three times more likely to be unemployed (Singh, 2000).

The construction industry's ability to attract new recruits from the BME communities has been particularly poor.

> The construction industry does not have a good record on recruiting and then developing employees from ethnic minorities. This is a missed opportunity, not only for people who may be well suited to a job in construction but also for the industry itself.
>
> (Cavill, 1999: 20)

The construction workforce is only 1.9 per cent Black and Asian, compared with 6.4 per cent of the working population as a whole. The industry employs 70 per cent fewer BME people than the UK industry average (CITB and Royal Holloway University of London, 1999). At the same time there are disparities in the representation of different BME groups in the construction industry. Particular BME groups are more likely to be found than others, especially African Caribbeans, Indians and those of mixed and other backgrounds, compared with those from the Pakistani and Bangladeshi communities.

In terms of construction professionals (such as architects, quantity surveyors and surveyors and town planners), it is estimated that only 2 per cent are ethnic minorities (Building E = Quality, 1996). It is difficult to be precise about the representation of ethnic minorities in each of the construction-allied professions since few of the trade and professional bodies routinely collect details of the ethnicity of their membership (Sodhi and Steele, 2000; CITB and Royal Holloway University of London, 1999). While there is an acknowledged severe under-representation of BME people in the construction industry at both trade and professional levels, this does not mean that these minority groups are not interested in employment within this sector, as evidenced by the high rates of participation in construction-related courses at colleges and universities. For example, the 1997 figures for ethnic minority entrants to degree courses in civil engineering and architecture were 11.2 and 12.5 per cent respectively (UCAS, 1998).

The question then is, why do so few ethnic minorities choose to enter the construction industry. Key to this are the barriers that exist for BME groups, relating primarily to the industry placing little emphasis on equality of opportunity. These barriers include the image of the industry, the prevalence of overt racism, the lack of career development opportunities and employment uncertainty, the lack of information available about the industry and the organizational culture of the industry. In this chapter, each of these issues will be discussed in turn. This will be followed by an examination of the initiatives that have been pursued to try to encourage the industry to embrace equal opportunities and seek to redress the imbalance in the number of ethnic minorities. The chapter will conclude by considering the lack of momentum of the industry at large in developing equal opportunities and the role that public sector clients/quasi public sector clients (i.e. housing associations and local authorities) can play in regulating the sector, to ensure that construction companies implement equal opportunities policies and monitor their outputs.

The image of the construction industry

Construction work in terms of the trades as opposed to professional roles is seen as being synonymous with 'lousy pay, dirty sites, cold and rain, no canteens, disgusting toilets, no pension, a macho culture and a poor safety record. It is no wonder teenagers dream of other things' (Latham, 2001). Hence, the construction industry is generally perceived as being hard manual work with little or no status. However, more than this, 'the industry is portrayed as being for white people only' (CITB and Royal Holloway University of London, 1999: 23).

This perceived lack of status of jobs within the construction industry is seen as a particular problem among some of the Asian community in terms of family and community pressure to pursue other professions, which are regarded as being of higher status. As one community leader noted:

> The Asian culture is to continue to better parents' achievements. Improve education prospects and earning potential...government posts in local councils are seen as high status jobs with reasonably secure salary and job. It has a cleaner environment and the job provides prestige and importance within the community...The status and image of the family matter very much in the Asian community. Children are encouraged to do well at school and continue with their education. Encouraged to take up professional jobs which will elevate the person's status in society.
>
> (CITB and Royal Holloway University of London, 1999: 30)

The industry also has a very limited number of BME staff in positions of authority. Hence, there is a lack of suitable role models for young ethnic minority people contemplating entering this sector. 'Role models of people from different cultural backgrounds who are outstanding achievers can...champion the cause and increase the perceived status of their respective groups' (Loosemore and Chau, 2002: 101). Recent studies within the social housing sector have found that the lack of senior BME role models can have a major impact on the BME communities' perception of the organization and particularly the extent to which the organization is characterized as being mono-culturalist (Somerville and Steele, 1998; Somerville et al., 2000). Even where ethnic minorities are in senior or professional roles, their position is often undermined by the view that, since they belong to an ethnic minority group, they must be in a manual or low status role. The President of the Society of Black Architects summed this up by suggesting that 'if you're Black, you're expected to be holding a shovel and digging, not in management' (Cavill, 1999).

Given this negative image of the construction industry, it is not surprising, that, as the Chief Executive of the Chartered Institute of Builders recently remarked, 'I don't think ethnic minorities value construction as a profession and would rather become, for example, a doctor or accountant before thinking about construction as a career' (Hampton, 2000: 7).

Racist attitudes and behaviour in the construction industry

Recent reports on the experiences of ethnic minorities in the construction industry have found that racism is rife and manifested itself in a number of ways (CITB and Royal Holloway University of London, 1999). Discrimination has been found to exist at the point of trying to enter the industry. Despite having the relevant qualifications, ethnic minorities often lose out to less qualified white people. These discriminatory practices have been explained by one construction industry manager as being due to the fact that 'the employer sees them [ethnic minorities] as being potentially problematic' (CITB and Royal Holloway University of London, 1999: 32).

The culture of the construction industry is widely regarded as one where jokes, banter and nicknames are commonplace. Jokes are often made about colour, race and stereotypes. Such behaviour validates the image of the industry as one where racism is prevalent. Recent research (e.g. CITB and Royal Holloway University of London, 1999) found that 39.4 per cent of ethnic minorities in the industry had experienced name-calling and one-quarter had experienced harassment, bullying and intimidation. Challenging such behaviour often leads to the victim being labelled as a troublemaker and can result in further victimization. The acceptance of this type of culture within the industry is illustrated by the comments of two construction workers from the Caribbean island of St. Vincent, one of whom was a chargehand and the other a site manager. Asked about racist graffiti on site, the pair admit there is racist graffiti in the toilets on site – their attitude is that it is only to be expected. 'Sure, there is some graffiti in there that says...is a Black bastard – it is normal...Graffiti can't hurt you, can it?' Senior managers are not seen as being sympathetic or willing to take action against the perpetrators, reinforcing the view that such behaviour is acceptable (CITB and Royal Holloway University of London, 1999).

The dominant culture within the industry is one of being part of a team and there is a great deal of pressure to conform, for example, drinking socially together, which may contravene ethnic minority, religious and cultural beliefs. However, not taking part in such activities can result in the individual Asian or Black person being ostracized or socially excluded and subjected to further victimization. In many instances then, ethnic minorities have to appear almost 'cultureless' and deny their cultural heritage and beliefs in order to succeed in the industry. There is also evidence that some ethnic minorities in the industry develop strategies for working with potential racists by avoiding situations where their ethnicity is obvious; for example, black managers often do the bulk of their selling work on the telephone or by post, to establish a relationship before the customer is aware of their ethnicity (Building, 1999; Sodhi and Steele, 2000). Similarly, a black employer, contemplating on his experiences, suggested that 'I think a lot of my success is because people think I am white...I don't have a clear Black accent or Asian accent' (CITB and Royal Holloway University of London, 1999: 37).

There is also compelling evidence that ethnic minorities are treated differently by employers. The CITB and Royal Holloway University of London (1999: 34) study found that 'just over one-third of respondents in the industry said that they would describe their training experience or working life as a Black/Asian person as different from white people'. Examples of being treated differently include being given different types of work, such as more menial tasks, compared with their white colleagues at the same level. This less favourable treatment is illustrated by the following comment: 'The types of jobs given to you to do. Best jobs, i.e. indoor, go to their own. Work like refurbishment of loos, Blacks and Asians do it.' Discrimination was also experienced in terms of access to training resources and level of pay.

Employment and career opportunities

A third barrier to the entry of BME people to the industry is the nature of the employment and career development opportunities. Continuity of employment within the construction industry is often precarious. The growth in labour-only subcontractors and self-employment has meant that permanent employment is difficult to sustain. Ethnic minorities are particularly at risk of being made redundant. Those wishing to establish their own business come up against additional barriers, which discriminate against BME Small and Medium Enterprises (SMEs). A recent study by Sodhi and Steele (2000) which looked at the contracting power of the UK housing association sector in relation to BME SMEs found evidence of discriminatory practices as well as wider societal barriers. Although virtually all associations maintained an approved list of contractors and consultants, the main methods used to invite companies to apply to join the list mitigated against the promotion of equality of opportunity. Particular emphasis was given to inviting 'known' or 'recommended' contractors and consultants (37 and 34 per cent respectively referred to this as their main method of attracting companies). This compares with just 20 per cent who advertised opportunities to join the approved list of contractors, although only a minority of these advertised in the BME press. As one BME contractor remarked, 'the old boys network is still very much alive and it's the accepted way in which things happen' (Sodhi and Steele, 2000: 35).

The process of completing the pre-qualification documentation was also found by many BME SMEs to be time-consuming and arduous. Reference was made to the amount of detail required which, in the opinion of some of the BME companies, far outweighed the value of the contracts available. Many of the BME SMEs lack the experience of dealing with the bureaucracy involved as well as having to deal with language barriers.

Discrimination was also evident in the awarding of contracts. While most of the BME SMEs were registered on at least one approved list with a housing association, few of them had been approached to tender for work. Furthermore, when contracts had been won, they tended to be for specific types of work, namely, the less lucrative repairs and maintenance work, as opposed to new development. Of those associations providing details of the value of the contracts let, it was estimated that a total of £414 m had been spent on contract work in the previous financial year and the vast majority of this (£251.8 m) was on new development, compared with £25.6 m on responsive maintenance work.

> There is a perception among housing associations that BME SMEs are unable to cope with big contracts. The reality for our company is that we could cope with them. The problem is trying to reverse this perception among the housing associations.
>
> (Sodhi and Steele, 2000: 35)

The study also found that only two-fifths of housing associations had written and agreed criteria for the awarding of contracts. Coupled with a lack of feedback on unsuccessful tenders, the view of many BME firms was that the whole process suffered from a lack of transparency and was open to abuse and discretion: bribes and 'back handers' were believed to be relatively commonplace.

> The move towards partnering and consortia approaches in the construction industry generally (Barlow *et al.*, 1997) has not benefited BME SMEs. Although four out of ten associations had established such practices, only a minority (15 per cent) had required the inclusion of

a BME sub-contractor. Even when this does occur, as one contractor commented the BME SME is often made to feel the subservient partner. In any case, partnerships tend to focus on big projects which absolutely excludes any role for small organizations. There is a closed shop culture – there is limited scope to be involved and in many circles we are not welcome at the hungry table.

(Sodhi and Steele, 2000: 36)

Other barriers which were found to be in operation revolved around the lack of flexibility by the commissioning body in relation to, for example, terms of payment, previous experience and insurance levels. At the same time, it should be borne in mind that, in addition to the barriers identified earlier, many BME SMEs are discriminated against in society generally, for example, in relation to securing credit from the lending institutions – a Black person may be unsuccessful in getting a bank loan to expand their business, proving that it is not just the construction industry that is prejudiced (Building, 1999). Access to leasing business premises can often prove difficult in the face of direct racial discrimination, with unnecessary and inappropriate questions being asked of Black-run businesses which are not asked for comparable white businesses (CITB and Royal Holloway University of London, 1999).

Increasing emphasis upon the 'packaging' of contracts to achieve economies of scale impacts on SMEs generally. Harding (2000) cites a growing trend in the public sector, reinforced by recent government policy, to roll a number of contracts together in favour of the major contractors. He quotes the example of the National Health Service, whereby

projects between £1 m and £10 m were to be bundled for the benefit of 'regional partners': only contracts below £1 m would be available for SMEs. The 550 health trusts, along with the 650 or so local authorities, are among the SME contractors' biggest clients. If all the best jobs are let to the major contractors and the £1 m–£10 m contracts are bundled for the regional contractors, the SME backbone of construction – 75 per cent of the industry – will be forced to cut their own throats in the discredited adversarial market.

(Harding, 2000: 25)

As regards the opportunities for career progression and promotion these are seen as being limited for BME workers in the construction industry. The study by the CITB and Royal Holloway University of London (1999: 36) found that 56.8 per cent of Black and Asian people felt that they did not have as good a chance of promotion as white people in the industry and that 'they were held back from pursuing their careers'. The construction industry is not unique in this respect. Recent studies of the career opportunities for BME staff within the social housing sector found evidence of discrimination in the availability of training, career advancement and promotional opportunities for BME staff (Somerville and Steele, 1998; Somerville et al., 2000). Staff from the African and Caribbean and Asian communities tended to be heavily concentrated at the bottom of the organizational hierarchy. For example, the four largest housing associations in the country, with over 1,000 staff each, between them employed a total of 275 BME staff, but only 8 of these were at a senior level. BME staff also tended to be found in specific roles, mainly support roles as opposed to technical (Somerville et al., 2000). Modood (1997: 43) refers to the 'glass ceiling' that affects all non-white men equally. Statistics from a variety of sources confirm this. Looking at the United Kingdom's top 100 firms, the proportion of BME staff is 1.75 per cent (Pandya, 1999). It is argued that the culture of racial indifference appears to be endemic in the housing world, as in other sections of British society (Somerville et al., 2002).

Lack of information on the industry

Another important barrier to the recruitment of BME people to the construction industry is the dearth of information available to these ethnic minority communities about the range of employment opportunities available. Levels of knowledge of the construction industry among the BME communities vary according to how much contact they have with employer bodies and the CITB. In general, there is a limited awareness of jobs in the industry. The CITB and Royal Holloway London University (1999) research found that information about opportunities in the construction industry and career paths were not reaching minority ethnic groups in the way that would produce improved representation of minority ethnic people. Where publicity information is available about the industry, it is often not available in community languages and almost exclusively portrays white employees. This latter point is particularly pertinent in relation to media coverage of the industry, given that the media is 'a very important tool for attracting people's attention' (CITB and Royal Holloway London University, 1999). In the absence of formal means of disseminating information to the BME communities about the construction industry, many potential entrants from the BME communities have to rely on informal means, family and friends being the main source of information. However, research elsewhere (Steele, 2001a,b) has highlighted the potential for inaccurate and misinformation to be conveyed by peers and family, further reinforcing their negative perception of the industry.

Organizational culture

Organizational culture has been defined as a set of fixed assumptions which are held by the organization, usually through informal networks which are quite different from the public structure (Davey *et al.*, 1999). In construction, white male values are the norm and are rewarded. This is illustrated by the attitude of senior white staff to the issue of equal opportunities. As a Bovis spokesman recently commented, surprised at being asked to provide details of the number of Black and Asian people that the company employs, 'racial discrimination is not perceived as a problem within Bovis' (Cavill, 1999).

Within the industry the principle of equal opportunities is not part of the culture. The vast majority of construction companies (94.1 per cent) were found to have an equal opportunities policy (CITB and Royal Holloway London University, 1999). However, these were very much characterized by what Tomlins (1994: 27) refers to as 'paper policies', in that such organizations are more concerned with having policies in place and 'being seen to be doing something, rather than actually doing it'. This is borne out by the findings from the CITB and Royal Holloway London University (1999) study. The proportion of companies whose equal opportunities policy addressed discrimination on the grounds of race was 93 per cent, while only 75 per cent stated that someone within the organization had been allocated responsibility for the implementation of the policy. A smaller percentage (64 per cent) admitted that they monitored the impact of their company's equal opportunities policy and only slightly more than half (56 per cent) had action plans to put their policy into practice. Hence, it can be seen that, although the vast majority of companies in the construction industry had an equal opportunities policy, few take the notion of equality of opportunity seriously.

The business case for equal opportunities

Despite the lack of progress on equal opportunities within the industry, there are important business arguments, beyond the issue of skills shortage, why the industry should embrace the philosophy of equality of opportunity. First, it is widely recognized that having a culturally diverse workforce can benefit the organization itself by harnessing those differences to improve production, creativity and decision-making processes (Kandola and Fullerton, 1994). Second, increasing the representation of the BME workforce can create new business opportunities. This is particularly relevant for regeneration work in inner city areas where the communities are multicultural. Rogers comments that

> customers will be attracted to an organization's services where they feel the staff have an understanding of their needs. Older people and lone parents may feel reassured to be offered a female electrician and ethnic minorities may request an electrician who understands a particular language.
>
> (2000)

This can only help to redress the white-dominated image of the company and encourage more ethnic minorities to seek employment in this sector.

A third business argument is that of pre-empting the possibility of companies being prosecuted for discrimination. Evans (2002) argues that in recent years there has been a dramatic growth in employment tribunal cases and an escalation in compensation claims. In the year 2000, 316 cases of discrimination were submitted to the tribunals, paying out £3.5 m in compensation. In addition, £1.2 m was paid out to one individual, whose case set a record for 'injured feelings'. Around 43 per cent of the total costs paid out by tribunals related to injury to feelings. It is interesting to note that awards are payable even if there is no reason to suppose that someone's feelings have been injured. The figure for 2000 represented a 38 per cent increase on the previous year. Given the number of out-of-court settlements, these figures represent the tip of the iceberg. Apart from the financial cost to the company, there is also the opportunity cost as customers, concerned about the image of the contractor and, by association, themselves will begin to source from equality-driven suppliers.

Positive action and the management of diversity

Despite the potential of positive action initiatives, which is an accepted approach within the relevant legislation in the United Kingdom to target recruitment at under-represented groups within the workforce, the CITB and Royal Holloway London University study found that three-fifths of the organizations consulted did not take any form of positive action towards race equality. This represents a major missed opportunity to develop a culturally and ethnically mixed workforce which would bring to the organization many of the benefits highlighted earlier. Although a number of specific schemes have been pursued in the last few years, they were often regarded as having failed to achieve their objective.

A number of initiatives have been developed through Government funded projects, City Challenge schemes and housing to promote equal access and participation of BME people in employment on the projects. However, several of these initiatives have failed both in terms of attracting the appropriate calibre of trainees and also in not being able to secure full-time employment for the trainees upon completion of the project or training programme. A further criticism is that they failed to employ sufficient numbers of local BME people, despite this being a requirement by the funding body, for example, regeneration funds or European funding. It has been suggested that, although the employment of local labour was included in contract specifications, this was rarely complied with or monitored (see, for example, Sodhi and Steele, 2000).

In the last few years, a range of other initiatives have been pursued to encourage ethnic minorities to enter and develop their career opportunities within the construction industry. For example, the CITB has advocated a positive action approach via the use of targets or quotas for the representation of BME staff within the industry. The CITB announced in the summer of 2001 that it wanted half of those entering construction to be women and from ethnic minorities and, more specifically, set itself the target of attracting 10 per cent more Black and Asian recruits to the industry by 2002. However, there has been some concern expressed by those in the industry about the establishment of quotas. The general sentiment is echoed by the following comment by the managing director of a brickwork specialist company employing 700 staff from a range of cultural backgrounds, 'I think you have to be careful with quotas. They are difficult to enforce and you always run the danger of diluting quality' (Akilade, 2000). A number of BME construction professionals have also expressed doubts about such an initiative, suggesting that this could be misconstrued as giving favourable treatment to particular community groups, highlighting another form of discrimination (Building, 2000).

In addition the equal opportunities pressure group Change the Face of Construction has launched a mentoring scheme. The scheme aims to support ethnic minorities in the construction industry by providing one-to-one sessions with a mentor to help with career development, and express hopes and fears. There is no information available about whether the scheme has succeeded in its aims; however, a similar pilot scheme for junior BME staff within the social housing sector has highlighted the general benefits of such a scheme for the mentees. In particular, the scheme enabled participants to share their experiences and network, broaden their horizons and knowledge, meet senior staff within the social housing sector in their mentoring role, build up their confidence and self esteem, become motivated and inspired and develop a sense of hope about the future – 'it reaffirms to me that there are people who are interested in minority groups being in housing. It has given me a hope that things will change and people are willing to assist in that change' (Miller, 2001: 15). The scheme within the construction industry has the support of some of the major key players who see mentoring as 'a good way of implementing change' (Construction Manager, 2000).

Conclusions

There is some scepticism about policies or initiatives which provide for special treatment of people by virtue only of their membership of a group – for example, policies of positive action and the adoption of racial equality targets. Kandola and Fullerton (1998: 34) suggest that 'positive action...is no better than applying a sticking plaster to a festering wound: it addresses the symptoms rather than the causes. It also provides activity without being purposeful'. In their view, the organization should be seeking answers to the question, 'Why are we failing to attract suitably qualified ethnic minority applicants in the first place?' Similarly, they regard minority targets as problematic in principle, not only because they single out groups for special attention, but because they are associated with a failure to examine the decision-making process and skills used by managers and also because they result in tokenism (Kandola and Fullerton, 1998: 141). Moreover, there is evidence that BME SMEs do not favour such an approach (Sodhi and Steele, 2000).

Rather than pursue positive action policies and initiatives, companies should be considering ways in which they can change the culture of their organization so that equality of opportunity becomes a core value. However, there is little evidence that companies in the construction industry are rising to this challenge. In the absence of self-regulation, changes are only likely to occur if there are pressing financial reasons for doing so – 'achieving equal opportunities is not the main priority, making profits is the main concern' (CITB and Royal Holloway University of London, 1999: 40). Here then, is an important role for clients and particularly public bodies who, under the Race Relations (Amendment Act) 2000 have a duty to promote race equality, eliminate unlawful discrimination and promote good relations between people of different racial groups. They also have the financial muscle to ensure that construction companies comply: those who do not may face an uncertain future. The Race Relations (Amendment Act) 2000 does not directly apply to housing associations, the main social housing developers, but the Housing Corporation which is their regulatory body is covered by the legislation and is promoting compliance. Clients can insist on a whole range of measures to ensure that BME people are fairly represented within the construction industry, perhaps the most important of which is ensuring that companies can demonstrate that they have an equal opportunities policy which provides more than lip service to ethnic minority communities.

Discussion questions

1   Why do so few Black and Minority Ethnic people choose to enter the construction industry?
2   What are the main barriers to employment progression within the construction industry for male and female, Black and Minority Ethnic workers?
3   What are the key features of successful initiatives addressing the imbalance of the number of Black and Minority Ethnic people in the construction industry?
4   What effective role can public sector and quasi public sector adverts play in improving the employment participation role for Black and Minority Ethnic groups in the construction industry?

References

Agapiou, A., Price, Andrew, D.F. and McCaffer, R. (1995) 'Planning future construction skill requirements: understanding labour resource issues'. *Construction Management and Economics*, 13: 2: 149–161.

Akilade, A. (2000) 'Race taskforce calls for mandatory quotas,' *Building*, 4 February, 13: 5: 17–18.

Ashworth, A. and Harvey, R.C. (1993) *The Construction Industry of Great Britain*, Butterworth-Heinemann, Oxford.

Barlow, J., Cohen, M., Jashapara, A. and Simpson, Y. (1997) *Towards Positive Partnering: Revealing the Realities in the Construction Industry*, Policy Press, London.

Building (1999) 'Minister slams industry record on minorities', *Building (Editorial)*, 3 December, 14.

Building (2000) 'Lipton offers to help black architects', *Building (Editorial)*, 24 November, 15.

Building (2001) 'CITB launches drive to hire minorities', *Building (Editorial)*, 7 September, 36.

Building E = Quality (1996) *A Discussion Document – Minority Ethnic Construction Professionals and Urban Regeneration*, RIBA.

Cavill, N. (1999) 'Purging the industry of racism', *Building*, 14 May, 20–22.

CITB (1991) *Technological Change and Construction Skills in the 1990s*, Research Document, Construction Industry Training Board, Bircham Newton, London.

CITB and Royal Holloway University of London (1999) *The Under-representation of Black and Asian People in Construction*, Construction Industry Training Board, Bircham Newton, London.

Construction Manager (2000) 'Construction managers lead on equal opportunities', *Construction Manager (Editorial)*, 21 September, 9.

Davey, C., Davidson, M., Gale, A. Hopley, A. and Rhys Jones, S. (1999) *Building Equality in Construction – Good Practice Guidelines for Building Contractors and Housing Associations*, Manchester, MSM Working paper, UMIST, No. 9901.

Druker, J. and White, G. (1996) *Managing People in Construction*, Institute of Personnel and Development, London.

Evans, D. (2002) 'Guarding against inequality', *Building*, June, 12.

Fellows, R., Gale, A.W., Hancock, M.R. and Langford, D. (1995) *Human Resources Management in Construction*, Longman, London.

Gruneberg, S.L. (1997) *Construction Economics – An Introduction*, Macmillan, London.

Hampton, J. (2000) 'Construction set to kick out racists', *Construction Manager*, September, 14–15.

Harding, C. (2000) 'Thanks for nothing', *Building*, 14 January, 25.

Kandola, R. and Fullerton, J. (1998) 'Diversity in action: managing the mosaic', Second edition, Institute of Personnel and Development, London.

Latham, M. (2001) 'Rules of attraction', *Building*, 23 March, 33.

Loosemore, M. and Chau, D.W. (2002) 'Racial discrimination towards Asian operatives in the Australian construction industry', *Construction Management and Economics*, 20: 91–102.

Mackenzie, S., Kilpatrick, A.R. and Akintoye, A. (2000) 'UK construction skills shortage response strategies and an analysis of industry perceptions', *Construction Management and Economics*, 18: 853–862.

Miller, S. (2001) *The Mentoring Master Class Evaluation*, The Places for People Group/University of Salford.

Modood, Y. (1997) 'Employment', in Modood, T. (ed.) *Diversity and Disadvantage: Ethnic Minorities in Britain*, Policy Press, London.

Morton, R. and Jagger, D. (1995) *Design and the Economics of Buildings*, Chapter 5, Clay Limited, England.

Pandya, N. (1999) 'Mentors for black and Asian managers', *The Guardian*, 26 June.

Rainbird, H. (1991) 'Labour force fragmentation and skills supply in the British construction industry', in Rainbird, H. and Syben, G. (eds) *Restructuring A Traditional Industry*, Berg, Oxford.

Rogers, D. (2002) 'Working Together', *Electrical and Mechanical Contractor*, October, 12, 16.

Singh, G. (2000) 'Why construction managers can't ignore ethnic minorities', *Construction Manager*, September, 6: 6: 11.

Sodhi, D. and Steele, A. (2000) *Contracts of Exclusion: A Study of Black and Minority Ethnic Outputs from Registered Social Landlords Contracting Power*, Equal Opportunities Federation, University of Salford/London.

Somerville, P. and Steele, A. (1998) *Career Opportunities for Ethnic Minorities*, University of Salford, London.

Somerville, P., Steele, A. and Sodhi, D. (2000) *A Question of Diversity: Black and Minority Ethnic Staff in the RSL Sector*, Housing Corporation, Research 43, Housing Corporation, London.

Somerville, P., Steele, A. and Sodhi, D. (2002) 'Black and minority ethnic employment in housing organisations', in Somerville, P. and Steele, A. (eds) *Race, Housing and Social Exclusion*, Jessica Kingsley, London.

Steele, A. (2001a) 'The housing and social care needs of the Asian community in Bury', Bury Metropolitan Borough Council/University of Salford.

Steele, A. (2001b) 'The housing and related needs of the Black and Minority Ethnic community in Tameside', Tameside Borough Council/University of Salford.

Thomas, J.C. (1996) 'Direct action', *Building*, 3 May, 36–38.

Tomlins, R. (1994) 'Housing associations and race equality: the report of the CRE into the housing association movement', *Housing Review*, 43: 2: 26–27.

UCAS (1998) *Annual Report*. Universities and Colleges Admissions Service, Cheltenham.

Chapter 11

Embracing diversity through the employment of disabled people

The missed opportunity?

Marcus Ormerod and Rita Newton

Introduction

This chapter examines the issues facing the construction industry in terms of enabling disabled people to participate fully in society. The construction industry be it through the design of buildings, or through the employment of people in those buildings, or through the employment of people within the construction process itself, has a responsibility to ensure that the environment it creates, is fully inclusive. An environment that differentiates or segregates leads to people not being able to fully participate in society, which in turn can lead to discrimination. An example of this can be seen through a survey undertaken by the UK Disability Rights Commission (2003a). They interviewed 305 young disabled people aged 16–24 about their experiences. They concluded that there are still many barriers in society preventing young disabled people from achieving their aspirations and having equal access to education, employment and social activities in their communities. 'This survey found that young disabled people often had to deal with prejudice and discrimination in many areas of their lives.'

The chapter initially describes models of disability that have emerged in recent years. The aim here is to establish our understanding of the nature of disability and our attitude towards it. The authors then suggest that there is a need to break down the barriers to disability through a twofold process of ensuring that we more effectively design buildings, and have employment practices that support social inclusion. However, this has to be set both within the context of the UK legislative framework (The 1995 Disability Discrimination Act), and the context of the construction industry itself. The chapter concludes by providing examples of opportunities available to employers in enabling and supporting disabled employees.

The nature of disability and our attitude towards it

Models of disability

Disability, like old age, is something that applies to other people but rarely do we admit that it is applicable to ourselves. Take a moment to consider the question: 'are you disabled?' The answer you give, more likely than not, will be 'not me'. This is strange when estimates for the numbers of disabled people in the United Kingdom vary from 10 to 20 per cent of the population. Nearly 1 in 5 people of working age reading this book should be a disabled person. But this is understandable, if the first situation is seen as an individual perspective and the second a statistical calculation. Individually we attempt to retain our uniqueness and resist attempts to be pigeon-holed into a category. Statisticians, on the other hand, much like government officials, use the seduction of numbers to argue their case and are comfortable seeing disabled people as a numerical value grouped together for the convenience of the analyst rather than the benefit of the user.

Traditionally disability has been considered to be a feature of certain individuals whose minds, bodies or senses appear to be different from a perceived norm due to impairments. People with impairments are viewed as being the source of the problem of disability; emphasis is placed on correcting this malfunctioning to bring the individual closer to a perceived normality. Impairments can be broadly seen as pertaining to mobility; sensory – mainly focusing on visual and hearing; cognition; psychological; size, shape and appearance and those acquired through the ageing process. This traditional perspective has been termed the individual or medical model of disability (Oliver, 1983). This approach results in a requirement for disabled people to be managed by professionals, who make assessments of individuals and administer services to compensate for an individual's inadequacies.

By contrast, through the result of discussion and debate, disabled people have identified that it is barriers in society that are the cause of disability (UPIAS, 1976) and have developed a social model of disability. The line of reasoning is that society is created and operates in a way that does not take into account those people who do not meet the perceived norm and so excludes and thereby disables them. In this definition 'impairment' relates to an individual's condition of mind, body or senses that results in an individual functional limitation; 'disability' is the limitations imposed by a society that takes no account of people with impairments. So people with impairments are 'disabled' by society. In this definition the term 'people with disabilities' makes no sense.

Using the social model of disability, the disabling barriers are not just physical barriers but are also attitudinal, systematic and institutional.

These barriers are present in areas as diverse as education, employment, leisure, transport, urban design and housing (Barnes, 1991). One individual can encounter disabling barriers several times a day in all these areas making disability a whole life experience.

The social model of disability allows us to recognize that the causes of disabling barriers in design are influenced by the broader social and cultural context. This enhances our insight into the ways in which disabled people are excluded by the built environment and poorly designed artefacts. In order to make best use of the social model of disability it is important to be clear about which limitations are caused by an individual's impairment (difference in mind, senses or body) and which barriers are created by society.

Applying the models of disability to the buildings we create and use

Between the two definitions of the individual model and social model of disability are other interpretations, such as Goldsmith (1997) who describes disability as being part of an individual (medically disabled), and that disability can be a transient experience according to the situation. He recognizes someone can be medically disabled if they have a physical impairment, but someone does not necessarily have to have an impairment to be disabled.

So, for example, someone can be architecturally disabled by stairs or narrow doorways when they are pushing a child in a pram, but that disability is removed when a ramp or lift is provided. However, Goldsmith does not seem to acknowledge that the same person would not be disabled if they returned to the same architectural feature without the pram. For Goldsmith this state of architectural disability is the same for people with impairments as it is for others who are disadvantaged by a building (the pram pusher, for example) – the state of being disabled is not constant or consistent. Goldsmith (1997) describes 'the disabled' as being those people who are architecturally disabled, medically disabled or socially disabled. Whilst this interpretation gives recognition to all people who are disadvantaged by poor design, it denies the extensive discrimination and exclusion experienced by those he describes as medically disabled.

In the construction industry, there is a tendency towards the assumption that we are creating environments 'for' other people rather than 'with' people for everyone to use. This is interestingly comparable to the individual or medical model of disability with architectural solutions being developed with minimal involvement of the diverse range of people who will use the environment when it is finished. In social housing developments the lessons have been learnt on the importance of community involvement in the design process, but this is much less likely to happen in other types of developments. The majority of architectural projects still miss the opportunity of embracing the knowledge and expertise of users at the design and planning stages. A quick glance at any of the architectural journals will confirm the lack of images of people using the building and resistance to include people even if requested by the designers (Lawson, 2001).

This lack of involvement extends into all aspects of the construction process. The false image of the construction industry being a harsh physical world not suited to disabled people is constantly put forward. Yet how is this industry meant to understand the design requirements of disabled people if they are not involved in the building process? There are many opportunities within the construction industry for disabled people to be involved. The interaction of non-disabled people with disabled people is that most likely to break down attitudinal barriers.

Breaking down the barriers to disability through effective building design

Increasingly, there is recognition that the built environment should encourage inclusion rather than create barriers (Imrie and Hall, 2001). Funders of developments are placing restrictions on what they will fund and requiring statements of accessibility to demonstrate that discrimination can be avoided (Ormerod and Newton, 2004). In existing buildings, an access audit can reveal the potential barriers and recommend a strategy for improvements that links with the maintenance programme. With new developments, an access appraisal can explore where the designer may be introducing barriers, but there is also scope for the designer to adopt an inclusive design approach from the outset.

Inclusive Design is a way of designing products and environments so that they are usable and appealing to everyone regardless of age, ability or circumstance by working with users to remove barriers in the social, technical, political and economic processes underpinning building and design. The process used to achieve an inclusive design solution is just as important as the finished product, or environment. A key element of this process is user participation. This is an aspect that the construction industry has in the main chosen to ignore. The client/designer relationship is recognized as important, but that between user and designer, or user and client, is given little, if any, consideration. John Zeisel (1984) shows this through a user gap model and illustrates in Figure 11.1.

Paying clients tend to assume that the designers they employ will be able to articulate the aspirations of the user client. At the same time, the briefing process used by the designer tends to assume that the paying client will explain how they deal with users in their business, or that this will be teased out by the designer. Unlike product design (Keates and Clarkson, 2004) the construction industry rarely uses focus groups, or user input, to shape the design process. If we want to make our environments work for a more

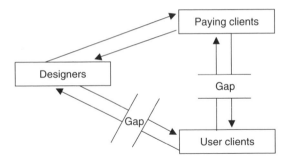

Figure 11.1 User gap model.
Source: Zeisel (1984).

diverse population then we need to develop tools that capture user information at the relevant stage of the design of the building.

Inclusive design also goes further than other comparable concepts. Accessible design provides solutions that make environments more usable, but often this is in a retrofit situation at extra cost to the client. Inclusive design starts at the outset of the initial planning and sets the scene for all design decisions made subsequently. Universal design (Ostroff and Preiser, 2001) uses an underlying seven principles to guide designers to inclusive solutions but focuses more on technical fixes and at times does not get to the root source of the exclusion. This could be argued to be a downfall of inclusive design in that it seeks where possible to eliminate the barriers to inclusion at the highest level, and at times might force the designer to regress back to a level at which they cannot make any changes. This is typified by the inclusive solution that embraces a wide range of users, but is expensive and therefore excludes unemployed disabled people on an economic basis.

The social, moral, legal and commercial benefits of inclusive design (DPTAC, 2003) form a powerful case for adopting a best practice approach in developing the built environment for the future. However, in adopting an inclusive design approach there are challenges that need to be addressed by all those involved in the process:

- Seeking inclusive design solutions rather than just additional accessible options.
- Involving disabled people in planning/design processes in a meaningful manner is not always a straightforward process.
- Balancing project requirements whilst increasing diversity requires a holistic view of the project.
- Inclusive design is an emerging area of practice and best practice is continually evolving.

- Legal issues such as reasonableness have yet to be tested, but duties in the DDA are anticipatory. New developments require long-planning periods and will have to anticipate the future position on accessibility standards.
- Historic properties, monuments and conservation areas can create a tension between preserving the significance of the site versus creating an accessible and inclusive experience.

It is easy to consider the designer as the person solely responsible for delivering inclusive design, when in fact, it requires the involvement of all parties in the construction process in order to be successful. Operatives on site are just as likely to be able to influence the inclusivity of a building project as the designer since they determine the actual positions of the many components that make up the finished product. For example, it is good practice to make accessible (commonly referred to as 'disabled') toilets differing hands on each floor of a multi-level building. This allows wheelchair users a choice if they can only transfer from one side. However it is common for the plumber to fit the flush handle on the same side of the cistern in both left and right hand toilets unaware that the handle trapped in the corner will be awkward to reach by a wheelchair user from the transfer side. Training of everyone involved needs to appreciate the wide range of user requirements and focus on what element of their input into the finished product is likely to engender inclusivity and remove potential barriers (Ormerod and Newton, 2003; SKILL, 2001).

Breaking down the barriers to disability within the construction industry

In order for the construction industry to respond positively to disabled people, either as employees, or as members of the wider community accessing the built environment, the whole industry has to take on board its responsibilities for social inclusion and look at ways to encourage change. Within the context of professional skills for example, within the construction sector, Fairclough (2002) highlights the current difficulties that the industry is facing and suggests possible remedies for this. Fairclough suggests that.

> there is considerable concern that the supply of professional skills required in the maintenance of the built environment has not matched the changing needs of the sector...the industry needs to find ways of attracting and retaining bright people – including the widening of its appeal to women and other under-represented groups.

Where a disabled person may not be able to directly work as a professional person on a construction site, by broadening the scope and role, a person may be easily accommodated within a wider job function. However, this can only be achieved if the industry takes a wider perspective on roles and responsibilities. Fairclough (2002) refers to the industry attempting to attract 'its new blood into the same old silos' since the industry is fragmented because of the many disciplines involved, and the 'desire of the professional institutions to maintain these silos as barriers to the much needed multi disciplinary approach to taking the industry forward'.

However, this approach to changing the nature of the 'silos' assumes that there will be sufficient people to fill this multi-disciplinary demand. In terms of disabled people, there are a significant number of disabled people who are capable and able to work, but they are unable to find work. Some 3.4 million disabled people were in employment in the United Kingdom in autumn 2001 which is only an increase of 2.7 per cent over a three-year period. The employment rate for disabled people is 48 per cent, compared to an employment rate of 81 per cent for non-disabled people (LMT, 2002). Clearly there is a problem. On the whole, disabled people feel that employers do not have fair recruitment practices, and that when they are in employment they are not treated fairly (DRC, 2003a). On the other hand, typically employers are unaware of the enabling and support mechanisms for disabled people when in employment. Also, employers need to have a greater understanding of their legal obligations in effectively employing disabled people, so there would be less employment tribunal cases on disability discrimination and employment.

The legal position on employing disabled people

Overview of the Disability Discrimination Act (1995) and related Codes of Practice

The Disability Discrimination Act 1995 (DDA) brought in measures to prevent discrimination against disabled people. Part 2 of the Act is based on the principle that disabled people should not be discriminated against in employment or when seeking employment. It also protects disabled people engaged in a range of occupations outside employment. Employers must comply with the duties set out in Part 2 of the Act, as must others to whom those duties apply. Except for the armed forces, these duties now apply (October 2004) to all employers (regardless of number of employees). In November 2000, the Council of Europe passed a Directive (2000/78/EC General Framework Directive for equal treatment in employment and occupation) requiring all European Union member states to introduce legislation prohibiting discrimination in employment on the basis of disability, age, religion and sexual orientation (race and gender discrimination is already prohibited by the EU). In relation to disability (and age), this must be done by 2006 at the latest. In response to this, the Disability Discrimination Act 1995 (Amendment) Regulations 2003 are enforced from 1 October 2004. The changes relate primarily to Part 2 of the DDA, although in respect of one particular issue (employment services) there will be changes to Part 3 of the Act as well. The Disability Rights Commission (DRC) has produced two new Codes of Practice (2003b,c) giving guidance on the operation of Part 2, and taking full account of the new Regulations. These replace the existing Codes of Practice. The DRC describe the status of the Codes (s51(3)-(5)),

> The Code does not impose legal obligations, nor is it an authoritative statement of the law – that is a matter for the courts and the tribunals. However, the Code can be used in evidence in legal proceedings under the Act [DDA]. Courts and employment tribunals must take into account any part of the Code that appears relevant to them to any question arising in those proceedings. If employers (and others who have duties under Part 2 of the Act) follow the guidance in the Code, it may help to avoid an adverse decision by a court or tribunal in such proceedings.

The European Directive applies to all areas of employment and occupation and, as a result, a large number of employment-type situations are covered by the DDA. Examples of these include:

- Every employer, regardless of number of employees is covered. This is a change because prior to the Regulations being issued, an employer with less than 15 employees was exempt.
- Additionally, all employers including the police, fire service, barristers, people in partnerships are covered.
- For people who have left their employment, post-termination discrimination is prohibited if it arises out of and is closely connected with the relevant relationship.

Exclusions are minimal, but include

- the armed services (the European directive permitted member states to have an exemption for the armed forces in relation to the disability and age provisions).

Types of discrimination

Under the DDA (1995, with 2005 amendments), there are three kinds of unlawful discrimination, namely

- less favourable treatment
  - direct discrimination on the basis of a persons' disability;

- direct discrimination on the basis of a disability-related reason (known as 'residual less favourable treatment');
- failure to comply with a duty to make reasonable adjustments;
- victimization.

The DDA says that an employer's treatment of a disabled person amounts to 'direct discrimination' if

- it is on the grounds of his or her disability;
- the treatment is less favourable than the way in which a person not having the disability would be treated;
- the relevant circumstances of the person with whom the comparison is made, are not materially different from the disabled person.

By contrast, 'residual less favourable treatment' concerns treatment which is for a 'disability-related reason' rather than 'on the grounds of disability'. With direct discrimination, the key to understanding the difference is to focus on the individual and to apply the three points given here. The DRC (2003b) provide an example of a disabled person with serious back pain who takes six months sick leave because of his disability and is dismissed by his employer. A non-disabled fellow employee (the non-disabled person to be compared against in point 2) also takes six months sick leave due to a broken leg but is not dismissed, hence direct discrimination on the basis of a persons' disability has occurred. With residual less favourable treatment, the comparison must be made with person to whom the disability-related reason does not apply (rather than a non-disabled person in the same circumstances).

The duty to make reasonable adjustments applied where 'a provision, criterion or practice' applied by an employer places a disabled person at a substantial disadvantage, compared to a non-disabled person. The employer has to take reasonable steps to prevent the disadvantage. Up to October 2004, there was no duty to make adjustments where the physical features comply with Part M of the building regulations. However, from 1 October 2004, any physical features may be subject to the reasonable adjustment duty, regardless of whether they meet the requirements of any building regulations. The DDA (s 18B(2)) gives examples of adjustments or 'steps' which employers may have to take and these have been described in Chapter 2.

Victimization is a special form of discrimination by which it is unlawful for one person to treat another (the victim) less favourably than he treats or would treat other people in the same circumstances (DDA s 55(1) and (2)). Additionally, the Regulations (2003) (DDA s 3B(1) and (2)) now define 'harassment' separately from discrimination, and prohibit harassment in relation to each category of person/employment or occupation situation. The position is a little less clear concerning 'constructive dismissal' in that in the case of Commissioner of Police of the Metropolis v Harley [2001]

IRLR 263 the Employment Appeal Tribunal (EAT) held that the phrase 'by dismissing him' does not include a constructive dismissal; in a subsequent case, Catherall v Michelin Tyres PLC [2003] IRLR 61, it was held that the DDA does cover constructive dismissal, thus leaving the situation somewhat unclear. The Regulations (regulation 4) would ensure that constructive dismissal situations are protected by the DDA.

The changes which have now been made to the Regulations (2003) do go some way towards reducing the basis on which employers can justify discrimination (quite how far depends on case law interpretation), although it remains possible to justify some treatment on the basis that it was for a 'material and substantial reason'. Justification will only apply to less favourable treatment; it will no longer be possible to justify a failure to make reasonable adjustments – either an adjustment will be reasonable or it will not. The Regulations say that treatment against a disabled person cannot be justified if it amounts to direct discrimination (DDA s 3A(4)). However, justification may be relevant in cases about 'residual less favourable treatment'.

The Act does not prohibit an employer from appointing the best person for the job, nor does it prevent employers from treating disabled people more favourably than those who are non-disabled people.

Proving discrimination and the burden of proof

Enforcement of rights in respect of discrimination in any of the circumstances contemplated by Part 2 of the DDA is by way of a claim to an employment tribunal (there is an exception for this in relation to qualifications bodies, in that where there is a statutory means of appeal relevant to that body, claims must be brought via that statutory route).

The Regulations alter the rules about the burden of proof in Part 2 claims. A person who brings a claim for unlawful discrimination under Part 2 must show that discrimination has occurred. The person must prove this on the balance of probabilities in order to succeed with a claim in an employment tribunal. However, if it is apparent from the facts of the case that there has been discriminatory conduct, the onus will shift to the employer to provide an adequate explanation for what has taken place. If the employer cannot provide such an explanation, the tribunal will uphold the claim. The DRC through the Code of Practice provide an example of this,

> A disabled employee scores very poorly in a redundancy selection process in comparison with other employees in the same position as himself, getting low marks for skill and performance in his job. The employee has always had good appraisals and no action was ever taken against him in respect of competence. In these circumstances, it would be for the employer to demonstrate that the reason for the low scores was not related to the employee's disability.
>
> (2003b)

Challenges for disabled people and employers

It would be easy for employers within the construction industry to fail to address their legal obligations in ensuring that policies and practices support the employment of disabled people. Within the context of the construction industry, some of the challenges that employers may face in so doing, are now described.

The previous exemption for small businesses

When the DDA was first introduced in 1995, the wording in section 7(1) was such that 'nothing in this Part applies in relation to an employer who has fewer than 20 employees'. This exemption was subsequently reduced to less than 15 employees. From 1 October 2004, this exemption is repealed. The implications of this are that the DDA employment provisions will now apply to any size of organization, and given that a significant number of construction companies would previously have been exempt this will have a significant impact in bringing all construction-related organizations into the scope of the DDA employment provision. Research by Roberts *et al.* (2004) has shown that smaller employers, in particular are unsure about the implications of the DDA for their organization.

Using health and safety as a reason to discriminate

The DRC Code of Practice (2003b) suggests that stereotypical assumptions about the health and safety implications of disability should be avoided. The Code further suggests that less favourable treatment which relates to such assumptions may count as direct discrimination – which is incapable of justification. Health and safety legislation does not require employers to remove all conceivable risks, and if there are concerns, it is appropriate to have a risk assessment carried out by an appropriately qualified person.

Research by the DRC and Health and Safety Executive (DRC, 2002b) has found evidence of organizations taking positive steps to overcome health and safety barriers to the recruitment or retention of people with a disability. It would seem that where a person acquires a disability whilst being employed, that the organization is generally keen to maintain the employment of the individual and to seek to make the necessary adjust-ments. This is supported by an evaluation of the government supported ONE programme (DWP, 2001) designed to increase employment opportu-nities for long-term unemployed people. It found that 62 per cent of employers said they would employ a disabled person.

However, equally there was evidence of a substantial number of organizations in the DRC survey who did not recruit a person because of their disability on the grounds of the health and safety risk. It was not possible from the research to determine whether or not these decisions were justified because individual case studies were not undertaken. However, if we refer to the previous evidence of the employment rate for disabled people being approximately that of non-disabled people (LMT, 2002), this is a real cause for concern. One of the main conclusions of the research therefore was that health and safety is frequently reported as being used as the rationale for the non-recruitment or dismissal of disabled people, and this in part would seem to be true. Clear guidance (though not currently available) is needed by employers on what constitutes a health and safety risk for a disabled people or those with poorer health.

Using medical information as an excuse

An example is provided (DRC, 2003b) of a labourer who some years ago was disabled by clinical depression but has since recovered now wants to work on a different building site. Although the labourer's past disability is covered by the DDA, the site manager refuses to accept him because of his medical history. It is likely that the contractor is acting unlawfully.

Discrimination against partners in firms

The DDA (s 6A(1)) says that it is unlawful for a firm, in relation to a position as a partner in a firm, to discriminate against a disabled person.

> A disabled person who uses a wheelchair as a result of a mobility impairment joins a firm of architects as partner, receiving 20 per cent of the firm's profits. He is asked to pay 20 per cent towards the cost of a lift which must be installed so that he can work on the premises. This is likely to be reasonable.

Work placements

The duty to make reasonable adjustments applies to a placement provider in the same way that it applies to an employer (DDA s 14D). Examples of this are:

- A student with learning disabilities on placement to a building-surveying firm is given personal instruction on health and safety procedure rather than written information.
- A disabled student on placement to a building control department has a personal support worker. The placement provider facilitates this by providing an extra work station for the support worker.

Opportunities for employers in enabling and supporting disabled employees

There are various opportunities for practical help and advice available to employers considering taking on a disabled person as an employee and research has shown that the proportion of employers employing disabled people or people with long-term health problems increased from 87 to 95 per cent in 2002 (Equal Opportunities Review, 2002). However, many employers are unaware of the extent of these opportunities and may unnecessarily exclude someone from a recruitment process due to the lack of this knowledge. This also leads to disabled people being naturally reluctant to complete application forms that ask for statements about their disability, even if the request is to genuinely help create an equal application process.

Access to Work is a scheme run by the Jobcentre Plus that primarily provides help to disabled people in getting employment through funding adaptations to premises and equipment; communication support at interviews; special aids and equipment; support workers and travel to work expenses for disabled people who incur extra costs due to their disability. The range of help extends from one-off interview adjustments through to long-term solutions for continuous employment situations. Employers must take into account that any benefits gained by other employees through the provision of adaptations for a disabled employee are likely, and as such Access to Work will only then fund a proportion of the costs. It is usual for Access to Work to fully fund items that only the disabled employee will use, such as a Braille embosser, but only a small percentage of a lift installation that everyone will then use. Another job support provided by Jobcentre Plus is WORKSTEP that seeks to provide assistance to disabled people who face more complex barriers to getting or keeping a job.

Jobcentre Plus recognizes employers who have agreed to meet commitments on recruitment, employment, retention and career development of disabled people by allowing them to use a Disability Symbol, commonly referred to as the two-tick scheme. The idea being to help disabled people identify organizations that are positive about disabled people. However, the construction industry has rarely taken up this opportunity to encourage disabled people with only 2 per cent of the list of Disability Symbol using organizations being in construction-related fields (Jobcentre Plus, 2003). There is also the New Deal for Disabled People that aims to help people on disability and health-related benefits to obtain and retain paid work through a network of job brokers.

The Employers' Forum on Disability provides advice to its members on a range of disabled employee issues, such as policy audits, DDA legal support and an extensive publication list of briefings and fact sheets. There are 36 Gold Card members of which only 1 is construction industry related and 382 organizations listed as ordinary members with 5 being construction based. In 1992, they conducted research to benchmark the extent of global inclusion and identified a 13-point check list against which companies should audit themselves. No companies in the research managed to report on disability in every benchmark criteria and whilst 28 per cent made reference to disability as a component of their equal opportunities policy only 12 per cent had disability as part of their strategic plan.

Employers are more likely to comply with their duties under the Disability Discrimination Act (1995) and to avoid the risk of legal action being taken against them, if they implement anti-discrimination policies. The proportion of employers with disability policies went up from two-thirds in 2001, to 90 per cent in 2002 (Equal opportunities Review, 2002). The Code of Practice (2003b) suggests the steps that an employer may undertake to

- establish a policy which aims to prevent discrimination against disabled people, and which is communicated to all employees and agents of the employer;
- provide disability awareness training to all employees;
- provide specialist training for people managing recruitment and training;
- inform employees that breaches of the policy are unacceptable, and respond quickly to any such occurrences;
- monitor the implementation and effectiveness of the policy, including consulting disabled employees about their experiences of working in the organization;
- design an effective complaints and grievance procedure to support the policy;
- keep clear records of decisions taken in respect of the aforementioned.

This is illustrated diagrammatically in Figure 11.2.

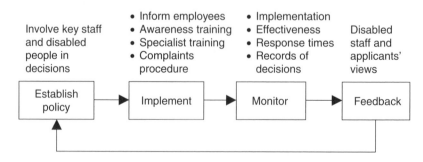

Figure 11.2 Implementing effective employment policies.

Conclusions

In the United Kingdom, disabled people are twice as likely as non-disabled people to have no qualifications, half as likely to go to University and, for those of working age, five times as likely to be out of work and claiming benefits (DRC, 2002a). In terms of employment, a DRC survey (2003a) found that 86 per cent of young people thought it was harder for disabled people to find jobs than non-disabled people. Thirteen per cent of interviewees had been turned down for a job because of their impairment; 18 per cent had been turned down for a job and told the reason was not related to their impairment, but they believe it was; 25 per cent of interviewees said they tend to apply for jobs that they did not really want. The British Government is seeking to overcome these barriers, and other barriers to employment experienced by disabled people through new legislation, and by providing guidance on policies and practices to support disabled people. The recent government paper on employment (DTI, 2003) sets out common values with European partners in 'putting high employment and social inclusion at the heart of our economic and social agenda'. Yet it recognizes that the 'challenge is making this happen'.

The construction industry accounts for a tenth of the United Kingdom's gross national product, and employs 1.4 million people (DTI, 2003). Further, the quality of the output from the construction industry provides the foundation for the work of other industries. The industry therefore has a responsibility to seek to make a difference in disabled people's lives by designing an environment that is inclusive, and by implementing employment practices that are non-discriminatory. Both of these responsibilities, if enacted, would lead to an improved quality of life for disabled people, and a society with a different attitude and approach supporting inclusion rather than exclusion; integration rather than segregation; opportunities for all rather than just for some. If this were to be achieved, then disabled people's lives would be significantly different since

> inclusion itself is a powerful way of changing non-disabled people's beliefs. When non-disabled people get to know disabled people, in a context in which disabled people are at least equal, and where positive messages are generated, this does influence belief systems.
>
> (Ralph, 1989)

Employment has a strong role to play in influencing our belief systems. Recent research by the DRC (2003a) finds that the group with highest disability discrimination awareness and the most inclusive attitudes about disability are people who work with someone who is disabled. They score higher than other disabled people themselves or relatives of disabled people. Employment opportunities for disabled people, and effective support for disabled people when in employment can both contribute to improving the skills and capacity of the UK workforce. Yet in the construction industry, there is very little focus on disability. If we think of diversity, we consider gender and ethnicity, and rarely disability. This chapter has shown that the industry has to change by

- developing a social model approach;
- generating inclusively designed buildings and environments;
- challenging discrimination;
- providing enabling policies and practices;
- supporting disabled people to, and in, employment;

There are currently very few statistics available on employment of disabled people within the UK construction industry (LMT, 2002), but the signs are that the industry is not proactively encouraging disabled people to consider employment in this area. Clients and other industries are well aware that disabled people, their friends and families are a significant purchaser of goods and services. A wake-up call is needed in the construction industry to develop an employment profile that is more representative of the society in which it operates. The image of the construction industry needs to change to one that is attractive to a more diverse population. If the construction industry is to create truly inclusive environments then it needs to be by working with, rather than for, a diverse society.

Discussion questions

1   Describe the key areas a training course seeking to cover disability issues to construction industry personnel would require. To what extent would the training course differ for different sectors of the workforce?
2   Assess the differences between accessibility, Universal Design and Inclusive Design with the context of construction projects by taking an example of a new build project and contrasting this to a refurbishment of an existing building. How would the design approaches differ between the two schemes in practice?
3   If the real challenge for employing disabled people is 'making it happen' (DTI, 2003) consider five possible ways in which the construction industry as a whole could seek to embrace a diverse workforce by employing more disabled people. Assess the extent to which your five ways are achievable, suggesting how a construction company may seek to implement your ideas.

References

Barnes, C. (1991) *Disabled People in Britain and Discrimination*. London: Hurst/ British Council of Disabled People.

Department for Work and Pensions (2001) *Recruiting Benefit Claimants: Qualitative Research with Employers in ONE Pilot Areas*. Research Paper Series 173. London: DWP.

DPTAC (2003) *Inclusive Projects*. London: DPTAC. http://www.dptac.gov.uk/inclusive/guide/02.htm#3

DRC (2002a) *Disability Briefing*. October. London: The DRC.

DRC (2002b) *Health, Safety and Disability: Are there Conflicts at Work?* London: The DRC.

DRC (2003a) *The DRC 2003 Attitudes and Awareness Survey*. London: The DRC.

DRC (2003b) *The DDA Employment and Occupation Code of Practice*. London: The DRC.

DRC (2003c) *The DDA Trade Organisations and Qualifications Bodies Code of Practice*. London: The DRC.

DTI (2003) *Employment Relations – Equality and Diversity: The Way Ahead*. http://www.dti.gov.uk/er.equality/

Employers' Forum on Disability (2002) *Global Inclusion Benchmark*. London: Employers' Forum on Disability.

Equal Opportunities Review (2002) 'The DDA after four years: part 1 – the meaning of disability'. *Equal Opportunities Review*. 94, 12–19.

Fairclough, Sir J. (2002) *Rethinking Construction Innovation and Research. A Review of Government R & D Policies and Practices*. London: HMSO.

Goldsmith, S. (1997) *Designing for the Disabled – The New Paradigm*. London: Butterworth.

Imrie, R. and Hall, P. (2001) *Inclsuive Design. Designing Accessible Environments*. London: SPON Press.

Jobcentre Plus (2003) *Disability Symbol User List of Major And National Organisations*. London: Jobcentre Plus.

Keates, S. and Clarkson, J. (2004) *Countering Design Exclusion. An Introduction to Inclusive Design*. London: Springer-Verlag.

Labour Market Trends (2002) *Labour Market Experiences of People with Disabilities*. London: Office for National Statistics.

Lawson, B. (2001) *The Language of Space*. Oxford: Architectural Press.

Oliver, M. (1983) *Social Work with Disabled People*. Basingstoke: Macmillan.

Ormerod, M. and Newton, R. (2003) *The Application of Research Theory to Provide Widened Access for Students with Disabilities Through a Virtual Learning Environment*. Best Practice in Building Education: HEFCE.

Ormerod, M. and Newton, R. (2004) 'Point of view: access statements'. *Access Journal*. 26–27.

Ostroff, E. and Preiser, W. (2001) *Universal Design Handbook*. New York: McGraw-Hill.

Ralph, S. (1989) 'Using video-tape in the modification of attitudes towards people with physical disabilities'. *Journal of the Multi-Handicapped Person*. 2, 223–235.

Roberts, S. Heaver, C., Hill, K., Rennison, J., Stafford, B., Howat, N., Kelly, G., Krishnan, S., Tapp, P. and Thomas, A. (2004) 'Disability in the workplace: employers and service providers responses to the Disability Discrimination Act in 2003 and preparation for 2004 changes'. Research Report No. 202. London: Department for Work and Pensions.

SKILL (2001) *Into Architecture*. London: SKILL.

Statutory Instrument (2003) No. 1673. The Disability Discrimination Act 1995 (Amendment) Regulations 2003. London: HMSO.

The Disability Discrimination Act (1995) London: HMSO.

UPIAS (1976) *The Fundamental Principles of Disability*. London: UPIAS / Disability Alliance.

Zeisel, J. (1984) *Inquiry by Design. Tools for Environment-Behaviour Research*. Cambridge: Cambridge University Press.

Part IV

Managing and implementing diversity

Chapter 12

Equality in the social housing sector

Sandi Rhys Jones

Introduction

The social housing sector has a better record in equal opportunities than the construction industry, which serves it. This is not surprising when seen in the context of the social housing commitment to meeting the diverse needs of local communities, regenerating local economies and giving opportunities to all sections to benefit from social housing building programmes.

In this chapter, the issues that are generated by this mismatch are examined together with examples of initiatives, research and case studies undertaken to combat it.

Social housing providers

Social housing in the United Kingdom is provided through two mechanisms, Registered Social Landlords (RSLs) and Housing Associations (HAs). Since the late 1980s almost all new social housing has been provided by RSLs, which are independent, not for profit private sector organizations. In addition, local authorities were given the option of transferring their stock to RSLs. One of the main reasons for this is that RSLs, being in the private sector, can raise private finance for new schemes and for investing in stock transferred from local authorities outside the constraints of Public Expenditure control and the Public Sector Borrowing Requirement.

The Housing Corporation is the Government agency responsible for regulating and investing in nearly 2,000 housing associations in England. Its current investment programme of £3.3 billion for 2004–06 is funding over 63,000 affordable homes. Sixteen thousand of these will be for key workers and 25 per cent will use some form of modern method of construction.

The Housing Corporation's mission statement is to work to improve people's quality of life through social housing. This is underpinned by five strategic aims:

- To help revitalize and maintain sustainable housing and neighbourhoods.
- Act in the interests of tenants and residents.

- Secure the effective stewardship of existing stock.
- Safeguard the public interest.
- Encourage innovation and the pursuit of Best Value.

RSLs submit bids for the Social Housing Grant to the Housing Corporation Regional Offices in a competitive bidding round. The Corporation approves schemes that meet local housing needs at affordable rents and that offer good value for money in terms of the public subsidy required and the quality of homes delivered.

Diversity in social housing provision

Social housing providers have a statutory and a moral obligation to operate and promote robust equal opportunities and diversity. The provision of social housing is not only a commitment to meeting the diverse needs of local communities, it is also a service for the most disadvantaged people in the community. Their need for housing provision may arise from direct or indirect discrimination on the basis of their race, gender, disability or sexual orientation, for example. Some social housing providers are committed to servicing the needs of these particular disadvantaged groups.

The commitment to diversity in service delivery is reflected in a commitment to diversity in the organizational structure of the Housing Corporation, the national body that regulates the UK social housing. In its *Equality and Diversity, Policies & Strategies report* (2003), the Housing Corporation clearly states its commitment to be an organization that

- develops services to achieve equality and diversity in all its activities;
- has a workforce generally reflecting the population;
- understands how valuing diversity can improve our ability to deliver better services;
- actively consults with all our customers to ensure the delivery of our goals;
- supports Registered Social Landlords, Local Authorities and other bodies in the achievement of equality and diversity;
- actively consults with different individuals and communities to ensure that services, which are provided, are responsive and reflect the diversity of need;
- provides all employees with the training and development they need to enable them to achieve organizational goals;
- provides a supportive open environment where all employees have the opportunity to reach their full potential;
- listens to its customers and involves them in the development of services that recognize and value diversity;
- believes that both customers and employees have important parts to play in making this happen.

Women make up 25 per cent of the management structure within RSLs and HAs. According to research carried out for the Housing Corporation by the University of Salford in 2000 (Somerville, *et al.*, 2000), the proportion of staff working for RSLs of black and minority ethnic (BME) origin is similar to that for the working population generally, namely 6.5 per cent. However at senior management level the proportion of staff in mainstream RSLs who came from BME background was 3.7 per cent, which compares well with the civil service but falls short of equality. In BME-led RSLs, BME and white staff were found in roughly equal proportions at most levels of the organization. Even at the top, nearly one in six of Chief Executives were white.

In contrast, the construction industry that services the social housing sector has a poor record in equal opportunities. It is a predominantly white male industry. The low representation of women, the disabled and people from BME backgrounds is documented in Chapters 10 and 11, but in summary the number of women in the industry as a whole is less than 9 per cent, very few women hold top positions, are sparsely represented at professional level (3.4 per cent) and at trade and craft level this figure falls to 1 per cent. Whilst the number of BME entrants to construction increased from 1.9 to 2.3 per cent in the four years up to 2001, this figure is well below the 6.5 per cent of the working population that comes from BME groups.

In addition to applying its equality and diversity policy in carrying out its statutory and corporate responsibilities, the Housing Corporation has an explicitly stated commitment 'to apply it to work undertaken for us by external consultants and contractors and to organizations receiving funding from us'.

Most RSLs have partnering arrangements or refer to Constructionline who act as an agent that has vetted all companies registered on its database. This includes providing two years audited accounts, references from clients, insurance and an equal opportunities policy. However, finding practical and effective ways of applying this policy in the white male-dominated construction industry servicing the social housing sector has been a major challenge for a number of years. It has been long recognized that a contractor simply dusting off an equal opportunities policy document to support the ticked box in an invitation to tender document is not a guarantee of active commitment.

Diversity and government

Following the Race Relations (Amendment) Act 2000, the Department of Transport, Local Government and the Regions (DTLR) established a Diversity and Equality Unit to help divisions deliver on both staffing and programme issues. In particular the DTLR aimed to ensure that working practices within the DTLR Housing Directorate were effective in bringing about better co-ordination of work on race and housing issues. As part of this activity, all divisions within the Housing Directorate participated in diversity training and a system established to monitor the proportion of ethnic minority staff employed by the Department at each grade.

In addition, the DTLR pledged by the end of 2001 to establish a programme for implementing race proofing of policies and research across the DTLR Housing Directorate, including:

- Assessing what information on impact and outcomes is necessary to 'race proof' housing policies, for example, in terms of data collection and analysis; feedback from BME groups, representative bodies, etc.
- Putting in place mechanisms for identifying high priorities and tackling them early on.
- Awareness raising and guidance for staff in Housing Directorate on BME issues.

To reinforce the process, a series of checks were established to ensure BME issues are given due consideration:

- When major policy options, or changes of direction, are put to Ministers.
- When research specifications are drawn up.
- When regular annual bidding rounds or similar exercises are carried out;
- Prior to publication of any external material on housing, from White Papers through to information leaflets.
- Through the inclusion of an objective on implementing race proofing in divisional level objectives.

Research and initiatives

Building positive action

Building Positive Action (BPA) was set up in February 1995 as a North West Housing Association initiative primarily to improve business opportunities for firms led by women, BME people and people with disabilities. Its other objectives were concerned with drawing mainstream contractors and consultants into a shared and purposeful working agenda on equal opportunities with Housing Associations.

Hosted by a small black Housing Association (Tung Sing (Orient) HA Ltd) BPA was funded and supported by nearly 40 HAs in the North West, together with the Housing Corporation, the National Housing Federation, Manchester Training and Enterprise Council and the Housing Associations' Charitable Trust. A key output from BPA was the production of a Directory of Minority-Led Firms, to help HAs identify such firms and to promote more and better planned use of minority firms in their day-to-day activities.

The dual role of social housing providers – moving beyond the provision of shelter for those in need to the developer and sustainer of communities – was explicitly highlighted by the introduction of the Housing Plus programme by the Housing Corporation in the 1990s. In late 1996, BPA undertook research into the extent of Housing Plus activities in the North West to include the use of minority contractors, local purchasing policies and the use of equal opportunities as a tool of sustainable economic development.

Funded by an Innovation and Good Practice grant from the Housing Corporation, the BPA produced two detailed reports in February 1998. The first, *Current Thinking on Housing Plus* (Smyth and Hopley, 1998) *was* intended to reflect strategic issues and contains useful research material and references. Containing more than 60 key findings, the report set out to provide a review of current thinking and the extent of Housing Plus activities. The second report, entitled *Approaches to Housing Plus in the North West* (Smyth and Hopley) is a collection of initiatives setting out to deliver the social and regeneration objectives of the Housing Plus programme.

The research revealed concerns about the financial support available to HAs in delivering the wider remit demanded by Housing Plus programme. The authors also predicted that the election of a new Government in May 1997 would impact significantly on the role of HAs.

Building Equality in Construction

Building Equality in Construction was an innovative project focusing on how to promote and maintain equal opportunities policies in the construction industry through developing a partnering methodology between housing associations and building contractors. The project was developed as a practical response to the call from Sir Michael Latham in his groundbreaking report *Constructing the Team* (1994) that the construction needed to address its poor record in equal opportunities. Led by University of Manchester Institute of Science and Technology (UMIST) in partnership with BPA (see earlier) and RhysJones Consultants, the project was funded by the Department of the Environment, Transport and the Regions under the Partners in Technology programme.

The project team set about bringing together housing associations and building contractors in the North West in a series of facilitated meetings and focus groups designed to encourage frank and constructive debate. Action-learning sets developed by University of Salford in collaboration with the UMIST project were also launched, with participation by more than 20 housing associations and building contractors from the focus groups. The project team produced its report and practical guidelines in August 1998.

In addition to the practical guidelines covering issues such as recruitment, retention and contractual arrangements, *Building Equality in Construction* (1998) generated a number of initiatives, including

- a housing association asked a contractor to work in partnership on a project;
- a contractor asked for help in recruiting women for a specific job opportunity, resulting in collaboration with other participants;
- a maintenance training proposal was developed between a contractor and a housing association;
- questions on equal opportunities are now included in an undergraduate questionnaire at UMIST.

Amongst the recommendations made by the Building Equality in Construction team was that the Housing Corporation and government should review the impact of policy and contractual matters on improving equality and diversity, for example, the emphasis on lowest price rather than overall and long-term benefit. It was also recommended that a central unit be established to co-ordinate and promote good practice and share knowledge. Overall, the project demonstrated that the emphasis on client-led change towards equality in construction resulted in a practical and effective joint-learning process for all participants, including the research team.

Race and sex equality in social housing

In 1999 the Department of Transport, Environment and the Regions funded a project to develop the work carried out by the UMIST-led team that delivered Building Equality in Construction. The two principal aims of the work were to

- disseminate and promote throughout Britain the Good Practice Guidelines on gender diversity for contractors and housing associations;
- to develop new good practice guidelines on race equality.

To achieve this, the project group was expanded to include teams from the Revans Centre for Action Learning and Research, the University of Salford, the University of Teesside and RhysJones Consultants. A key element was the setting up of three action-learning sets in Salford, Teesside and London, bringing together social housing providers and building contractors to generate practical ideas for change. (*Race and Sex Equality in the Construction Industry.*)

In addition to achieving its two principal aims, the major outputs of the project included

- research by the University of Salford in collaboration with the Housing Corporation into race equality;
- development of integrated proactive industry, community and academic networks in the North East and South East.

The action-learning set methodology demonstrated sustained commitment of participants from client, contractor and academic organizations in a non-adversarial collective style. A number of the relationships established in the project have continued, demonstrating its value as a sustainable and replicable approach.

Case studies and examples were generated for use by the Construction Best Practice Programme, underpinning the requirement for practical proposals required to implement the recommendations by the Latham and Egan reports. The work was relevant to issues associated with New Deal and in engaging Small and Medium Enterprises in the construction industry.

The report entitled *Race and Sex Equality in the Construction Industry* contained a large number of proposals for improving equality and diversity and commented on the fact that current literature contained no shortage of suggestions for change. In addition, the researchers reiterated the need for policy changes at national level and the establishment of a co-ordinating function to gather and promote good practice.

Equality in the social housing sector

A research project into gender equality in social housing in the North West was carried out in 2001 by a team at Manchester School of Management (Dawe *et al.*, 2001). The objectives were to identify the factors inhibiting the representation of women, both in the social housing sector and the wider construction industry, in order to recommend policy changes.

Overall, the researchers were disappointed by the level of response to their written questionnaire, which had been developed in collaboration with one of the major RSLs. All 240 RSLs and HAs in the North West were contacted for possible interviews and sent questionnaires. Because of the structure of the sector, the 27 replies to the questionnaires accounted for 86 of the 240 RSLs contacted, representing 35.8 per cent of the total population of RSLs in the North West. This low response is at odds with the stated commitment of the sector to equality and inclusion. However, some of the research team had already experienced this apparent lack of interest, when conducting work in developing practical guidelines for great equality in the social housing sector. The researchers working on the *Building Equality in Construction* guidelines project noted that it was initially difficult to engage representatives from the social housing sector (Davey *et al.*, 1998). Even some of those who expressed interest in the project often failed to attend meetings and as a result were outnumbered by building contractors. In contrast, building contractors reacted very positively to telephone contact, regularly attending focus groups and making significant contributions.

Researchers for Equality in the Social Housing Sector found that time pressure was the usual reason for the low-response rate. Two of the largest RSLs were contacted at the grassroots level by the project worker at conferences and invited to participate in the project, which they willingly accepted. However, the director of the RSLs had different views about participating in the project. In his written response he commented that the organization received so many requests to participate in research projects, that unless there was a direct bearing on its work he did not wish to add burdens to over-stretched staff.

In-depth interviews were conducted with a cross-section of the housing industry sector representatives, in order to provide rich qualitative data and to form the basis of a wider questionnaire survey. The interview groups included construction companies, RSLs, women's training groups, the Construction Industry Training Board, and schools and colleges.

Most regeneration programmes require that local authority and housing association contracts use local labour. In order to benefit financially from the regeneration activities in their environment, social housing providers must also adhere to the principle of 'Best Value'. To fulfil these obligations there is increasing recognition that the diverse nature of the population has to be reflected within industry's workforce, and equality of opportunity has to be implemented. There is also an increasing recognition that social inclusion through the use of local labour has numerous benefits, with reduced vandalism and fewer problems on site.

Case study

Arawak Walton Housing Association: BME Residents Training Programme

Arawak Walton Housing Association is a BME RSL based in Manchester. Like many RSLs, the association estimates that 25–50 per cent of its day-to-day maintenance expenditure represented invoices of no more than £50. The Housing Association had a pool of available labour among its tenants, the majority of whom were unemployed and women with children of pre-school and school age. It was a logical conclusion that with proper training, the HA could solve both problems by training residents to carry out basic maintenance work themselves.

Initially it was felt that the main problem would be how to finance the project, but through involvement in the 'Building Equality in Construction' programme Arawak Walton was offered funding for the project from a contractor. Support then came from two more contractors who attended the programme workshop, in the form of training facilities and further funding. The nation wide retail chain B&Q also provided toolkits free of change for tenants who successfully completed their training.

However, despite the fact that many areas currently under regeneration have extremely high levels of unemployment, construction companies have difficulties in recruiting the local population. As a result, labour is often imported into the area. A spokeswoman for a major construction company, which works on regeneration activities for housing associations, commented to researchers about the problem of motivating people on estates where between 80 and 90 per cent of the residents were unemployed.

The company had tried several ways to aim their employment promotions at local residents living in areas of regeneration. These included meeting directly with the local population on Saturday mornings in places such as libraries, which are safe and have no religious connections. But such efforts to encourage local people into employment generated little enthusiasm. The researchers concluded that this is an area where in-depth research may help to identify some of the barriers currently being experienced by those who are socially excluded from entering into local economic activities.

Furthermore, some HAs had held open days to encourage direct recruitment from housing estates, yet these have not been effective. It was suggested that problems of numeracy and literacy were a serious issue, but it is the benefits trap that prevents any real progress being made, particularly for the many women on such estates.

Contracts of exclusion

In February 2000, a study into the barriers experienced by BME-led construction enterprises in the social housing sector in London was published (Sodhi and Steele, 2000). Funded by the Housing Corporation, the work was commissioned by the London Equal Opportunities Federation, an organization that maintains a register of BME, female and disabled contractors/consultants for utilization by RSLs in the London area. The researchers Dianne Sodhi and Andy Steele, from the Salford Housing and Urban Studies Unit, carried out extensive research which led to a number of challenging findings.

For example, it was suggested that although RSLs felt it was important to use BME contractors and consultants, they did not generally promote equality of opportunity or social inclusion through their policies. Standards were not regularly enforced, one in ten indicated that they had targets but a tiny minority actually specified them. Few RSLs could provide examples of good practice and little monitoring was carried out as well.

The research also indicated that RSLs were much more likely to employ BME consultants and contractors than females, but it was not clear whether this is a reflection of their representation in the industry generally, or whether gender inequalities were greater than race. On a more positive note, the report highlighted examples of exemplary practices such as Haringey Council's seminars on meeting selection criteria for BME companies and the training provided for BME companies by Laing Partnership Housing.

In summary, the authors concluded that, 'It is clear that, despite the rhetoric, there is a general lack of appreciation of why the employment of BME small and medium enterprises is important either from an equal opportunities perspective or from the point of view of supporting and encouraging community sustainability'.

Case study

Riverside Housing and Shokoya-Eleshin in Liverpool

Ways of encouraging black and ethnic minority people were described in the report, including the successful construction events held by the Society of Black Architects working with the London Skills Council. Another example of this is the successful Estate Renewal Challenge funding bid submitted by Riverside Housing in Liverpool involved Career Development Services and Shokoya-Eleshin, a local black-led construction company. Ethnic minorities, young and old unemployed men and women have been trained in construction, construction management, business management and IT as part of the project. Trainees varied in age between 16 and 68, more than 70 per cent were from ethnic minority groups and 17 per cent were female.

The report contains a wide ranging and comprehensive set of recommendations for action by RSLs, the Housing Corporation, professional institutes and trade federations and agencies maintaining registers of contractors/consultants working in the construction industry.

Housing provision for the disabled

A report into the current and future needs of disabled people, *Where Do You Think You're Going* was published in 2003 by the charity John Grooms (Ackroyd, 2003). The charity is the sister organization of the John Grooms Housing Association, the largest specialist provider of wheelchair accessible housing in the country. The objective of the inquiry was to make recommendations concerning the fundamental strategic direction of the charity's services.

Although the introduction of Building Regulations Part M, Lifetime Homes Criteria and Supporting People are encouraging moves towards long-term accessibility, the report draws attention to the national shortage of accessible housing for the disabled. John Grooms Housing Association estimates a need for up to 300,000 new or adapted wheelchair accessible homes. The inquiry also drew attention to the fact that information on the supply and ownership of accessible housing is fragmented and incomplete, with no standard measures or definitions used. The Habinteg Housing Association with the Papworth Trust has now developed a planning model based on information gathered in a study carried out by Pathways Research in 1999/2000 (*Pathways to Accessible Housing*).

The Housing Corporation and diversity

How do the equality and diversity requirements of the corporation and social housing providers impact on the construction industry? The obvious starting point is the regulatory code 2.7 and its guidance, which includes performance of contractors/supplies/consultants in its list of areas to set performance targets.

Clause 2.7 of the regulatory code states that:

> Housing associations must demonstrate when carrying out all their functions their commitment to equal opportunities. They must work towards the elimination of discrimination and demonstrate an equitable approach to the rights and responsibilities of all individuals. They must promote good relations between people of different racial groups.

Alongside this code the regulatory guidance identifies a number of areas where performance targets should be set to assess continuous improvement, including tenants satisfaction, dealing with racial harassment, governing body membership, staffing, tenants/residence representation and the performance of contractors/supplies/consultants.

Good Practice Note number 4 shifts focus from equal opportunities to equality and diversity, and includes an outline timetable of action required of HA first published in the Race and Housing Inquiry Challenge report.

This report challenged HAs and RSLs to

- integrate race equality into business planning;
- develop action plans with milestones and targets;
- involve tenants, staff and BME organizations;
- ensure measurable outcomes for BME communities;
- review, evaluate and monitor plans.

The Housing Corporation's commitment to equality and diversity is expressed in its June 2003 report Equality and Diversity: Policy and Strategies. The organization has funded many pieces of research and supported Innovation and Good Practice Grants (IGP) in all areas of Equality and diversity.

For example Career Opportunities for Ethnic Minorities (COFEM) was set up following Corporation funded research carried out by Salford University to improve the career opportunities of Black Minority Ethnic (BME) staff in the housing sector. Commissioned by the National Housing Federation (North West) and funded by the Housing Corporation and the Ethnic Minorities Housing Trust (North West), the research identified a number of barriers faced by BME housing staff including the lack of networking opportunities, access to information and career development.

COFEM was initially piloted in the North West of England where local authority housing departments and housing associations signed up to take part in the scheme. COFEM – NW oversaw the production of a mentoring guidance and training pack to help organisations who wished to set up mentoring schemes as well as provide contact details on where further help was available.

COFEM – NW also piloted a Master Class Project which enabled 58 BME housing staff from housing associations and local authorities, to attend four sets of classes run by six mentors, The mentors included some of the most senior housing directors in the region and gave mentees the opportunity to network and meet with managers who they may not have usually been in contact with. Since then similar steering groups have been developed in London, Leeds and Birmingham.

At the end of 2005 the COFEM initiative was handed over to the Housing Diversity Network (HDN), a not for profit organization set up in May 2002 to provide information and practical assistance to the housing sector on all aspects of equality and diversity. Based in Huddersfield, HDN is fully self-financing, generating income from its consultancy and training activities and its corporate membership network.

In August 2005, the Housing Corporation published The BME Action Plan 2005-2008, outlining how the Corporation will work with housing associations and key stakeholders to ensure people from all ethnic groups have equal access to services and are represented at all employment levels within and across the housing sector. It also aims to ensure BME communities are fairly represented on providers' governing bodies and suppliers, contractors and consultants reflect the communities they work in.

Case study

Bramall Construction: award winning diversity programme

It was due to requests from clients and housing associations for data on their employees that John Hammond from Bramall Construction Limited recognized there was a need for change. As a result, the company has implemented an equal opportunities initiative that addresses long-term problems in skill shortage areas, ensuring best practice is employed in all aspects of management including recruitment. The company sought help in developing these practices from outside agencies, including the employment advisory service, employment service, race relations' employment advisory service and Training and Enterprise Centre. A multi disciplinary cross sectional steering group was established, from a cross section of staff from different working locations, jobs, ages, gender and race. This group took the lead in driving the company's equal opportunities programme and re-educating the work force in appropriate behaviour. Bramall has changed from being a white, macho, male dominated workplace to being a welcoming and pleasant working environment. For their efforts, they have been awarded with a Kick-Start Equality Award and a Disability Award.

The Housing Forum

The Housing Forum was set up in April 1999 to promote radical change and continuous improvement in the housing construction industry, in response to the recommendations of the Rethinking Construction report of Sir John Egan. One of the major achievements of the Housing Forum was that it established itself as the only housing organization whose membership spans all sectors of housing construction. This provided opportunities for cross sector learning and networking at all levels amongst organizations and individuals who are leading the movement for change. Originally established to operate for three years, the Housing Forum continued beyond this timescale as part of the Rethinking Construction programme. Whilst working alongside the Movement for Innovation (M4I) and the Local Government Task Force, the Housing Forum continued to maintain its separate board and membership in order to formulate the policy for housing and to formulate action until March 2004.

Key outputs of the Housing Forum from 1999–2002 included:

- building a portfolio of demonstration projects promoting innovation and good practice;
- setting up the National Customer Satisfaction Survey, the regular measure of progress for speculative house builders and canvassing opinion of some 10,000 new house buyers;
- establishing a Benchmarking Club for members to measure their performance with peer organizations;
- developing Housing Sector Key Performance Indicators for refurbishment, repairs and maintenance works;
- setting up Working Groups to examine specific topics.

It is through such a Working Group that The Housing Forum addressed equality and diversity. The Recruitment, Retention and Respect for People (3Rs) Working Group of the Housing Forum was set up in 2000, chaired by Charmaine Young CBE, Director of housebuilder and developer St George Regeneration. The Working Group saw shortage of skills and labour as the main challenge to the construction industry and pointed out that a significant proportion of the potential workforce – women, BME people and the disabled – were being ignored.

A survey of employment by the House Builders Federation (2001), found that

- 35 per cent of house builders were experiencing shortages of site manager/supervisors and design/technical staff;
- 73 per cent of companies reported that the shortage of suitably skilled site operatives I already impacting on their activities;
- 57 per cent noted that they are also affected by a shortage of professional staff.

The survey also showed that whilst the house building sector employed far more women than the construction industry as a whole (28 per cent compared with 8.6 per cent in the industry) only 0.5 per cent were from black and ethnic minorities and only 0.1 per cent were registered disabled.

To begin to address the challenge, the 3Rs Working Group set out to

- gather examples of current initiatives in the housing sector;
- pilot a series of recruitment roadshows to universities, using male and female role models from a variety of disciplines and backgrounds;
- identify and publish a series of housing role models;
- produce a Work in Occupied Premises toolkit for both workers and residents.

In addition, the Working Group gathered information and case studies through a series of open meetings, workshops and a correspondence network. The results were disseminated through a report that identified 20 ideas for delivering the 3Rs, including examples of good practice and initiatives (The Housing Forum, 2002).

Case study

Southern Housing Group in partnership

In Hackney, East London, the Southern Housing Group (SHG) is working in partnership with Mansells, Wates Construction and Durkan

Construction to refurbish homes whilst tenants remain in occupation. Emphasis is on providing both service and opportunity for residents. To achieve this

- sHG, the contractors and the Residents Liaison officer hold weekly surgeries on site on four different estates;
- monthly residents' progress meetings are held with contractors;
- landscaping design meetings are held separately for young people.

Training in construction initiatives have also been established, aimed at including women and young people, particularly from BME groups who are under-represented in construction. Wates have introduced schoolchildren to site skills, Mansells are aiming at NVQ training for 16–18-year olds and Durkan is employing women on site in a number of trades.

For example, the 3Rs Working Group report called on the housing industry to 'think outside the box' when recruiting staff, in order to attract a larger number of recruits and a wider range of skills. Labour-only subcontractors tend to favour white males, as they are seen as more easily integrated into a gang. The report highlighted Notting Hill Housing Trust's Local Initiatives Fund, which is supporting 15 local unemployed residents to achieve a highly innovative training programme marketed to tenants and residents.

The 3Rs report also gave examples of how to encourage more women into construction, including the innovative Building Work for Women project based in London. It addressed the problem of trained women who have failed to gain employment in the industry by working with employers to provide supported site experience leading to full-time employment. Contractors Durkan, Willmott Dixon and Llewellyn provided work experience and employment for tradeswomen on a variety of social housing projects in the London area.

Case study

Chantry Builders Ltd, based in Rotherham, is a member of the South Yorkshire Employers Coalition, which is a subgroup of the National Employment Panel, headed by Gordon Brown. Targets for this group are to get people working within construction and to encourage disabled people, ethnic minorities and women into work.

Chantry has many public sector clients and is currently working on a kitchen and bathroom refurbishment programme with Northern Counties Housing Association in Sheffield and Rotherham. They won the project partly because they could demonstrate diverse workforce.

Denis Jolly is Financial Director and Company Secretary at Chantry and is a strong advocate of the company's commitment to employing women and ethnic minorities. He comments,

> It makes good business sense because our customers require that we
> have equal opportunities written into our policies and we win contracts
> based on the fact that we have a diverse workforce. For example,
> women of certain religions are only allowed to have women in their
> homes to work and the fact that we have eight women employees
> enables us to accommodate this.

Chantry is currently working with schools in the South Yorkshire area
to encourage young women into the industry by taking in their female staff
in to speak to them. The company is supported by the CITB-Construction
Skills Yorkshire and Humber Office with STEP and other services.

Conclusions

A review of progress in the social housing sector since 1994, when the
issues of equality and diversity in construction were first highlighted at
national level by Sir Michael Latham, presents a mixed picture. It is encouraging to see the issues of equality and diversity in social housing being
addressed regularly. It is encouraging to see action taking place on some of
the policy changes identified by a number of research projects. It is encouraging to see the positive approaches being taken by some contractors and
organizations delivering services to RSLs.

However, the pace of change is slow. There is government commitment to
delivering social and affordable housing. Whilst the challenge of developing
a more diverse construction industry remains a major one, housing is the
sector that touches people more directly than any other and which offers the
most accessible route for change. There is no shortage of research data,
recommendations, examples of good practice or practical guidelines, as
demonstrated in this chapter. The requirement for co-ordination, effective
dissemination and compliance remains. What is also needed is to progress
more quickly from rhetoric to commitment and from policy to practice.

For housing associations, a lack of forward thinking may have culmi-
nated in ownership of properties in hard to let areas, sheltered schemes that
are under used – a mismatch between the growing number of customers and
available housing stock. Against this background, housing associations
need to develop partnerships with forward thinking and diverse contractors
and consultants, businesses that have a view about the future and do not
carry on simply developing and designing 'the norm'. This is the added
value of a diverse workforce, which simply by being part of a business can
contribute to business thinking and delivery.

Women are the biggest client group within the social housing sector
and more than half of all tenants are women. Businesses that provide

gas and electricity have recognized for a long time that if a small child is in a household, repair needs should be seen as a priority. Housing associations try to do the same but in most cases rely on their maintenance contractors. A housing association is more likely to use a contractor who has thought about the needs of women and children and has a diverse workforce that can meet customer needs, as well as service needs. A contractor who, by sending a woman plumber or by making sure that staff understand the cultural sensitivity of some customers, can demonstrate that they have really thought through the diversity issues of their service. The Housing Corporation's 2003 Race and Housing conference – D3 Design Diversity Dialogue – set the scene for bringing together leading architects and housing associations to look how design can transform affordable housing and improve the lives of all sections of the community.

Housing associations need progressive partners. If the building industry starts now to consider diversity in all its activities whether design, employment, training or service provision it will be able to provide the skills and performance needs of housing associations in the future.

Discussion questions

1 An obligation to provide a significant percentage of affordable housing as part of private housing developments is one of the ways that government is seeking to address the housing shortage. How might private sector housebuilders, perceived to be one of the most profitable sectors of the construction industry, also contribute to creating a more equitable and diverse industry?

2 Many young people are concerned about environmental and humanitarian issues, of which the provision of good housing is an essential element. Yet the construction industry, which provides housing, sanitation and the related infrastructure, has difficulty in recruiting sufficient numbers of the right men and women. How might the housing sector do to resolve this issue?

3 What are the reasons that inhibit opportunities for smaller and minority-led firms and companies from obtaining work in the social housing sector? How might this be addressed without triggering the resentment that sometimes arises from what is perceived to be preferential treatment?

References

Ackroyd, J. (2003) *Where Do You Think You're Going?* London, John Grooms.

CIB (2002) *Recruitment, Retention and Respect for People: 20 Ideas for Delivering the 3Rs*, Rethinking Construction, London, Constructing Excellence in the Built Environment.

Davey, C., Davidson, M.J., Gale, A.W., Hopley, A. and Rhys Jones, S. (1998) *Building Equality in Construction: Good Practice Guidelines for Building Contractors and Housing Associations.* Manchester School of Management, Working Paper Series 9901, ISBN: 1-86115-050-4.

Dawe, C., Fielden, S., Gale, A.W. and Cartwright, S. (2001) *Equality in the Social Housing Sector,* funded by ESF, Manchester School of Management.

HMSO (1994) *Constructing the Team,* London, HMSO.

Housing Corporation (2003) *Equality & Diversity, Policy & Strategies,* June.

The Housing Forum (2002) *Recruitment, Rentention and Respect: 20 Ideas for Delivering the Three Rs,* March.

Pathways to Accessible Housing for Wheelchair Users, Habinet Housing Association Ltd.

Smyth, K. and Hopley, A. (1998a) *Approaches to Housing Plus in the North West,* Building Positive Action.

Smyth, K. and Hopley, A. (1998b) *Current Thinking on Housing Plus: A Report on the Findings of a North West Survey,* Building Positive Action.

Sodhi, D. and Steele, A. (2002) *Contracts of Exclusion: A Study of Black and Minority Ethnic Outputs from Registered Social Landlords Contracting Power.* Salford Housing and Urban Studies Unit, ISBN 1-902496-15-9.

Somerville, P., Steele, A. and Sodhi, D. (1998) *A Question of Diversity: Black and Minority Ethnic Staff in the RSL Sector,* Housing Corporation Research Source Findings No. 43, Housing Corporation.

Somerville, P., Davey, C., Sodhi, D., Steele, A., Gale, A.W., Davidson, M.J. and Rhys Jones, S. (2000) *Race and Sex Equality in Construction Good Practice Guidelines for Building Contractors and Housing Associations.* Manchester School of Management, Working Paper Series 2002, No. 0201 (UMIST).

Chapter 13

Formulating and implementing diversity in a contractor's organisation

A case study

Chrissie Chadney

Introduction

This case study is examines the strategy employed by one national construction contractor, to develop its workforce in such a way that by December 2005 the percentage of both female and ethnic minority members of staff was double the average percentage of other contractors. It traces developments over the past five years, which have included a culture change programme and a major emphasis on recruiting, retaining, promoting, developing and rewarding the very best staff. It sets the evolution of a more diverse workforce in the context of good working practices, which put meeting the needs of the organisation as paramount, along with the appreciation that both the customer and its own people are its most important asset. The aim of this chapter is to explore the development of culture change, perceptions and actions in relation to diversity within the context of a contractor's company. It has four major objectives. First, to explore the benefits of a 'low key' approach to policy and practice change; second, to identify the key activities, processes and influences which have brought about change; third, to consider issues of both recruitment and retention in sustaining diversity and fourth, it explores external and other influences which detract from, or support change.

Setting the scene and context

Willmott Dixon is a thriving construction contractor member with a turnover of £400 million in 2005 and a workforce of just over 900 people. It operates primarily in the management of construction with most of the 'hands-on' site work being sub-contracted. Eighty-two per cent of the workforce are male and 18 per cent female overall, but this percentage includes all members of the largely female administrative and support staff, together with all the male operatives. The staff can be considered as three sub-groups. Eleven per cent are administrative and support staff and 7 per cent are operatives. Therefore, 82 per cent are construction professionals, mainly

Quantity Surveyors and Building Managers and Design Managers and those in senior management positions with a construction or allied background. It is in the latter group, those involved in management positions that the staff composition has changed over the past five years to more accurately reflect the diverse culture in which we live.

The Willmott Dixon Group has a number of unique features as a contractor within the Construction Industry in the twenty-first century. It remains a family business with a very small main board, consisting of shareholder, chairman and directors and a nationally known advocate of the Construction Industry as non-executive deputy chairman. It has shown a profile of consistent controlled growth during its 153-year history. It is a constantly evolving business, presently structured in differing operational companies and business units. Two major divisions cover contracting activities in Construction and Housing, and there are two smaller development business.There is a very strong belief that change is both a necessary and welcome feature of business today and all staff are encouraged to understand and 'buy in' to this reality. Strategic plans for the business are fully shared and in 2000 the whole company underwent a development programme, which introduced the notion of an empowered workforce, one where each individual and thus the whole business could strive to reach their potential.

The Group has always had a major commitment to training, both in the recruitment of management trainees and in the development of all its staff through comprehensive training plans. This commitment starts from the very top of the organisation when each new member of staff attends an Induction at Group Headquarters run by members of the Main Board. It is also a major advocate and practitioner of a partnering approach to construction, which again was introduced to every individual member of staff in a development programme which ran throughout the Group in the late 1990s. Through its training programmes, policies and procedures together with the sharing of strategic objectives, the Group attempts to ensure that it has a distinct ethos and a clear culture to which it aspires. In common with many companies it would submit that 'people are its greatest asset' but it tries to constantly address this in a live way. One of the ways it does this is by keeping 'people issues' high on the agenda of its companies. 'Life Style Balance' has been a frequently re-assessed agenda item and the company has a good record of flexibility in approach. The issue of 'Corporate Social Responsibility' has become increasingly important since the start of the new century and so looking at the company and what it does from the view points of enviornmental, economic and social responsibility is actively encouraged. There is always a tension between the aspiration to be considerate and look after staff, and the need to meet shareholder aspirations, but the company has managed to accommodate these potentially divergent wants and needs.

There are standard procedures for resource planning, recruitment, performance review, training and development, and also an annual 'People Asset Presentation'. Managing Directors are required to draw up a review of each member of their staff and define their performance standard and development needs, along with staff succession planning and an overall assessment of the 'people assets' of the business. There is a very small central Human Resources team providing expert, advice, support and strategic initiatives. In the divisions a further small team deal primarily with human resources processes. However, in each Business Unit the Managing Director is expected to take a major responsibility for human resource management, recruiting and retaining personally knowing and being personally known by each member of staff.

As a company, Willmott Dixon is in the vanguard of change in several aspects of building innovation, as well as ways of working with both customers and its own staff. The winning of several national awards over the past five years including those for Partnering, Training and for Building Manager of the Year, illustrate the commitment of the Group to excellence within the Construction Industry. In 2005, Willmott Dilon was named as 'Best Company to work for in Construction' as a result of the recommendations and voting form its own satff and this indicated that the focus on 'people' was more than rhetoric.

Genesis of awareness of the issue of diversity

In common with most companies, Willmott Dixon has had an equal opportunity policy for many years, but in order to trace the development of a greater awareness of diversity and subsequent actions, it is necessary to go back to the early 1990s. The then Chairman of the Company was a major player in the Construction Industry as past president of the CIOB and second chairman of the Construction Industry Board. He contributed to the debates which led to the 1994 publication of the Government commissioned document 'Constructing the Team', known as the 'Latham Report' (Latham, 1994). As a long-time advocate of the value of women in construction, the Chairman was particularly involved in Working Group Eight – 'Tomorrow's Team: Women and Men in Construction' (1996) which followed the Latham Report and recommended a number of actions. These were under the headings of 'Raising Awareness'; 'Education and Training'; 'Recruiting and Retaining'; 'Changing the Culture'. The findings and recommendations of this report influenced the way in which Willmott Dixon then began to seek to bring about a change in relation to women in the company.

At this point it would be fair to say that there was little by way of ethnic diversity awareness evident in construction in the public domain, nor in the thinking of the company itself. It had a typical workforce with a few members of ethnic minorities in the staff group, rather more in the support roles than in the professional roles and very dependent on geographical location.

Informal policy formulation

'Tomorrow's Team' (1996) put much emphasis on the business case for equal opportunities and in particular, the employment of women. In 1996 the Chairman decided to more actively encourage the recruitment of women building professionals into Willmott Dixon by setting a target that 25 per cent of the management trainee intake each year should be female. Thus, the first and only formal strand of any policy element in a cultural change towards diversity was set.

The changes in Willmott Dixon have all derived from a longer-term viewpoint, which identifies the strategic aim for controllable growth and sustainable profit and then addresses the strategic objectives through which these will be achieved. This way of approaching the issue of change and its accompanying behavioural and attitudinal components, derived from the work of the Main Board and particularly the leadership of the Chief Executive from the late 1990s. In formulating the long-term objectives, it became clear that in order to improve performance and meet these two aims of controllable growth and sustainable profit, it was necessary to engage both intellect and emotion of staff, to really get 'buy in', to ensure that we attracted and kept the very highest quality of staff from the whole population. Thus, one of the six objectives was to 'recruit, retain and reward the very best people'. All of the objectives focussed on business processes and for each of the six, specific targets were set and shorter-term objectives and implementation plans were made. In order to progress the overall strategy and to particularly meet aspects of the 'recruit, retain, reward' objectives, a major training and development programme was devised called 'Achieving Potential'. This focussed on developing a real understanding of the company's strategy and culture and specifically addressed the concept and practice of empowerment as the way to realise potential. All staff attended this programme, run over 18 months by senior directors, and through it a very clear idea of the culture to which Willmott Dixon aspires was promulgated. At each event it was one of the two shareholder directors who stood in front of the 20 participants and stated their vision for the business and what this would both give to, and expect from its staff.

The cultural change programme made no specific reference to diversity in itself. Rather, it chose to set the scene and articulate a culture, which was inclusive of those who wanted to be part of an ethical, hard work/hard play, changing organisation. Its policies supported training and development and equal opportunities as undisputed requirements, and in this way the Company was, on one level, simply aspiring to best practice.

Prior to the formal inception of the cultural change programme, the Personnel, Training and Development Team were working on the updating of personnel-related policies and the implementation of systems to better track staff statistics. The head of the team, who had long experience of equality issues and training and development in this area in the public sector, introduced a new recruitment and selection course, which addressed equal opportunities more clearly. Further work was done in examining existing policies and practice guidelines in this area. It was a deliberate decision not to have many separate policy documents about equal opportunities, but to embrace all the issues in one short concise document, and ensure that such matters as sexual and racial harassment were clearly stated as gross misconduct issues in the Disciplinary Procedures of the Terms and Conditions of Employment.

Willmott Dixon had always had good links with other construction industry organisations and again took a pragmatic approach to this. Such links raise the profile of a company but at the same time expertise can be shared and support given to the wider industry. Hence the Head of Personnel, Training and Development was pleased to build relationships with 'women in construction' organisations and Willmott Dixon actively supported, in a matched funding relationship, the Oxford Women Skills Centre and contributed to discussions with the CITB. Links were established with other organisations, which aimed to get more women into the industry and the company was involved in research projects looking both at women and ethnic minorities in construction. All of these activities helped to enhance awareness of the issues and were reported through the Executive reports to the Divisional and Business Unit management teams.

Implementation

The initial policy in relation to the numbers of female management trainees proved a very difficult target to meet, but drawing on the few women graduates available from the universities, on internal networks, daughters, nieces of existing managers and following up speculative enquiries, this target was reached each year for five years. Therefore, by the first years of twenty-first century the management trainee group, the numbering about 70 graduates and non-graduates was about 20 per cent female. Also by this point, earlier female trainees had moved up the business to assistant manager and manager positions. The distribution of the women trainees and managers was not evenly spread. Some roles and some types of business within the Group proved more attractive than others. Whilst several women, particularly undergraduates on a four-year programme, joined the company planning to be site managers, the majority moved into design and build or quantity surveying. The quantity surveying women, particularly in the Social Housing Contracting Business Units, grew exponentially so that, in one Business Unit, half of the Quantity Surveyors were women by 2004. Quantity Surveyors are site based and there is no doubt that the number of women now present on site has subtly altered the acceptable site culture, to one which appears less traditionally aggressively masculine.

As the actual number of women building professionals has grown, so also has the variety and level of roles they are undertaking and the business itself appears to be perceived as more attractive to qualified women. There are now a few established women site managers, assistant site managers and final year management trainees working on major construction projects such as colleges, hospitals and office blocks, as well as in social housing. There have also been promotions into senior management positions of building professional women from a variety of construction backgrounds and whilst none has yet reached a technical director position, this is no longer something which would be considered inappropriate. Indeed the first female participant has now been invited to attend the company's Management Development Level Four programme for new and potential directors starting in early 2006 and this is anticipated to lead to a director-ship within two years.

The women engaged as management trainees are frequently above aver-age within the management trainee group as a whole. This is not due to the criteria being different, rather that those women who do consider con-struction have already had to assert themselves and tackle a number of hur-dles before they join us. They will probably have taken non-female traditional qualifications and have actively chosen a non-traditional career where they will be a minority and potentially under greater scrutiny just because they are still somewhat unique. Within the Group they have proved their competence in the workplace and this has fed a continuing process whereby the Business Units are keen to take on more women. The need for women frequently to be not just good, but very good, is no different here from the experience of any women entering hitherto male-dominated organ-isations. The high quality of these women has been illustrated in several ways. Academically, proportionally more women professional builders in the Group have gained first class honours degrees than their male equiva-lents, and in the past seven years, women trainees have won the company's prestigious 'Trainee of the Year' award on three occasions.

It remains very difficult to reach the target the company has set for itself for the number of women trainees and new ideas for engaging the interest of woman constantly have to be sought. The active targeting of women to the management training scheme has clearly supported the first step of recruiting more women into the company. Recognising that whilst the women would be competent they would be entering an environment unused to them and one where they might be quite isolated, certain retention activ-ities were undertaken. In 1998, the Head of Personnel, Training and Development started the Willmott Dixon Constructive Women Group for the 28 professional female staff. Its terms of reference describe it as a 'a cross-section of construction and allied women professionals in Willmott Dixon, meeting bi-annually to discuss issues, provide mutual support and act as a forum for positive action'. The meetings always include either a guest speaker or a short training session, as well as the opportunity to share knowledge, experiences and discuss current issues in both the company and the wider construction community. Speakers have included the project manager of the 'Respect for People' initiative and the chairperson of 'Building Work for Women' and the 2003 'Business Woman of the Year'. Training has included 'Lateral Thinking'; 'Understanding Self through Psychometrics' and 'Assertion Skills' and 'Neuro-linguistic programming'. It has provided a unique opportunity, fully supported by senior management, for the growing number of women professionals to come together. By 2003 the group had grown to over 60 members and was too large to meet as one, but it continues to meet once a year in two geographically split sub-groups. As a result of the existence of 'Constructive Women' a number of informal mentoring relationships have developed between the participants and a 'women trainees focus group' was organised for a short period of time.

As for most organisations, the issue of women and maternity leave is not without its problems. However, Willmott Dixon has been very clear that having invested a great deal in the training and development of women building professionals, it actively seeks that they should return to work in some capacity after maternity leave. The actual leave arrangements are the statutory ones and there is no enhancement, but there is maximum flexibility when a woman wishes to return. Thus there are examples of reduction of hours, working from home arrangements or project work, which can be fitted around childcare needs. Some roles, such as quantity surveying lend themselves more readily to this flexibility than others. Site management, for instance, clearly requires the presence of a site manager for the whole period that a site is open and the nature of the work does not lend itself to job share arrangements. In all cases, the business need is considered first and Willmott Dixon consider that this priority is the right one. Given the business need, we match the skills and availability of the employee and negotiate the best way forward. This has proved a successful way of working, on one level requiring only common sense and willingness on both parts to meet a common goal. In several cases women have now continued their professional cereers through two periods of maternity leave and the return to work has been helped since 2004 by the company's adoption of the Government's Chidcare Voucher Scheme'.

The outcome of efforts made over the past five years to address the in-balance in this area of diversity has led to a significant number of female professionals in Willmott Dixon as Table 13.1 illustrates.

Diversity in relation to ethnicity has come about through an incremental process of looking in the market place for those able and willing to work

Table 13.1 Professional staff male/female composition 2004

Willmott Dixon Professional Managers	Total	Male	Female	Female as percentage
Directors	42	40	2	4.8
Senior managers	90	79	11	12.2
Managers	482	448	34	7.0
Management trainees	68	59	9	13.2
Totals	682	626	56	8.2

Note
The group includes all qualified construction professionals and the small group of professionally qualified staff in functions such as accountancy, human resources and IT.

hard and deliver to a high standard within our company culture. There has been no numerical targeting to increase ethnic minority building staff, but there has been an openness and willingness to recruit and assess all potential candidates. This has naturally included both ethnic minority construction professionals from Britain and increasingly, driven by skill shortages, from overseas. The management trainee group, whilst primarily consisting of construction undergraduates or graduates, has in the past three years included IT graduates who work in setting up and maintaining technology on the construction sites of each Business Unit. The normal university recruiting efforts took place. However, a high percentage of IT graduates from universities are from ethnic minorities and this was then mirrored in the membership of the group which was appointed. A number of graduate and undergraduate trainee production managers are from ethnic minorities particularly in the London area reflecting the student composition of the London universities. The severe lack of trainee or qualified quantity surveyors has led to this role in particular, being filled by non-UK citizens, often from Asia, and this has added to the numbers of ethnic minority professionals.

Willmott Dixon has always had a significant percentage of staff from 'non-visible' ethnic minority groups, primarily of course those from Ireland, and they are found at every level in the business, including the Business Unit Directorates. In addition to the increased number of trainees from visible ethnic minority backgrounds, there is a small but significant number of ethnic minority professionals joining the company at middle management and then progressing. It is not that Willmott Dixon makes any particular efforts to make this happen. Rather, it appears to be an outcome of the overall culture we have tried to establish and perhaps by 'word of mouth'.

Whilst there has never been a major policy thrust nor required practices to increase diversity in Willmott Dixon there have been supporting activities, which have perhaps contributed to the changes in the profile of the workforce. The general level of awareness in relation to diversity has been enhanced by bi-annual printed 'Employment Briefings' from the Human Resources team. These address changes in the law, in relation to diversity and other areas, with particular attention given to the implications for the business and actions which should be taken. A new course 'Diversity Issues for Tendering' commenced in 2004 and addressed the ways in which we, as a contractor, could assist clients in meeting their diversity targets through good practice in our own company. A further course 'Diversity Awareness' has been designed as part of a suite of management development courses open to all staff. This course primarily explores cultural awareness and the effective management of diversity within the team. There has been a focus in the recruitment and selection training, which middle and senior managers undertake, to discuss diversity and the value of a diverse workforce. There has been a stress on the business benefits, as these have to be the drivers. This has included the particular business benefits of having, for instance, members of ethnic minorities on the team when working in a multi-racial area. It has included an appreciation of the value of the perspective and presence of women at middle to senior level, particularly when working with our public sector clients. In this sector there are already more women at senior levels themselves and the balance of the workforce in the contractor's organisation is noticed. The Human Resources team have provided information and training on cultural diversity issues to site teams and individuals with advice as appropriate. Tangible support has been given to site teams by the provision of multi-lingual posters depicting construction equipment and common phrases. In 2005 the company's 'Equal Opportunities' policy was updated to better reflect current practice and re-named the 'Equality and Diversity' Policy.

In common with issues of gender, the company chooses to link with external organisations in pursuing its own understanding and involvement in issues of ethnic diversity. It has contributed to two CABE (Commmission for Architecture and the Built Environment) research projects into the representation of black minority ethnic students and professionals in the built environment industry; this project also looked at the evidence and experiences of barriers to entry and progression. The company has chosen to be registered with the Department of Trade and Industry's 'Constructionline' which requires the demonstration of positive action in areas of good race relations and equality and diversity practices.

It is only recently that the company has collected detailed statistics on ethnicity but Tables 13.2 and Figure 13.1 illustrate the staff composition in the autumn of 2005. As can be seen, the percentage of ethnic minority staff, both building professionals and administrative staff represent 4.6 per cent of the workforce against a construction industry average of 1.9 per cent.

Table 13.2 Ethnic composition of staff 2005

	No. of staff	Percentage
Unspecified	31	3.4
African (includes North Africa & African/English)	6	0.7
Asian (includes Indian, Asian/English, Chinese, Malay)	26	2.9
Caribbean (includes Caribbean/English)	9	1.0
European (includes English/European)	11	1.2
English	503	55.8
Irish (includes English/Irish & Scottish/Irish)	19	2.1
Welsh (includes English/Welsh)	88	9.8
Scottish (includes English/Scottish)	17	1.9
Other	191	21.2
Total staff numbers	901	
Total known ethnic minority	41	4.6

□ White male
■ Ethnic male
□ Ethnic female
■ White female

Figure 13.1 Gender and ethnicity breakdown of all staff 2004

Notes
The percentage of visible ethnic minority staff is 4.6 per cent and if we include Europeans this rises to 6.2 per cent. The Group operates across England and South Wales and hence the significant percentage of Welsh staff. Some staff who identify as Irish, Scottish or Welsh do consider themselves as in ethnic minorities, and the combined figures here would give an ethnic minority population of more than 19 per cent.

Recommendations

In recommending ways in which diversity in a contractor's organisation can be improved, Willmott Dixon's experience suggests the following three underlying premises, which in a traditional equal opportunities environment may be seen by some as almost heretical:

Premise 1 – Always put the business need first;
Premise 2 – The ideological imperative is the secondary consideration and should not be made more of an issue than is necessary;
Premise 3 – Basic good practice, common sense and a willingness to be open-minded can bring about real progress.

Using these three premises as a backdrop there are some pre-requisites without which change is unlikely to happen, namely

- commitment from the top;
- realistic and pragmatic approach;
- people focus with an acceptance of change as a major feature of company culture, and diversity as just one aspect of a broader picture;
- supporting behaviours across a range of activities;
- involvement with external agencies.

Commitment from the top

In Willmott Dixon we were fortunate that the Chairman recognised the sense in attracting more women into construction. This was based not so much on an ideological stance but on sound business sense. Previous failures to tap into a potential workforce were recognised for what they were; which was very foolish. The fact that some of our customers in particular would value the presence of women, both on site and in the company hierarchy was acknowledged as a business reality, which should be addressed. The contribution that women could make in moving the construction industry away from the traditional aggressive male environment, so often displayed, to one more conducive to good working relations and a better public image, was recognised and welcomed as much needed. The initial action by the Chairman in making a small step to change the shape of our management trainee intake was fully supported by the Board. In subsequent years, ideas to support the retention of women staff, through activities such as 'Constructive Women' and through flexible post maternity arrangements for women professionals were fully supported at the top of the organisation. The promotion of two women, the Head of Personnel, Training and Development and the Company Secretary/Lawyer, to the Executive of the Group in 1999 also helped in showing that those running the business were not only supportive but willing to include women at the top of the organisation.

Realistic and pragmatic approach

The 'business need must come first' philosophy is fundamental in this approach to implementing diversity. Having a realistic and pragmatic approach to addressing issues of diversity is rooted in the primary need for the business to succeed and for the needs of the organisation to be always served before the needs of individuals. It could be argued that the organisation is only the sum of its people and that where there has been discrimination and in-balance in the past, positive discrimination is needed. The approach recommended is more subtle than this. In any change, and the inclusion of a significant number of women and members of ethnic minorities is a change, a full analysis of the positive benefits for the business needs to be articulated and embraced by all. There does need to be intervention, for without intervention the status quo will remain, but that intervention should be very carefully considered and the dis-benefits explored as well as the benefits. For some staff, the very fact that we were putting a target on the number of women trainees was problematic in itself. Some had to have the difference between target and quota explained several times and similarly there were the usual cries of 'why a Constructive Women Group, why not a Constructive Men Group?' which arose when we started that activity, and indeed still sometimes occur today. Careful explanation of the facts and openly addressing the emotions, which inevitably seem to be engaged for some, is necessary. Fortunately, the business benefits and the simple sense and justice of being an inclusive organisation are usually obvious to all but the most narrow-minded.

People focus with change as a major feature of company culture, and diversity as just one aspect of a broader picture

The fact that a company actively embraces change is a major benefit when addressing issues of diversity. If a company shares with its staff its strategic objectives and helps them to recognise that it really believes the maxim 'people are our greatest assets,' then it should follow that changes away from the traditional workforce to a more diverse one should not pose a threat. Changes in the workforce composition will occur, as will changes in methods of procurement; changes in the materials we use to build; changes in the way we use IT and the way we reward staff and so on. Whilst changing the profile of a workforce is factually very different from changing the way we procure work as a contractor, there are parallels that can be drawn both in the intended end product and the processes required on the way.

Indeed a comparison could be made between the change towards a more people focussed culture and diverse workforce, and the change in the Construction Industry towards a Partnering culture. In the latter case the focus for the industry is on shared objectives, openness and honesty, continuous improvement and the challenging of assumptions and of the old adversarial way of working. To bring about the change, an organisation and its individual staff have to be really aware of the limitations of the old ways and of the negative outcomes for all parties. People have to be carefully taken through an awakening process so that they can see how a change can happen, perhaps incrementally, perhaps more radically. They need to see how it is in the individual's interest as well as the organisation's interest. They need to examine what would get in the way of the proposed change and learn how to address these 'rocks in the road'. They need to be trained and skilled in the behaviours, which will support the change. As a consequence of all of the aforementioned, there is a reasonable chance that you will get 'buy in'. The processes suggested do not need to be through formal 'teaching' but can occur through role modelling, through discussion, through some key people challenging assumptions and by keeping 'people issues' on the agenda. This is in exactly the same way that commercial and production issues are traditionally and appropriately addressed all the time in the running of a construction contractor organisation.

Supporting behaviours across a range of activities

Whilst 'taught courses in diversity issues' are not in themselves going to bring about change, training and development does need to be one aspect of a repertoire of supporting activities. At the very least, there needs to be a clear emphasis on equality and diversity issues within Recruitment and Selection training. There could also be value in designing very customised training activities, which address manifest business needs in this area. There can, for example, be very real issues for a contractor's site staff working on predominantly Asian housing estates with a high Muslim population. Site staff do not want to appear rude but can unintentionally insult or act inappropriately through ignorance of other cultures. In such a situation 'cultural awareness' training for those working in culturally diverse settings makes sound business, customer satisfaction and equality and diversity sense.

In relation both to ethnic diversity and gender there is value in the business encouraging open-mindedness and the challenging of assumptions through discussion. In the construction industry a 'politically correct' way of doing this is almost always entirely detrimental. If a heavy-handed approach is taken, then a reversion to what those in the industry would recognise, albeit sometimes reluctantly, as an old 'white males together' defensiveness, can still manifest itself. All involved in construction know about the 'craic'. This form of banter, chat and humour can sometimes be a vehicle for the display of sexist and racist attitudes but in itself is one of the joys of construction. It can be such a positive contribution to communication and the oiling of interpersonal relationships and it is perfectly possible to use the 'craic' to develop debates and challenge assumptions.

Targeting of numbers in a group is an established way of attempting positive action to address imbalance. An advantage of targeting is the way it keeps the issue of diversity near the front of the recruiter's mind and ensures that all efforts are made to tap into the whole potential pool of applicants in an effort to reach the target, even if sometimes this proves impossible. It also encourages recruiters to 'think outside the box' in terms of person specifications and sources and types of contract. It challenges the tendency towards 'cloning' or 'like will appoint like' danger which is ever present in recruitment. Providing the overarching intention remains that the company appoints 'the best person for the job', and then targeting can be useful in focussing minds on the importance of balance in the workforce. The role of personnel, training and development here is in training and coaching, but also in acting as 'guardians' of best practice. By ensuring that at short-listing stage an open mind is kept, the pool of applicants for eventual interview can be broader and more diverse.

Within the organisation there can be value in looking at different forms of support and development activities, which can help to ensure the retaining of under-represented groups once they do join a company. These could include 'think tanks', support groups, mentoring schemes, but in all cases attendance should be by invitation, not a compulsory activity. There is a danger in making individuals feel more different than they wish, by imposing a difference, which might not meet individual needs. Other supporting activities can include the active development, as now required by law, of more 'family friendly' policies and flexible working arrangements which will tend to have a greater effect on women in the business than men. The principle of business needs first may seem to conflict with this flexibility, but with proper dialogue and a focus on common objectives, this can be overcome.

Involvement with external agencies

Whilst this may not be seen as essential, it can be very helpful for a company tackling any change, to tap into the myriad organisations which have real expertise in the change area, in this case, that of diversity. Important bodies in Construction such as the 'Rethinking Construction – Movement for Innovation', the Construction Industry Training Board, the professional bodies and the Major Contractors Group all have material and research projects. These help to move things forward in Construction and to record and disseminate best practice. The Movement for Innovation's (M4I) (Rethinking Construction Movement for Innovation, 2000) 'Respect for People' initiative (3), with its 'tool kit' can be usefully used in training, development and monitoring. There is value in a company being involved in groups, which are focussed on developing opportunities for women and ethnic minorities. Membership of such groups builds the knowledge of the company about the issue and it also means that a clear employer perspective can be added to deliberations.

Conclusions

This case study has traced the development of change in an organisation, which has led to one contractor having a more balanced workforce than it did five years ago. Its experience advocates not radical actions but a more modest process, working on several fronts at once. The commitment to change must be based on business needs and on the valuing of different contributions of the workforce. The proof of any success must lie in outcomes and not policies or set practices. Over six years the number of women and ethnic minorities has increased, as has their presence at increasingly higher levels of the organisation. This change is becoming embedded in the organisation and the value of diversity increasingly appreciated by staff at all levels and also by our customers.

Discussion and review questions

1 The case study illustrates a stated 'subtle' approach to implementing diversity, which 'may be deemed heretical'. Why is this asserted and what could be the shortfalls of this approach?
2 How might a contractor organisation assist a client organisation in implementing diversity policies and practices?
3 Consider the importance of a company being both people focussed and embracing change as a major feature of company culture, in formulating and implementing diversity in an organisation.

References

Construction Industry Board Working Group 8 (1996) *Tomorrow's Team: Women and Men in Construction*, London, Thomas Telford.

Latham, Sir Michael (1994) *Constructing the Team – Final report of the Government/Industry review of Procurement and Contractual Arrangements in the UK Construction Industry*, London, HMSO.

Rethinking Construction – Movement for Innovation(2000) *Working Group on 'Respect for People' – Toolkit*, London, Department of the Environment, Transport and the Regions.

Chapter 14

Government initiatives
and tool kits

Sandi Rhys Jones

Introduction

Government is a regulator of the construction industry, through policy making and legislation, and a major client. Government, both national and local, is also a significant construction employer, through the provision of professional services such as planning, architecture and building control and also through direct labour organisations of tradespeople. In terms of equality of opportunity and diversity, the public sector is generally perceived to be a more active and effective employer, compared with the private sector in general and the construction industry in particular.

In recent years, government responsibility for the construction industry has been subject to considerable change, as a result of departmental re-structuring. Currently, housing comes under the remit of the Office of the Deputy Prime Minister and construction in general sits with the Department for Trade and Industry. This chapter includes a review of the plethora of government initiatives for change within the construction industry, with comments on their relative effectiveness and some suggestions for further action.

Appendix I and II in this book provide good practice guidelines on gender and race equality in the construction industry in more detail – including tool kits.

Latham – the trigger for change

The first time the low representation of women and black and minority people in construction was highlighted at a national level was in 1994, as a result of the government's determination to bring about improvements to the industry as a whole. Sir Michael Latham was asked by Government to investigate the causes of (and possible solutions to) the poor performance of construction. In his far-reaching report *Constructing the Team*,[1] Sir Michael painted a robust picture of the construction industry's reputation for expensive, poor quality products and services, its shortcomings in training and development of its workforce and its macho, dangerous and adversarial culture. Calling upon the industry to increase the very low numbers of women in construction, he recommended that, 'Equal opportunities be vigorously pursued by the industry, with encouragement from Government.'

The call for greater diversity was both long overdue and challenging. Historically, a number of government initiatives for change in construction had impacted negatively on efforts to improve equal opportunities in the industry. Contract compliance, for example, the mechanism developed in the early 1980s to foster greater equality in awarding local authority work, left a long-term legacy of resistance to positive action. The well-intentioned but misguided application of the contract compliance rules led to a number of patently discriminatory incidents, fuelling an understandable antagonism to political correctness. This undermined efforts to improve diversity in the construction industry, particularly in changing attitudes towards greater gender and racial equality.

Compulsory competitive tendering, a government initiative designed to achieve better value, also had a dramatic negative effect on equality, particularly at site and operative level. Many local authority employers axed their direct labour workforces, the traditional stronghold of training and employment of women 'on the tools'. Virtually overnight, the number of tradeswomen in construction, the least representative group in the industry, was decimated.

Moreover, there was cynicism over the real impact of government reports and initiatives on the industry in general. Many people recalled the Banwell Report (1964)[2] of 30 years earlier, which inquired into reforming construction relationships in the long post-war boom period. It recommended replacing open tendering by select approved lists and favoured system building and package deals. Thirty years earlier, the report was seen as an excellent document but did not prove to be a mechanism for significant change.

Against this background, the impact should not be under-estimated of Sir Michael Latham's report on improving construction performance in general and the drive for greater diversity in particular. Not only was the issue clearly identified in a seminal report that triggered real change in the industry as a whole, further credibility was given by the appointment of leading construction industry figure Sir Ian Dixon as chairman of Working Group 8 for the first six months of its formation. A dynamic and committed supporter of equality, Sir Ian continued to be an active champion when he took over from Sir Michael Latham as chairman of the Construction Industry Board (CIB).

Equal opportunities and diversity had been clearly identified as an issue that could no longer be left in the 'too difficult tray' in the construction industry, although actually bringing about change in attitude and behaviour remained a major challenge.

Implementing Latham

The influence of the Latham Report was such that the widely fragmented construction industry came together for the first time to drive forward change. The mechanism for achieving this was the CIB, the first single, truly representative body to be established for the construction industry.

The mission of the CIB was to provide strategic leadership and guidance for the development and active promotion of the UK construction industry, its clients and government, in order to improve efficiency and effectiveness throughout the construction procurement process. In particular, the CIB aimed to secure a culture of co-operation, teamwork and continuous improvement.

An indication of the scale of the task in gathering consensus to carry forward the Latham recommendations can be gauged from the fact that there were 30 bodies representing professional institutions, 23 organisations in the Building Materials Producers body and 11 different representative groups in the confederations representing contractors.

The membership and funding for the CIB came from the following six groups:

- Government in Partnership (DoE, DTI, The Scottish Office and the Health and Safety Executive)
- Construction Industry Council (the construction professional bodies)
- Constructors Liaison Group (Building Structures Group, Specialist Engineering Contractors Group, National Specialist Contractors Council)
- The Alliance (Alliance of Construction Product Suppliers, National Council of Building Material Producers)
- Construction Clients Forum
- Construction Industry Employers Council (Major Contractors Group, Federation of Civil Engineering Contractors, Federation of Master Builders and the Building Employers Confederation).

The first task of the CIB was to form Working Groups to address the main issues of the Latham Report, with representatives nominated by each of the six member bodies. Seizing the opportunity created by the official acknowledgement of the lack of diversity in construction, a number of individuals pressed for the issue to be specifically addressed in this process. As a result, one of the thirteen Working Groups set up by the CIB was dedicated to investigating the poor equal opportunities record of the industry and recommend ways of bringing about improvement.

Mainstreaming diversity

Working Group 8 (WG8) of the CIB, set up to address equal opportunities, consisted of a cross-section of men and women from various parts of the construction industry and government. It was chaired by Sandi Rhys Jones, who took over as chair from Sir Ian Dixon when he became Chairman of the CIB. The working group recognised the need to review representation and attitudes towards gender, race, disability and sexual orientation, but in order to ensure that practical recommendations emerged from their work in the short timescale allotted, it was agreed to focus attention on women, as representing 50 per cent of the population.

A fundamental premise of the group was that the changes needed to recruit and retain a more representative workforce – better working conditions, both physical and social, improved training and career planning – are the same changes that would improve the construction industry in general. A key aim of WG8 was therefore to mainstream the issues of diversity and equal opportunities in the sea change sweeping through the UK construction industry. As part of this, WG8 made a point of reviewing and commenting on the reports of all the CIB Working Groups, to ensure that at the very least, the language used acknowledged that the industry is not entirely white and male and that women take a role in the process.

WG8 members believed that it was important to draw from the experience of other industries and professions that had achieved significant change in equal opportunities. As Sir Ian Dixon commented in his foreword to *Tomorrow's Team*, 'At least by being one of the last industries to change, the construction industry can learn from the experience of others.'

A seminar was held at which representatives from accountancy, the health service, television, manufacturing and government outlined how they had increased the representation of women. It became clear that whilst construction does have particular challenges, it is not as unique as many people believe. Innovation in retaining and recruiting women in the IT, accountancy and media sectors provided useful guidance and ideas.

The Working Group also carried out a small research project amongst 95 major construction clients to identify attitudes and practice relating to diversity and equal opportunities. All responding organisations had equal opportunities policy statements in place, but significantly most also had a comprehensive set of measures to implement the policies. This is in marked contrast to the construction industry, where more than 90 per cent have a policy but only 37 per cent have an equal opportunities plan (*Black and Ethnic Minority People in Construction*, 2002).[3]

Tomorrow's Team: Women and Men in Construction

The Working Group's Report, *Tomorrow's Team: Women and Men in Construction* (1996),[4] consisted of three parts:

- Part 1: The value of diversity, put forward the business case for greater diversity, the need for culture change and guidelines on implementing equal opportunities in the workplace.

- Part 2: Developing people addressed education and training, recruitment and retention and the influence and attitudes of clients towards diversity.
- Part 3: Facts, figures and information, provided statistics, legal background and the methodology applied by the Working Group.

A number of recommendations were also proposed, clustered under the headings of 'Raising awareness', 'Education and training', 'Recruiting and retaining' and 'Changing the culture'. These included calling upon the professional institutions to demonstrate commitment to equal opportunities, to increasing support for women-only trade training schemes in the voluntary sector, to improving opportunities for site experience, developing technical skills for appropriate administrative staff and instituting a cross-disciplinary mentoring scheme.

The practical guidelines on encouraging greater diversity were designed as a pullout section of the main report and were devised not only for employers but also for individual women in the construction industry. The Construction Industry Training Board (CITB) sponsored the printing of additional copies of the guidelines as a stand-alone document for wide distribution.

WG8, like all the others in the CIB, was disbanded in 1997, a few months after delivering its final report. The task of taking forward the recommendations on equal opportunities and diversity was left to individual companies or organisations in construction industry. Direct and indirect government support was on offer, either through applying for DETR (Department of Environment Transport and the Regions) funding under its Construction Research and Innovation Programme or by calling upon advice and support from the Construction Best Practice Programme. However, it became clear that significant efforts were still required to bring about change, and that only lip service, rather than commitment and action, was being given to the issues of equal opportunities and diversity.

Egan – rethinking construction

Meanwhile, the drive for improved performance in construction was sustained, despite a change of government. In 1997, the Conservative Government that had instigated the Latham report and funded the establishment of the CIB and its Working Groups was replaced by a Labour administration. Contrary to the expectations of many, when Labour took power after 17 years of being in opposition, the momentum for change in construction was maintained.

In fact, and perhaps unsurprisingly, another report into construction performance was commissioned, led by Sir John Egan as chair of the Construction Task Force. While some regarded the exercise simply as the need for a new government to make its own mark, others welcomed the new report as client driven, rather than government/industry-led. Be that as it may, the Egan Report *Rethinking Construction* produced in 1998 acknowledged and took forward the issues highlighted by Latham, set out in five drivers for change: committed leadership, focus on the customer, product team integration, quality-driven agenda and commitment to people.

The Egan report emphasised the need for mutual respect, good human resource management and health and safety, but to the disappointment of those committed to improving equal opportunities, there was no explicit reference to the continuing need for greater diversity. However, there were some encouraging signs of support. The new Minister for Construction Nick Raynsford had already taken on the mantle of government champion. Speaking at the CIB Conference, he stated,

> I do not think it is acceptable that construction has next to no commitment to equal opportunities... the industry is doing itself damage by this approach in that it is turning its back on a huge potential source of recruits. Not only that but I have to say that a failure to offer equal opportunities is simply not consistent with this Government's view of a business which involves all of the community.
>
> (Nick Raynsford, June 1997)

The movement for innovation

Within weeks of the launch of *Rethinking Construction* in 1998, there was a proliferation of Government supported programmes and projects designed to improve construction through the Egan agenda. A new body, the Movement for Innovation (M4I), was established to deliver practical demonstration projects measured against Key Performance Indicators (KPIs). The CIB continued to operate, sustaining its Latham agenda through the Construction Best Practice Programme and the Construction Research and Innovation Strategy Panel (CRISP) whilst supporting the implementation of the Egan agenda.

Another vocal and committed diversity champion appeared in the person of Alan Crane, appointed by Nick Raynsford in 1998 to chair the M4I. The commitment to people issues, including diversity, was reflected in the decision to form the Respect for People (RfP) Working Group of the M4I established in response to Nick Raynsford's challenge,

> The industry will only be able to rise to the challenge of Egan if it can recruit, retain and develop the talent and skills it needs. Companies who do not treat their people well, who cannot retain the talent they need, will lose competitiveness, lose margins and eventually lose their business entirely.

These issues of training, recruiting and retaining the construction workforce had become of major importance, as the acute skills shortages in the construction industry had become evident. Not only had many workers left the industry during the ravages of the recession of the early 1990s, but also fewer young people showed interest in taking up construction as a career, particularly in the face of competition from other sectors. For example, engineers were wooed by law and accountancy firms (professions which had achieved parity in male/female student numbers and high levels in the workplace) and potential operatives saw IT as a better paid, cleaner and more exciting opportunity.

Diversity toolkits

In November 2000, under the chairmanship of Alan Crane, the Respect for People Working Group produced a report *A Commitment to People 'Our biggest Asset'*, identifying a range of issues for employers to address in order to improve working conditions and job satisfaction. Diversity was one of the issues included, together with employee satisfaction, health, safety and career development. A set of eight checklist type toolkits was produced, together with pan-industry performance measures.

A trial programme was established, in which more than 100 companies in all sectors throughout the United Kingdom participated, providing feedback that enabled appropriate refinements to be carried out before publication of the toolkits. The objective was to validate the recommendations, toolkits, performance measures and business case before contributing to a strategy for full industry implementation of the RfP agenda.

It was a source of considerable disappointment that the diversity toolkit scored the lowest in terms of perceived relevance, effectiveness, usability and impact. Not only did the RfP group identify the low take-up of the diversity toolkit, there were difficulties in identifying robust KPIs. As a consequence, the KPI Handbook (2002) has no headline KPI for diversity.[5]

To analyse the reasons and identify improvement, a report was commissioned by Rethinking Construction and the CITB, carried out by the University of Loughborough.[6] In the conclusions of the report, the researchers comment,

> Unlike some of the other Respect for people focus areas, diversity is a difficult and contentious issue to tackle. Thus, the unwillingness of construction companies to embrace the diversity agenda may reflect a failure of the industry to accept the arguments for workforce diversification, rather than a direct criticism of the toolkit itself.... Hence a key challenge is to convince construction companies to embed diversity as an integral part of good employment practice.

The researchers also tested five alternative toolkits and documentation on diversity amongst the responding companies, all of which received positive feedback, including three toolkits produced by Change the face of construction (discussed in the following section). It was recommended that good practice from these toolkits could be taken on board in further development on the RfP diversity toolkit.

Change the face of construction

In parallel with the RfP initiative, there was a drive to sustain the Latham agenda on diversity within the CIB. In 1999, the DETR provided funds to devise a programme of cultural change to assist employers in recruiting and retaining under-represented groups such as women, black and ethnic minorities and people with disabilities. Under the four-month long contract, Sandi Rhys Jones OBE and Helen Stone OBE, devised Change the face of construction. The three core tasks of the pilot programme were to

- establish a communications mechanism to reach minorities in construction,
- create a cohesive visual identity and devise and distribute a newsletter,
- begin a targeted communications campaign.

A further six-month contract followed in 2000, funded by the CIB and CITB. Its objectives were to formulate a self-sustaining programme, with funds contributed by industry, and to begin to deliver a practical toolkit to companies to help improve their recruitment and retention of under-represented groups. Practical outputs included

- an information hub, based on a Web site providing information, news and events,
- meetings and consultation with organisations around the UK to identify level of support required and to promote the business benefits of diversity,
- a database of speakers and role models,
- toolkits on Recruitment, Retention and Satisfying the Customer,
- practical case studies,
- a model mentoring programme,
- numerous articles and features, presentations at conferences and exhibitions,
- liaison with diversity groups and organisations, including Opportunity Now, Commission for Racial Equality, Race for Opportunity, London Equal Opportunities Forum.

The programme attracted significant interest, primarily as a much-needed resource on diversity issues, particularly from the media, students and academics. However, continuing government commitment was dependent on financial support from the construction industry, which was not forthcoming. The reasons for the lack of financial support appeared to be a combination of initiative fatigue, a belief that training-related issues should be covered through existing levy-related contributions and a low priority on diversity. Since then, the Change the face of construction Web site has been sustained on a voluntary basis by Sandi Rhys Jones.

The Housing Forum

Another of the new bodies set up by government and industry to implement Egan recommendations for improvement and innovation was The Housing Forum, under the chairmanship of Sir Michael Pickard. The construction sector most familiar and relevant to the general public and the individual consumer, housing is regularly exposed in the media for dubious practice and poor workmanship. Whether a major housebuilder, jobbing plumber or registered social landlord, housing bears the brunt of the macho 'cowboy builder' image.

Determined to bring about change, The Housing Forum took up the same KPIs as the M4I and launched the annual National Customer Satisfaction Survey to find out what 10,000 buyers think. The Housing Forum also established its own RfP Working Group in 2000, chaired by Charmaine Young of major housebuilder St George Regeneration Ltd. The Working Group set about establishing an extensive network of members, gathering examples of good practice and holding events for debate and exchange of experience. The result of this activity was the collection of big ideas and simple ones, some based on the Housing Forum demonstration projects and others based on the knowledge and experience of members of the group. The material was published as a series of case studies and practical suggestions in a guidance document and wall poster entitled 'Recruitment, Retention and Respect for People: 20 ideas for delivering the 3Rs.'

The RfP Working Group of the Housing Forum also worked with the House Builders Federation, the CITB and the Change the face of construction team to deliver a series of roadshows to universities and colleges. This programme gathered together young men and women from a variety of backgrounds, disciplines and companies to describe in an open and informal way their reasons for choosing housebuilding, the challenges and the benefits. Before and after questionnaires revealed that the roadshows changed perceptions and raised interest levels amongst students.

Egan – Accelerating Change

In May 2001, the CIB was wound down, five years after it had been set up to take forward to recommendations of the Latham report. It was replaced by the Strategic Forum for Construction, again with Sir John Egan at the helm. Rethinking Construction took over the task of umbrella organisation to carry forward the change process. This time, rather than consisting of industry representative bodies, Rethinking Construction Ltd was set up as a not-for-profit company generating best practice and innovative ideas for improvement across three areas of activity, working in partnership with the Construction Best Practice Programme:

- M4I, leading on non-housing construction in the private sector.
- The Housing Forum, leading on housebuilding, refurbishment and repairs and maintenance in the public and private sectors.
- Local Government Task Force (LGTF) established to help local authorities meet best value obligations, learn from demonstration projects and maximise efficiency and effectiveness.

In 2002 The Strategic Forum produced the report *Accelerating Change*, which set the following strategic targets:

- By the end of 2004, 20 per cent of construction projects by value should be undertaken by integrated teams and supply chains to increase to 50 per cent by 2007.
- 20 per cent of client activity by value should embrace the principles of the Clients' Charter, rising to 50 per cent by 2007,
- Strategic Forum members will develop and implement strategies which will enable the industry to recruit and retain 300,000 qualified people by the end of 2006, resulting in a 50 per cent increase in suitable applications to build environment higher and further education courses by 2007.

Welcome though this explosion of activity was, it is not surprising that the range and inter-relationships of activities led to a combination of confusion, overload and apathy.

Initiative fatigue began to appear, acknowledged in an interim report on the toolkit trials produced in November 2001 by the RfP group (Interim Findings, 2001)[7] and also by the People Working Group of the Strategic Forum. A low level of awareness of Latham and Egan generally also became apparent. For example, in a survey commissioned by Rethinking Construction in 2002, unprompted recognition rates were very low – 13 per cent for the Egan report 3 per cent for Rethinking Construction and less than 1 per cent for KPIs and Accelerating Change.

Interestingly, awareness levels amongst clients was considerably higher than amongst contractors and manufacturers – mirroring the difference in attitude and practice on diversity by clients that had been identified by the CIB WG8 on Equal Opportunities five years earlier. This difference was reflected across a range of issues, for example, only 22 per cent of contractors and manufacturers could name one Egan initiative without prompting. In contrast, 50 per cent of public sector clients could do so. Similarly, 34 per cent of clients value an integrated supply chain whereas only 20 per cent of suppliers think they do. Suppliers also underestimate the importance clients attach to sustainable construction techniques and good safety records.

As part of the effort to raise awareness of the Rethinking Construction agenda, a network of ten regional groups was developed during 2002–2003 bringing together and disseminating the activities of the M4I, the Housing Forum, the LGTF and the Construction Best Practice Programme. In addition, a National Strategy Panel was established to promote the benefits of demonstration projects generated by the various partners throughout these regions. The impetus to have robust programmes in place was driven by the knowledge that government funding for the programme was scheduled to end in March 2004.

The National Strategy Panel produced a report on the demonstrations programme (*Demonstrating Success through Rethinking Construction*, 2003),[8] which claimed that if one-third of the industry applied the principles of Rethinking Construction to their projects

- client construction costs could be reduced by £1.4 billion,
- the cost of accidents could be reduced by £1.2 billion,
- organisations could deliver an additional £446 million in profit.

In his foreword to the report, Chair Professor David Gann of Imperial College highlighted the need to transfer lessons from successful projects across processes and within organisations and in particular to co-ordinate with the work being carried out by organisations using the RfP toolkits.

Rogers – Constructing Excellence

In February 2003, another re-structuring of the delivery of the Egan agenda took place, this time explicitly to co-ordinate the various programmes and initiatives under way. It was announced that Rethinking Construction (and its constituent groups M4I, The Housing Forum and the LGTF) was to be combined with the Construction Best Practice Programme. The new body, *Constructing Excellence*, is chaired by Peter Rogers of developer Stanhope, Chair also of the Strategic Forum for Construction.

The key objective of the new organisation is to deliver an effective and streamlined approach to the delivery of construction industry reform and offer the industry a single port of call. Its remit is to continue delivering the Rethinking Construction initiative through a network of best practice clubs, regional cluster groups and demonstration projects.

The good news for those committed to diversity is that Peter Rogers has unequivocally stated that the construction industry needs to be more diverse and in particular to attract and keep more women at all levels.

A wider view

The Greenfield report

Whilst diversity in construction was being addressed as part of the Rethinking Construction/RfP initiatives, a gender specific initiative in a wider arena was being developed by government. For some years the DTI had been funding Set4Women, a gender unit for science, engineering and technology, staffed by five people and with the resources to run a Web site, produce materials and support individual initiatives. When construction responsibility was moved to the DTI from the DETR, it was hoped that the sector would benefit from a similar level of resource and focus on diversity.

In 2002 the opportunity presented itself, when the Patricia Hewitt, Secretary of State at the DTI appointed Baroness Greenfield to advise on a strategic approach to increasing the participation of women in science, engineering and technology (SET). A number of women from the construction participated in this review, concerned to ensure that the industry was recognised as a specific and also to maintain a balance between academia and industry.

Following a full-day workshop to which a wide range of women were invited to contribute views, three teams were set up to address barriers and opportunities to women at three stages:

- Early career development and re-entry after a break (chaired by Professor Teresa Rees).
- Mid-career management (chaired by Dr Nancy Lane OBE).
- Getting the culture right for women to make it to the top (chaired by Dr Gill Samuels CBE).

Many of the recommendations in the report (*Set Fair*, 2002)[9] were similar to those identified in the Latham WG8 report *Tomorrow's Team: Women and Men in Construction*. For example the establishment of a centre to provide an infrastructure for women in SET, organisations, networks and initiatives, particularly to 'reduce duplication of activities and act as a focus for the media, head-hunters, government, industry and professional societies'. This was also a need that the Change the face of construction initiative hoped to achieve. Other key proposals included

- the provision of a returners' scheme,
- raising the status of diversity by including gender on the existing R&D scoreboard, requiring formal reporting on the diversity measures adopted by SET employers,
- establishing a part-time/job-share programme with advantages for employers such as tax breaks/funding.

The Equal Opportunities Commission

The Equal Opportunities Commission (EOC) is an independent, non-departmental public body funded primarily by government, dealing only with sex discrimination. In 2003, the EOC embarked on a general formal investigation (GFI) into areas where there are currently shortages of trained staff and an imbalance in the concentration of women and men. These areas include engineering, construction, plumbing and ICT, which are male-dominated careers, and childcare, which is female-dominated.

The two year long GFI took place in two phases. From May 2003 to December 2003 the EOC took a national overview of gender or occupational segregation (the concentration of women and men in different kinds of job) and inequality in Modern Apprenticeships. The Phase 1 Report, *Plugging Britain's Skills Gap: Challenging Gender Segregation* looks at the case for change, including productivity, competitiveness and choice; current policy and practice; Modern Apprenticeships and their opportunities and challenges.

Phase 2 of the GFI took place from January 2004 to March 2005 and explored barriers and solutions through LSC (Learning and Skills Council) case studies and focus groups with employers, young people and women in 'non-traditional' work like construction and engineering. The resulting report *Free to Choose: Tackling Gender Barriers to Better Jobs*, confirmed what many activists had been saying for some time, namely that stereotypes begin young and are reinforced at school age and that careers advice is too ineffective to challenge these stereotypes. Young people, especially girls from lower socio-economic groups, are being channelled into jobs traditional to their sex and are not being given access to careers advice, work-experience and training opportunities.

In parallel with the GFI, the EOC also embarked on a specific campaign called *Know Your Place, Encouraging Women* to consider a career in the construction, engineering, IT and plumbing industries, aims to encourage women. Working with JIVE partners, who create change for women and girls in engineering, construction and technology, the EOC produced posters featuring careers in construction, engineering careers, IT and plumbing together with a Web site and leaflet with advice and information.

The Construction Industry Training Board

The CITB's stated strategic approach for diversity is to focus on engaging employers in the business case for diversity, as much as on widening participation in education and training. The CITB has a network of diversity advisors working nationally, co-ordinated by two diversity managers at its Head Office in Norfolk.

A specific CITB initiative is the STEP Into Construction programme, which support to employers for recruitment of job-ready black minority ethnic people and woman aged 16 plus. The programme includes support for a six-week work trial and the guarantee of a job interview at the end. Other activities in the CITB diversity plan for 2004 includes continuing its advertising and public relations campaigns, aiming to support 100 Positive Action Education Projects aimed at ethnic minorities and women and the setting of apprenticeship targets for ethnic minorities and women aged 16–25.

A key element of the CITB approach to diversity is the development of strategic alliances. These include the LSCs, to enhance mechanisms that support employers to improve their recruitment practices, and the Construction Industry Trust for Youth partnership, to support ethnic minorities and women in finding work experience and job opportunities.

CITB-Construction Skills sits with the Engineering and Training Board (ETB) on the Steering Group of Women into Science and Engineering. In 2003, the CITB-ConstructionSkills Construction Ambassador scheme was aligned with the SET Ambassador scheme run by SETPOINT with ETB funding. CITB-ConstructionSkills is also a member of the Department for Trade and Industry's Office of Science and Technology Women in SET Implementation Group.

The UK Resource Centre for Women in SET

The UK Resource Centre for Women in Science, Engineering and Technology was set up in 2004 in response to the Greenfield *Set Fair* report, and complements the Government's 10 year investment framework for Science and Innovation. Based in Bradford, the UKRC is being developed by the JIVE consortium of Bradford College, Sheffield Hallam, Open and Cambridge Universities and has a budget of over £4 million to be spent over a three year period. The JIVE partnership has particular expertise in construction trades and IT skills.

The aim of the UKRC is to establish a dynamic centre that provides accessible, high quality information and advisory services to industry, academia, professional institutes, education and Research Councils within Science, Engineering and Technology (SET) and the built environment professions, whilst supporting women entering and progressing in SET careers. The UKRC is working closely with other organisations in the field, including the EOC and also the Women and Equality Unit, based at the DTI and which supports the work of the two Ministers for Women.

In addition to a comprehensive website, good practice guidelines, mentoring schemes and employer networking events and award schemes, the UKRC has also launched a programme to attract and support women returning to SET after a career break for motherhood or other reasons. The UKRC also has funds available to pump prime and support relevant initiatives and activites, including bursaries.

Progress and barriers

So what currently has been the effect of government initiatives on those committed to bringing about and managing greater diversity in construction? There is little doubt that the topic of diversity has been taken out of the 'too difficult tray', and that many working in the area have been encouraged to continue their efforts. However, the multiplicity of programmes and activities-generated initiative overload, and promotion of the active advocates and activists has tended to create a 'usual suspects' attitude. There is also growing concern amongst employers in general about the burden of compliance with regulation and legislation, combined with huge difficulties in finding young people with the commitment to training for a career in the crafts and trades.

Whilst awareness has increased, behavioural change remains slow. The majority of small and medium enterprises that make up the majority of employers in the industry see little relevance in change initiatives relating to diversity, and resist anything that may add to administrative load. Informal recruitment methods and self-employment at trade and operative levels present particular barriers, not only to women, black and minority ethnic and disabled entrants, but also to the development of higher standards of behaviour, training and working practices. There is also a reluctance to contribute financially, particularly to training and education programmes, for which many employers believe they already contribute sufficiently through levy to CITB and other providers. Compared with other sectors of industry and commerce, the profit margins of many companies, even in prosperous times, are relatively low.

Examples of good practice in diversity that have taken place are usually driven by voluntary sector groups working in partnership with industry – voluntary sectors who spend their time on the perpetual search for funding. The enormity of this challenge was reflected in 2004 when the highly successful and well-regarded Women's Education in Building (WEB) organisation went into liquidation. Fortunately much of the expertise and skill gathered during the 20 years that WEB operated are to be harnessed through the CITB and some committed employers. Employers committed to increasing diversity are generally major companies, with appropriate human and financial resources, and medium-sized private or family-owned businesses with management commitment and relative freedom from stock market pressures to innovate.

There is also a wider image question to be addressed. Traditionally the construction industry has tended to debate issues and promote achievement within its own sphere. There have been few speakers or presenters representing construction at general business conferences and events on culture change and diversity. The fragmentation of the industry and its supply chain has also inhibited change.

However there are encouraging signs that the long awaited joined up thinking to address these problems is beginning to take place between Government departments, educational establishments and the private sector. The UKRC is the valuable hub that has been demanded for so long by those committed to change, and is bringing together the various public, private and voluntary sector groups such the DfES, the DTI, the EOC, Learning and Skills Councils and employer groups such as the Sector Skills Councils.

The Women and Equality Unit, based at the DTI and supporting the work of the two Ministers for Women, is also working with the DfES and the UKRC and its partners to address the occupational segregation highlighted by the EOC and which is a key issue for gender diversity in construction.

Conclusions

In summary, the construction industry's reputation for being dirty, dangerous, sexist and racist is seen to be the major barrier to attracting a more diverse workforce. To overcome this, there is a need to illustrate the benefits of a more diverse workforce and to promote the business case more persuasively. It makes sense to sustain the various initiatives already begun, modifying where necessary, such as the RfP toolkits, the Construction Best Practice Programme and the Rethinking Construction Demonstration projects. This is where the UK Resource Centre for Women in SET will play a valuable role.

In addition there are benefits in bringing construction together with the science, engineering and technology sectors under the UKRC umbrella. Some of these organisations and businesses are further ahead in gender diversity than their construction counterparts – and indeed, many are construction industry customers. This combination creates a network and exchange of valuable experience and strengthens representation to government. The increased profile of committed organisations and individual women working in the construction sector is also helping to encourage change, supported by the recognition by the three main political parties of the role of women in the workplace at all levels.

As for the future, it will be interesting to watch the impact of latest proposal for more joined-up government thinking, namely the single equality commission planned for October 2007. The result of the biggest review of equality institutions in twenty-five years, the new Commission for Equality and Human Rights (CEHR) will bring together the work of three existing equality commissions – the Commission for Racial Equality, the Equal Opportunities Commission and the Disability Rights Commission.

Led by the Women and Equality Unit, the review proposed that one body will ensure that work on equality is co-ordinated across Government as a whole. The official view is that gender is a main stream issue, rather than a special case. 'In a world where individual identities are becoming increasingly important, we need to see people as a whole – not put them into boxes marked race, religion, gender, sexual orientation, disability,' said the Rt Hon Patricia Hewitt MP, speaking as Secretary of State for Trade and Industry and Cabinet Minister for Women at the launch of the Government White Paper. The question is whether the construction industry is ready for this, or still requires positive and clearly defined action to bring about greater gender diversity in order to contribute to the DTI target of 40 per cent female representation on Science, Engineering and Technology related boards and councils by 2008.

Discussion and review questions

1 What practical help can be given to small builders to encourage them to take on apprentices, particularly women and ethnic minorities?
2 An increasing number of tradespeople from Eastern Europe, particularly those countries newly admitted to the European Union, are coming to the United Kingdom in response to the building boom. What are the issues raised by this trend?
3 The countries described in question 2 above have a tradition for educating and training significant numbers of women in engineering, construction and technology. What might the United Kingdom learn from this?

Notes

1 *Constructing the Team*, HMSO, 1994.
2 Banwell, H. (1964) Report of the Committee on the placing and management of contracts.
3 *Black and Minority Ethnic People in Construction*, Royal Holloway College and CITB, 2002.
4 *Tomorrow's Team: Women and Men in Construction*, Thomas Telford Ltd, 1996.
5 *Key Performance Indicators Handbook*, produced by Respect for People Working Group, 2002.
6 *An Evalutation and Development of the 'Diversity in the Workplace' Toolkit and Key Performance Indicators as Mechanisms for Improving Business Performance in the UK Construction Sector*. Department of Civil and Building Engineering and Department of Social Sciences, Loughborough University, April 2002.
7 Report of Interim Findings for the Respect for People Toolkit and Performance Measures Trials, produced by Rethinking Construction, November 2001.
8 *Demonstrating Success through Rethinking Construction*, Rethinking Construction Ltd, 2003.
9 *Set Fair: The Greenfield Report on Women in Science, Engineering and Technology*, Department of Trade and Industry, November 2002.

Building equality in construction good practice guidelines – Part I

For building contractors and housing associations

Davey, C., Davidson, M.J., Gale, A.W., Hopley, A. and Rhys Jones, S. (1999)

Source: Building Equality in Construction Good Practice Guidelines for Building contractors and housing associations', Working Paper No: 9901, Manchester School of Management Working Paper Series, ISBN: 1–86115–050–4, 30 pages.

Introduction

This is a short report following a research project into how housing associations and building contractors can work together to improve equal opportunities in the construction industry and to meet the diverse needs of local communities.

The Building Equality in Construction research project was funded by the Department of Environment, Transport and the Regions under the 1996 Partners in Technology programme. It was carried out by UMIST in partnership with Building Positive Action and RhysJones Consultants Ltd.

The report was produced as a 'live' document, for debate at the industry forum held on 6 May 1998 to mark the completion of the Building Equality in Construction project. Response was extremely positive, with the Equal Opportunities Commission (EOC) requesting to use parts of the document as an example of good practice in its information packs.

Additional information provided by the Construction Industry Training Board (CITB), North West about its successful girls-only training courses has been incorporated in this version. Comment was also received from the Commission for Racial Equality, suggesting that racial discrimination could be addressed in more depth. This issue has been incorporated in the team's proposal to expand the project nationally.

The project team gratefully acknowledges the contribution, support and interest of many people during the progress of the work, which ran between May 1997 and May 1998, in particular the financial contribution and assistance of the Construction Sponsorship Directorate of the DETR (Department of Environment Transport and the Regions).

Summary

The project

Building Equality in Construction is an innovative project focusing on how to promote and maintain equal opportunities policies in the construction industry, through developing a partnering methodology between clients (housing associations) and the building contractors they employ. The project has developed a management mechanism and good practice guidelines for building contractors and housing associations.

The project was carried out by UMIST, Building Positive Action (representing around 40 housing associations in the North West) and RhysJones Consultants Ltd. The Department of the Environment, Transport and the Regions provided matched funds. under the 1996 Partners in Technology programme.

The problem

The construction industry is under pressure from housing associations to improve its performance in equal opportunities. The need for improvement is increased by a social and commercial environment in which women in particular are taking a growing role. Add to that the significant role of women in purchasing decisions as well as being major users of the finished project and it is clear that the construction industry needs to catch up with other industries in recognising and involving women.

How can a male-dominated industry known for its poor reputation and working conditions, tight margins and entrenched attitudes, improve its performance in equal opportunities? This project builds upon the report produced by the Construction Industry Board, Working Group 8, 'Tomorrow's Team: women and men in construction' (1996) by highlighting possibilities for change specific to building contractors and housing associations involved in social house building programmes.

The players

The housing association sector has a better record in equal opportunities than the construction industry which serves it. This is perhaps not surprising when seen in the context of the housing association commitment to meeting the diverse needs of local communities, regenerating local economies and giving opportunities to all sections to benefit from social housing building programmes.

Women represent 25 per cent of housing association management, compared to 3 per cent of professional women in construction (the Chartered Institute of Building had 0.7 per cent female corporate members in 1998) and a mere 1 per cent at trade and craft level. This lack of representation and the poor record of equal opportunities in construction causes concern to housing associations.

Building contractors recognise the importance of the housing associations as a client group – in the North West the sector spend is some £1 billion over five years. Despite a degree of cynicism about the effectiveness of monitoring compliance with equal opportunity requirements of housing associations, many are nevertheless concerned to improve their equal opportunities performance and widen the recruitment pool.

Triggering change

The innovative approach of the project, focusing on client-led change towards equality in construction, resulted in a joint learning process for all participants, including the research team. Moreover, the research process itself has achieved some progress. Before the project was finished, a number of initiatives were undertaken, including:

- a housing association asked a contractor to work in partnership on a project.
- a contractor offered to sponsor an undergraduate at UMIST.
- a contractor asked for help in recruiting women for a specific job opportunity, resulting in collaboration with other participants.
- a housing association offered to reward contractors active in equal opportunities through promotion in its annual report.
- a maintenance training proposal is being developed between a contractor and the housing association.
- a contact person on equal opportunity issues in construction was established at UMIST, providing a link on job incentives, sponsorship and lecture programmes.
- questions on equal opportunities are now included on undergraduate satisfaction questionnaires at UMIST.
- the Construction Industry Training Board and the Professional Management Liaison Group approached the project team to discuss ways of working in partnership.
- a housing association and a contractor instituted an exit monitoring system to track trainees.

In addition, the action learning sets developed by James Powell OBE and Donna Vick (Revans Scholar) at the University of Salford in collaboration with the UMIST project team, were successfully launched in March this year with the participation of more than 20 housing associations and building contractors from the focus groups.

Changing perceptions

Doubts about building contractors' commitment to equal opportunities appeared at first to be confirmed by the poor response to postal invitations to participate at the beginning of the project. However, when the researchers spoke to contractors on the telephone, there was a very positive response.

In contrast, housing associations often failed to attend meetings and as a result, were outnumbered by building contractors, not only in attendance but also in terms of contribution. The genuine enthusiasm shown by building contractors at the focus groups surprised and impressed researchers and participants alike.

Contrary to popular belief, some contractors are committed to equality on the grounds of social justice rather than business interest alone. Housing associations, more predictably, react badly to equality efforts driven by financial motives alone. Nevertheless, housing associations recognise that reward mechanisms for good practice are more effective and practical than penalties for apparent failure.

However, housing associations appear very anxious at the prospect of letting contracts using criteria other than price. Lowest price is a key driver for poor employment practices and a conservative attitude towards equality and indeed change. This is a fundamental issue with respect to public accountability in relation to 'value for money'.

Fear of failure

Many contractors believe that setting specific targets, say 30 per cent of the workforce to be female or ethnic minority groups, is a demotivator. If they cannot achieve these targets, despite commitment and much time knocking on doors, they fear being seen by the client as failing. Some other mechanism is required, over which the contractor has more control.

Recognising that reward mechanisms for good practice were potentially more effective and practical than penalties for apparent failure, housing associations reacted favourably to other mechanisms for improving equal opportunities. These included reviewing project deadlines to allow flexible working, agreeing education and training schemes as part of the package and including in annual reports contractors who promote equality. Housing associations were aware of the need for improvements in monitoring and enforcement of equal opportunities criteria, but, due to perceived practical difficulties, were reluctant to develop specific proposals further without support from the Housing Corporation.

There was considerable debate about the effectiveness and practicality of contractual agreements to promote equality, with no obvious division in opinion between contractors and housing associations. Some thought contractual agreements for equality ensured fairness, enforceability and an opportunity to get paid for extra work, others anticipated a potentially adversarial and litigious situation.

Climate of co-operation

The project was launched by a seminar, attracting representatives including 21 housing associations and 9 construction companies, face-to-face interviews were conducted with 16 representatives of the industry and a series of focus groups were run with housing associations and building contractors.

The friendly, relaxed and informed atmosphere of meetings was a key factor in securing the confidence of people to work in partnership. The process stimulated the formation of a very effective network. Following feedback from the first focus group, the sessions were extended from two hours to three, reflecting the wish by participants to devote more time to debating the issues.

Moving forward

It became clear during the project that there is value in extending and promoting a number of existing initiatives designed to improve equal opportunities. For example, the CITB would like more housing association involvement in the curriculum centre programme, building contractors would like to see more programmes such as Women as Role Models (WARM) going into primary schools. The Contractor Code for housing associations and their contractors, published by the National Housing Federation, also contains useful guidelines on a range of topics including equal opportunities.

There are some policy issues however, requiring input from the Housing Corporation and government, which relate to financial implications of some of the proposals put forward during the course of the Building Equality in Construction project. For example, selection criteria based on lowest price prevent the implementation of some of the innovative and inclusive packages that could be developed. Loss of housing benefit for tenants offered training or work opportunities also present difficulties.

In order to investigate ways of reviewing policy and to sustain the enthusiasm and momentum generated by the project, the project team has identified that there would be support for the formation of an 'exchange'. The tasks would be to act as a co-ordinating body, particularly in widening existing initiatives, monitoring progress and perhaps linking with a charter scheme for the industry. Several individuals were willing to investigate the possibility of providing support for a secretariat function for the exchange.

Most significantly, the project team believes that the experience of this regional project could be replicated on a national level throughout England and Wales. Based on the Housing Corporation regions, the structure would develop the Building Equality in Construction model, linking an independent third party team (university, training provider and consultant) with local housing associations and building contractors in order to develop similar partnership projects and initiatives to improve equal opportunities.

Guidelines for building contractors

Introduction

These good practice guidelines offer suggestions on how building contractors employed on social housebuilding programmes can set about promoting equality, particularly in the manual trades. They are a development of the Construction Industry Board Working Group 8 Report 'Tomorrow's Team: women and men in construction'.

The guidelines deal with how to attract more recruits into construction, particularly female, provide training and development opportunities for women, operate fair recruitment procedures, contribute towards career development and improve retention. Ways of identifying, measuring and benefiting from practical steps to promote equality are included.

The guidelines are divided into the following sections:

- Why bother with equal opportunities?
- Attracting young people into the industry
- Widening the recruitment net
- Innovative approaches to training
- Keeping employees
- Starting the process.

Why bother with equal opportunities?

The construction industry operates in a changing social and commercial environment in which women in particular are taking a growing role. Add to that the significant role of women in purchasing decisions, as well as being major users of the finished project, and it is clear that the construction industry needs to catch up with other industries in recognising and involving women and other under-represented groups.

- Women are increasingly responsible for house purchasing decisions. A survey by Barclays Bank showed that 17 per cent of mortgages were taken out by women – double the percentage for 1983.

- In the United States, one in four people is employed by a woman – an indication of the growth of female entrepreneurs and business people in all sectors of trade and industry.
- Housing associations are a significant client group – in the North West the sector spend is some £1 billion over 5 years. Housing associations are concerned to improve their equal opportunities performance and widen the recruitment pool – despite a degree of cynicism about the effectiveness of monitoring equal opportunities compliance with equal opportunities requirements.
- Women represent 25 per cent of housing association management, compared to 3 per cent of professional women in construction and a mere 1 per cent at trade and craft level. This lack of representation and the poor record of equal opportunities in construction encourages scepticism amongst clients and undermines relationships.
- Equal opportunities is a significant component of construction projects offered by housing associations and local authorities. Action by contractors to promote equality has the potential to improve performance in construction projects stipulating local labour, apprenticeship schemes, tenant satisfaction and involvement and the regeneration of the local economy.
- Action to promote equality can increase applications from quality applicants, especially women and other under-represented groups, and help find local labour and apprentices as well as overcome skills shortages, especially in the manual trades. It can also help the company attract and develop women in more traditional occupations.
- A positive response to issues of equality will help to ensure the position of construction companies that are at the forefront of change – change necessary designed to enhance the long term competitiveness of the construction industry and promote equality for men and women.

'Women have to make such an effort to succeed in construction that well qualified women are an excellent resource under the right conditions that is, flexible conditions, stability of employment, crèche etc. They like the work and can do it well because they are interested in doing a good job. They are also interested in green issues. A clever employer would specialise in this and employ tradeswomen.'

Attracting young people into the industry

Schools and colleges

The building industry needs to attract more good people. The current skills shortage in many areas has emphasised the difficulties of recruiting people with the basic skills, knowledge and motivation to build a competent workforce. In particular, there is concern that very few women are attracted to building as a career, at a time when the number of women entering other non-traditional careers is growing.

There is therefore a need to raise awareness of the opportunities offered by the construction industry, to improve conditions and to present an image and a reputation that will attract and keep high quality recruits – both men and women. The number and quality of applications for training and careers in the construction industry needs to be increased, especially from women. The participants who took part in the project Building Equality in Construction 'Building Equality in Construction' were very keen for information about construction to be provided for girls at a young age. Contractors also felt they had a key role in making girls aware of opportunities in construction and wanted detailed information about how to go about this.

Visits to schools and visits by schools to sites and production facilities are an excellent way of raising interest and improving understanding of the industry, not only amongst children and young people but also amongst the people who influence them, their parents and teachers. Showing the real process of housebuilding can be fascinating and eye-opening experience for children and young people. After all, early learning for most young children is playing with bricks and sand and the traditional role of women is as home-makers, so why not home-builders?

When a contractor in Lancashire was asked by a client, a local school, to run a careers event for the pupils, he set the children the task of quantifying a wall. The girls enjoyed the exercise even more than the boys and appeared very capable.

School visits

Visits to co-educational and girls primary schools in the local area are an effective method of encouraging interest in construction and improving the construction industry's image. It is also important to remember that whilst there is a need to warn of the dangers inherent on a construction site, this should be balanced by creating a sense of excitement and interest in the process itself. It is important to define the objectives of visits and also to ensure that boys and girls have equal opportunities to participate.

- Highlight interesting and positive aspects of the construction industry.
- Site visits by health and safety officers focus on the dangers of the industry.
- Encourage boy *and* girls to consider opportunities within the industry by providing information about the wide range of jobs. Avoid statements which explicitly refer to jobs that girls can do.
- Make the most of women role models – Ideally, women from the contracting company who are employed as engineers, quantity surveyors

and trades people (Word of caution: Make sure that women employees are not taken away from essential activities, overworked, take on too much or attract criticism from male colleagues).

- Encourage women working in the more traditional jobs (e.g. secretaries) to get involved too. Women working in building companies on the administrative or support side can also take part in school visits and careers events. Existing female employees are well placed to provide information about construction and their feedback can help the company identify and overcome any problems in encouraging girls to consider working within construction.
- Offer appropriate training to employees interested in attending careers events and visiting schools to help them to talk confidently and appropriately about a range of opportunities within construction. This training could be provided in-house or through the CITB. Training and school visits will help female employees develop their own careers and may also encourage some even to consider working in non-traditional fields.
- Arrange for a female role model through London Women and Manual Trades (LWAMT) and Women as Role Models (WARM) to take part in careers events, host visits and attend speech days.

A personal assistant/PR manager working for a medium-sized construction company gained useful experience by visiting schools. She subsequently helped her boss find applicants for management/professional positions within the company and attended activities organised for women working in construction and housing.

Careers events

These are often organised by schools and colleges and are an opportunity for contractors to encourage new recruits. Such events are also opportunities to meet potential clients – schools and universities frequently commission building work.

- Encourage interest from girls by ensuring information is presented in a form which appeals to boys and girls. Design material using photographs of women on site or articles from the media about the women in construction.
- Ensure that men and women on exhibition or career stands react positively when girls ask about opportunities in construction and follow up their interest in an encouragingly way.
- Provide participative activities of interest and relevance which to children can take part in, with appropriate guidance.

- Present a positive image of the industry by highlighting impressive construction projects such as bridges, the Channel Tunnel, the Millennium Dome, company buildings or local amenities such as housing and school buildings.
- Talk about the challenges of building safely, for example reinforcement against earthquakes, increasing personal security and building for the high risk industries.
- Encourage children to discuss the design of a building in terms of effective use of space, image and maintenance. For example, ask the children to work in small groups to redesign the school or playground, with particular emphasis upon the needs of children, teachers or parents. Research shows that school students are attracted by the concept of constructing a 'socially useful' building.

Tarmac Bachy hosted an open day for four local schools in an embryonic Jubilee Line Extension at Canary Wharf comprising stalls to represent all the allied construction disciplines. The company also provided instructors to enable young boys and girls to sample jobs such as laying bricks and pouring concrete (Alexander, 1997; Building, 31.10.97).

Curriculum centres

The CITB Curriculum Centres enable schools, colleges, local enterprise councils and education business partnerships to work together to develop construction as a context for learning. The CITB is involved in organising site visits, in-service training for teachers, school–college partnerships, curriculum developments, visits to primary and secondary schools, the provision of mentors, the delivery of NVQs for 14–19 year olds and university liaison.

- Get involved with CITB curriculum centres to demonstrate commitment to the long term success of the industry, gain contacts with members of WARM and meet potential clients from the housing associations and schools.
- Encourage more male and female employees to get involved in the centres, don't just leave it to senior managers.

The CITB welcomes involvement by contractors and, in return, offers participants courses in presentation skills – an excellent way to promote the industry.

- University sponsorship, placement and liaison.
- Contractors interested in forging links should contact their local universities and colleges.

- Provide sponsorships for women at university and ensure women work with the company during their course and probably afterwards.
- Consider giving guest lectures, getting involved in research projects and advertising jobs at the firm to students and university staff.

Following involvement in the Building Equality in Construction project, one large contractor contacted UMIST to discuss a marketing placement/ job for a male or female undergraduate in particular and in forging links with the university in general.

On site

Site visits

Site visits not only stimulate children's interest in construction, but also provide opportunities for parents and teachers to learn about construction and to thus offer more informed guidance. Site visits can be made to new buildings, rehabilitation or specialist construction projects. Contractors can also arrange visits to manufacturers and suppliers.

- Provide lively information about a range of jobs – both professional and manual – give a short talk and encouraging interaction with employees on site.
- If possible, provide opportunities for children to see female employees working on site. This can be achieved by arranging visits to sites where women bricklayers, quantity surveyors, safety officers and site managers are based.
- Consider arranging for the visit to be hosted by a female office based employee who can provide information about women and construction.
- Organise a short, interactive visit to a local factory which is of relevance to the pupils and teachers.

One contractor successfully organised a visit to a brick factory where girls and boys were allowed to manufacture bricks inscribed with the school badge. The bricks were subsequently sold to raise money for the school.

Take your son or daughter to work

Research shows that children whose parents work in the construction industry are more likely to join the industry themselves and that 25 per cent of women in construction have parents in the industry. Contractors could provide opportunities for the daughters of their employees to find out about the industry.

- Try to ensure that girls are not restricted to office-based activities traditionally undertaken by women and/or boys to site-based activities dominated by men.

- Provide information about a range of occupations/activities in construction for girls and boys.
- Prepare lively, easy-to-read information for parents, staff and children, as well as ensure suitable clothing and safety procedures.
- Ensure that children are accompanied by an appropriate adult, where possible designate a female member of staff for site visits, otherwise girls will gain the impression that only men work on site.

Doing painting and decorating has worked wonderfully for me because now I can get work. I want to specialise more, which I can do through further training. I had no idea that there was so much work out there. Me and my partner have had to turn work down. It's amazing.

(Female trainee, WEB, 1996)

Shadowing and work placements

Placements are an opportunity for the contractor to learn more about potential new recruits as well as to encourage men and women to enter the industry.

- Provide young women with the opportunity to shadow staff for two or three days. This is an effective way of highlighting the range of occupations on site, the challenging nature of the industry and its suitability for women.
- Offer work experience for young women before they choose their A levels, preferably in the summer after their GCSEs.
- Consider paid holiday work for A level students.
- Provide short placements for teachers, careers advisors and youth leaders so that they can find out about the industry and advise children appropriately.

Tarmac invited eight sixth form students from a local girls school to spend a day shadowing female staff on four of its sites in London. The girls commented:

Our visit was a real eye opener. We were amazed at the diversity of professional jobs available, and how these were not gender-biased. Everyone on the sites seemed to work together as a team towards a set goal, and each team had a good skills mix.

Widening the recruitment net

Targeting new sources

There is a need for the construction industry to target women in recruitment procedures and ensure fair recruitment policies. Target new sources such as the daughters and wives of existing employees, female employees currently working in traditional occupations, women from housing associations and tenants.

- Think about attracting a wide range of people when designing advertisements, through the use of pictures, profiles of women and other minority groups.
- Include statements such as 'working towards equal opportunities' and 'we welcome applications from under-represented groups' in advertisements, brochures and so on.
- Emphasize job skills traditionally associated with women such as communication skills, ability to work in partnership, language skills and teamwork. Put less emphasis, where possible, on qualities traditionally associated with men such as mechanical knowledge, technical skills, mathematical abilities and salesmanship.
- Highlight the need for experience in management, project management, consultancy gained working for other industries, housing associations or local authorities rather than focusing solely on construction.
- Highlight possibilities for job share, flexible hours part-time work and training.
- If mentoring and support networks are in place, talk about them.
- Consider flood lighting site for evening visits, as happens in France.
- Make sure that commitment to equal opportunities policies and procedures is fully understood and applied by employees and subcontractors.
- Provide training to help management address negative attitudes towards women and take practical steps to increase the representation of women.

The potential to widen the recruitment net was illustrated by a large contractor whose company recently employed a single mother aged 29 years and a man aged 43 years. This apparently surprised some people in his company. A medium sized contractor has looked into the possibility of recruiting staff from the housing associations.

Selection criteria

When recruiting, pay particular attention to criteria regarding age, construction experience, professional qualifications and technical skill. Use a structured procedure, with consistent questioning and reporting methods.

- Ensure that selection criteria do not discriminate unfairly.
- Accept equivalent experience and qualifications, where possible.
- Use the formal recruitment methods described by the CIB's Working Group 8.
- Use recruitment panels comprising both men and women.

Some women who currently work in construction have become interested in the industry through DIY (Do it Yourself). One who worked as a carpenter with Liverpool City Council and, at the age of 22 years, won the National Training Award for her commitment to personal development, said that her interest in carpentry developed at a young age when she used to help her father with DIY jobs.

Encouraging minority groups into the manual trades

Innovative ideas are needed to reach under-represented groups and encourage them to consider employment and training in building, particularly in the manual trades. One effective method is to develop 'dynamic duos', women and men working in pairs, who appreciate the need to attract under-represented groups, can visit target areas and who are able to answer questions about the training and the company. These people should be in a position to offer practical and financial help for anyone interested in attending any meetings and/or arrange a home visit. They should also be present at subsequent meetings so that potential new recruits are reassured that a friendly face will be present.

- Design documentation to appeal to women as well as men in order to overcome reluctance to apply for training in non-traditional occupations.
- Find appropriate channels to communicate to a wide range of groups.
- Extend the distribution of leaflets to housing estates, local amenities such as launderettes, the city centre, women's sports classes, careers events, girls sixth form colleges, housing association properties, DIY courses for women, gay and lesbian organisations, music events, night club queues, especially those popular with black and ethnic minorities.
- Send leaflets to people prepared to recruit under-represented groups such as the contractor's own employees, housing association representatives, people from the directory of minority led firms, Moss Side and Hulme Women's Action (MOSHWAF) participants in 'New Deal' and others involved in equal opportunities initiatives.
- Consider an incentive scheme to encourage people to either recommend potential recruits and/or hand out leaflets. The RAC and the Co-operative Bank offer vouchers, payment and free gifts.
- Set up a task force within the company specifically charged with the recruitment of under-represented people.
- Encourage female apprentices to tell their friends about opportunities in construction.
- Give a realistic, but positive, account of training and future employment prospects during the recruitment phase.
- Publicise training schemes for female apprentices in the press.

While advertisements in papers often do not produce many applications, WEB found that an article in the Sainsbury in-store magazine was more successful. There are also magazines such as 'Your Home', part of Prima Magazine, which may be appropriate; and other women's and general interest magazines and newspapers which are interested in producing editorial on non-traditional topics.

Directory of minority-led firms

As well as recruiting individuals to the industry, building contractors can help improve equal opportunities by giving minority-led firms, the opportunity to obtain sub-contract work. For example, there is evidence to suggest that many more women are setting up their own businesses, partly as a result of being unable to progress in traditional employment in the construction industry. To help in identifying minority-led firms the Directory of minority-led firms has been compiled by Building Positive Action and is currently held by North British Housing Association.

* Consider sub-contracting work to firms from the directory, especially when providing housing or services for women and under-represented groups.
* Consider providing financial support for the directory of minority-led firms.

Supporting the directory of minority-led firms demonstrates commitment to equality, gains publicity for the company and potentially comes into contact with housing associations. The directory is supported by nearly 40 housing associations who meet regularly and are committed to equal opportunities.

Taster courses

Short or taster courses designed to provide basic skills have proved an effective method of attracting women on to courses and give them some experience of construction. A short course also gives trainees a good start in a trade of their choice and enables established tradeswomen to become multi-skilled.

* Offer one-day a week courses in electrical installation, painting and decorating, plastering or carpentry, lasting for approximately ten weeks.
* Give help with finding training and/or employment in construction once the course has finished. This would be in the form of passing on information about job opportunities, help with job searchers or advice about curriculum vitae.

- Put on displays and exhibitions of the work done and the achievements of those taking part, to give others an opportunity to view women's work.
- Offer evening courses one evening per week, again in a 10 week block or taster courses lasting for six weeks, with trainees spending two weeks on each trade.
- Provide women-only courses, as these are an effective method of attracting women and countering the male-dominated culture.
- Mixed course are satisfactory as long as there are a reasonable number of women on the course and sexist and racist behaviour is not tolerated.
- Provide a female instructor from your own organisation or from the directory of minority-led firms.
- Examine the costs and opportunities for funding taster courses with housing associations.

The CITB Construction Curriculum Centre in Manchester, with CITB, MANCAT and Manchester TEC, has run two girls-only summer schools and taster programmes for Key Stage 4 pupils (14–16 years). The first summer school (2 days a week for 6 weeks) attracted 11 students. Eight of them completed the course and three of them were destined to join the construction industry the following summer. Since then, 19 Key Stage 4 pupils from North Manchester High School for Girls have embarked on a one afternoon/week, two-year course in NVQ Level 1 Building Craft Occupations.

Innovative approaches to training

Providing attractive packages

The construction industry needs to increase the quantity and quality of training, especially for women. This will help the industry provide qualified professional employees and tradespeople, improve opportunities for local people, reduce vandalism on site, reduce property mistreatment, increase tenant involvement and improve building design. It will also enable contractors and housing associations to work together in partnership and gain a competitive edge for future contracts.

Whilst opportunities for women should be available in the full range of manual trades, contractors could start by focusing upon apprenticeships for painters, carpenters and electricians, as they are a popular option for women.

- Lift age restrictions and open up courses to a whole range of women including women returning to work after maternity leave, single mothers, black and ethnic minority women.

- Offer supplementary courses in maths, English and communications/ negotiation skills for work in the construction industry; English as a second language; and computing and multimedia studies.
- Investigate the possibility of developing and using material about eco-technology, electrical installation and plumbing in training (and employment) schemes, as green issues have been found to be of particular interest to women and clients.
- Highlight the provision of training opportunities and good links with local training providers in promotional material or discussions with female applicants, as these issues are important to tradeswomen.
- Draw attention to possibilities for tradeswomen to attend college, take courses offered by the CITB and/or supervise apprentices.
- Ensure that women are not isolated by recruiting other women, offering women a female mentor from WARM or encouraging networking with other women in the organisation.
- Consider supporting the setting up of a training centre in Manchester, similar to the London-based Women's Education in Building (WEB).

WEB offers courses exclusively for women, taught by qualified tradeswomen and has successfully trained over a thousand women since its formation in 1984. Its trainees report enjoying the course, finding employment and earning a decent wage; which though only £150 for a trainee, rises to about £250 per week after two years.

- Provide opportunities for apprentices to gain qualifications which will make them employable and competent. Ideally these should be City and Guilds or National Vocational Qualifications in the manual trades at level III – not just I and II.
- Monitor the training programme by the contractor, client and/or CITB by regular, personal contact in addition to forms and questionnaires.
- Monitor the cost and quality of training by setting aside houses specifically for training purposes and nominating a training supervisor.
- Display the work produced by trainees to senior managers in the company, the client, partners and visitors such as potential new recruits, local residents and school children.
- Consider flexible hours and taster/short courses.

One female change agent says, 'Learning a trade is strange and some women are unwilling to commit themselves for a year. This year we are running short courses and hope that women will stay on. We are looking to recruit women returners.'

DIY and maintenance for tenants

DIY courses run with housing associations enable contractors to work in partnership with housing associations, develop a pool of women potentially interested in working in construction and gain contact with people listed in the directory of minority led firms. Funding could initially be provided through 'Housing Plus' or 'New Deal' or Innovation Grants, but the savings in maintenance costs arising from fewer unnecessary call outs and less property mistreatment would make the courses self-financing in the longer term.

- Consider working with housing associations to provide DIY and maintenance courses for women.
- Investigate the costs and benefits of DIY. Housing associations estimate the cost to be between £2,000–£3,000.
- Highlight the potential for courses to increase the quality of the service provided for tenants by increasing levels of tenant involvement, training and potential employment.

DIY and maintenance courses were run successfully in Rochdale in a predominantly Asian area and in France. Two other housing associations are planning to provide DIY training for tenants, one intends to use a woman from the directory of minority-led firms and the other a mainstream male contractor.

Keeping employees

Building contractors should aim to not only build a good workforce by recruiting women, but also encourage them to stay by helping them to develop their careers – good people are a valuable resource.

Preparing for apprentices

An apprentice may find it daunting to arrive on site for the first time and have little idea about what to expect. When women arrive at a site, supervisors, foremen, and co-workers may react with varying degrees of enthusiasm, resentment, surprise, indifference or embarrassment.

- Involve and inform the workforce by outlining the organisation's commitment to hiring women, the practical steps which have been taken to ensure equality and the company's belief in the ability of women and other minority groups to do the job.
- Provide opportunities for new apprentices to meet other apprentices and to learn about the company, the type of work and the working conditions before starting on site.
- Produce a trainee's handbook with introductory information, an outline of training and health and safety structures.

- Ask the workforce to call women apprentices by their name, rather than 'sweetie' or 'luvvy'.
- Ask the supervisor to ensure follow-up of any complaints of harassment.
- Start training on site during the spring or summer, rather than the winter.
- Provide apprentices with a telephone number for emergency calls from, for example, child minders and schools.
- Ensure that any posters, calendars and reading material that could cause offence are removed and/or stored out of sight.

A contractor in the Manchester area worked with the CITB to provide training for young local bricklayers and joiners which encouraged team spirit, reduced vandalism (e.g. fewer windows broken) and increased the respect shown by local people. He would like to see the scheme extended to people up to the age of 35. Other contractors worked closely with housing associations and local authorities to provide apprenticeship schemes.

Keeping apprentices

Building contractors often experience a high drop-out rate amongst apprentices, both male and female, for various reasons. Although some apprentices, particularly women, recognise that the manual trades provide good employment opportunities and wages in the longer term, building contractors have to attempt to satisfy the short-term expectations, needs and aspirations of apprentices. To reduce wasted time and effort in recruitment, it is important to find out why apprentices leave and try to resolve problems where appropriate.

- Carry out exit monitoring to find out why apprentices leave and whether they have been lost to the industry – recognising that it might be necessary to keep some reasons confidential.
- Provide apprentices with a reasonable wage/allowance and if possible contribute to travel costs, equipment, overalls and luncheon vouchers.
- Where appropriate, help with the cost of childcare.
- Offer apprentices opportunities to supplement their income by doing paid work for the company.
- Identify benefits of training already provided and costs of leaving.
- Remedy any problems revealed by the exit monitoring.

An apprentice leaving does not necessarily represent a failure, especially where a contractor introduced professional recruitment methods and/or the individual has benefited from the training. One contractor based near Stockport suspects that apprentices have been poached by construction companies prepared to pay higher wages. He intends to introduce exit monitoring to find out if this is the case or if there are problems with training and/or recruitment.

Improving working conditions

The world of work is changing dramatically. More people are rejecting poor conditions and long hours, and technology is offering different ways of carrying out traditional tasks. Learn from the manufacturing process, where for example, the off-site production of components has meant that work is conducted indoors and flexible hours can be provided more easily. Also compare what other industries and professions are providing in terms of working hours and support mechanisms.

- Provide flexible and/or shorter hours, as this is a real benefit to male and female employees who care for children and/or elderly relatives.
- Consider flexible starting and finishing times, part-time employment, shift system, temporary work during the school holidays, job share and home working.
- Adapt shifts and piece-work to suit family commitments.
- Maintain a dialogue with housing associations about lengthening project deadlines in order to provide flexible or shorter hours for employees.
- Provide and maintain good toilet and washing facilities – the two-plus-one toilet block has alleviated some of the practical problems traditionally associated with employing women.
- Investigate the possibility of providing childcare facilities.
- Examine the possibility of helping with the costs of emergency child-care cover for work required outside of normal office hours and/or at short notice.
- Support women and minority groups by encouraging them to participate in networks or mentoring schemes. These may be specifically for women such as WARM, for those who wish to meet and support other women, or mixed organisations such the unions and professional bodies.

One contractor already employs a quantity surveyor who works flexible hours and another provides flexible hours for office staff on an informal basis. He has family responsibilities himself and therefore appreciates the difficulties. 'I understand the difficulties because I have to pick up the children. The nurseries say that you have to pick up your children by 6pm, otherwise there is a £1 per minute fine. It creates a lot of stress' (Male contractor).

Starting the process

Running a pilot project

Achieving results and bringing about change demand that actions are measured against targets. Identify activities proposed in these guidelines that are feasible and appropriate to undertake.

- Develop a pilot project, likely to attract media attention, and a series of supporting actions.
- Meet with other managers to discuss possible pilot projects in more detail and identify further action.
- Extend meetings to potential clients and/or partners.
- Encourage open discussion, welcoming ideas and minimising criticism, to encourage innovation and change.
- Sustain momentum by arranging follow-up actions and meetings.
- Once a pilot project and supporting practical steps have been agreed, define objectives, specific activities, timetable for completion, outcomes and benefits.

A number of building contractors identified pilot projects during the Building Equality in Construction project, based upon developments in their own companies and/or suggestions from housing associations. Others already had ideas about practical steps.

Developing and using measurement tools

Develop appropriate measurement tools through discussions with participants. This involves defining parameters, identifying practical steps, developing specific measures, testing and incorporating the results into overall performance measures.

- Establish a baseline, against which progress can be measured.
- Identify ways of measuring outcomes, for example numbers of applications from women, number of apprenticeships; and number of careers events etc.
- Keep a qualitative record of innovations, successes and problems. The records are useful for illustrating accounts during discussions and presentations to clients, the media and work colleagues.
- Use the measurement tools to gain feedback about the successes and costs of action to promote equality.
- Identify and build upon successes.
- Establish learning points and take remedial action.
- Encourage people to admit to problems and to learn from difficulties.
- Increase motivation by recognising progress, giving recognition, giving praise where due and avoiding criticism.
- Encourage a sense of shared responsibility, especially when problems are encountered. Avoid allocating blame.
- Communicate successes to employees, clients and potential new recruits.
- Keep material to publicise interest and action such as photographs, articles in the building press and letters of recommendation.

Promoting the results

Getting involved in equal opportunities initiatives and activities enables contractors to gain insight into current thinking and behaviour within housing associations, contributing informed decisions about future business and marketing activities. Taking practical steps to promote equality enables building contractors to demonstrate commitment and initiative to housing associations, other clients and the public.

- Contact housing associations, Building Positive Action and those involved in Housing Plus, New Deal and Local Labour schemes to ask about working in partnership.
- Inform potential partners about action taken to promote equality.
- Show measurement tools and supporting evidence to demonstrate commitment.
- Improve the industry's image by highlighting opportunities for women.
- Try to obtain media coverage to promote the company and the industry, especially when involved in specific projects.

Guidelines for housing associations

Introduction

Housing associations are expected to ensure equality and high performance not only within their own organisations, but also within the construction industry which serves them. Government bodies put pressure on housing associations to fulfil social needs, regenerate local economies and improve service delivery. At the same time, housing associations are held accountable for reducing long terms costs.

How does a client work with the construction industry to achieve equality? A range of difficulties present themselves – the sometimes indirect relationship between the client and the contractor, a complex and restrictive legal system, competing demands and pressures.

The Guidelines are broken down into the following sections:

- Why bother with equal opportunities?
- Contractual mechanisms
- Promoting good practice
- Working in partnership.

Why bother with equal opportunities?

Local authorities have a duty to ensure that tax payers' money is used to gain quality products and value for money from contractors. Whilst this has traditionally been evaluated in terms of cost, technical competence and financial soundness, the Commission for Racial Equality argues that equal opportunities policies are increasingly being seen as part of good management practice.

*   The Housing Corporation expects greater readiness amongst social landlords to employ tenants to manage estates and undertake maintenance tasks such as repairs and painting (the United Kingdom currently lags behind France).
*   There is growing commitment to the idea that tenants – male and female – should be involved in designing and managing their homes.
*   Housing Plus aims to encourage registered social landlords to create and maintain sustainable social housing; obtain added value from housing management and investment; and build partnerships with stakeholders in communities.
*   Through social housing, community action plans can provide a coherent approach to dealing with the social exclusion, multiple deprivation and managing the mix of communities.

> I just cannot image a situation where the very people that we are providing housing for are not involved in the whole process – of course they should be.
>
> (Baroness Dean)

Contractual mechanisms

It is argued by some that promoting change through contractual mechanisms shows that housing associations are serious about the standards for equal opportunities and encourages compliance by contractors. It is one way in which housing associations can demonstrate that they meet the standards laid down by the Housing Corporation.

Contract clauses

Of course, housing associations and building contractors are obligated to conform to statutory obligations, but the importance of doing so can be highlighted in contract documentation. More detailed information about promoting equality through contracts is available from the Commission for Racial Equality.

*   Remind building contractors in the contract to abide by their statutory obligations under the Sex Discrimination and Race Relations Act for example, in order to demonstrate reasonable steps to ensure the widest response from all sections of the community to employment opportunities

within the undertaking and that all potential applicants receive equal encouragement to apply.

- Consider including a clause on discrimination in employment, for example: 'The housing association has adopted equal opportunities policies in its working practice and is opposed to any discrimination on the basis of gender, colour, race, ethnic origin or disability by contractors or subcontractors'.

- Consider including a clause on neighbourly relations. For example: 'The housing association wishes to promote neighbourly relations with the areas in which it is working. Contractors must ensure that their employees and sub-contractors do not cause abuse or harassment while engaged by the association. Contractors are required to take prompt and firm action against any employee causing abuse or harassment. The association requires that pornographic material likely to embarrass women or men must not be displayed within its building sites.'

- Remind building contractors of their obligations under the Disability Discrimination Act (1996), which requires employers operating in the UK with 20 or more employees to employ a quota of registered disabled people, currently 3 per cent. Employers who are 'below quota' are required to recruit only suitable registered disabled people when vacancies arise, or to obtain a permit allowing it to recruit other than disabled people.

> Associations should develop and adopt criteria for checking that contractors and consultants invited to tender for their work operate non-discriminatory employment practices.
>
> (The Housing Corporation)

Approved lists

Housing associations should evaluate building contractors' ability to meet the criteria for entry on to approved lists using questions such as, 'Do you comply with the Sex Discrimination Act and the Equal Pay Act?' Open questions will elicit more useful responses than simple yes or no questions.

- Emphasise the importance of equal opportunities by providing guidance to contractors about the approved questions. Ideally, a face to face meeting should be held with contractors but the guidance can be in written form.

- Suggest contractors refer to guides and articles such as: 'Racial Equality and Council Contractors' by the Commission for Racial Equality (1995) and 'Equal Opportunities in Development' by George Barlow (1989).

- Ask for documentary evidence to support claims of meeting the criteria, such as recruitment manuals, recruitment advertisements, promotional brochures, flyers about training opportunities, application forms, relevant extracts from disciplinary and grievance procedures and equal opportunities policies and statements.

- Seek confirmation that a senior person from the firm is responsible for equal opportunities policy and its operation.
- Check that the text and illustration of documentation shows that people are welcomed from different backgrounds, race and gender.
- Identify good practice and areas for improvement and feed the information back to contractor, either individually or collectively.
- Follow up the forms and the documentary evidence submitted for entry to approved lists by checking a number of contractors.

> Ethnic minority contractors should be encouraged to join the associations approved list for tenders.
>
> (The Housing Corporation)

Checking compliance

It is easy for building contractors to show apparent compliance with equal opportunities criteria by simply ticking a box saying 'yes' and producing appropriate documentation. Housing associations should aim to ensure that building contractors are checked regularly and effectively and are given help to meet the standards where appropriate. Check to see whether contractors have recently been involved in any industrial tribunals by contacting the Commission for Racial Equality and Equal Opportunities Commission.

- Talk to people responsible for equal opportunities, recruitment, training and front-line management to evaluate successful implementation of the policy.
- Find out about meetings and training sessions to discuss equality and specific steps to ensure the fair representation of women and ethnic minority employees.
- Examine figures broken down by gender for recruitment, training and employment.

A housing association took children from local schools to see a roof being made out of straw bales and found that many children's mothers out of interest came along too.

Annual review of approved lists

An annual review gives housing associations the chance to demonstrate their equal opportunities objectives and goals. It also provides an opportunity to ask contractors for a commitment to continue to meet the housing association criteria as a condition of entry or retention on the approved list.

A meeting for all the contractors on the association's approved list is an effective method of motivating and guiding action, rather than issuing a written questionnaire, and gives contractors an opportunity to ask questions. An association might consider a joint meeting organised by several housing associations in order to minimise costs and maximise impact. Whether by interview or by mail, a standardised questionnaire ensures consistency.

- Review building contractors on the approved lists using a questionnaire designed by the housing association or a standard questionnaire published by the Joint Consultative Committee for Building.
- Ask 'how' rather than 'if' the contractor complies with requirements such as the Sex Discrimination Act to encourage contractors to engage with the issues and provide more specific details about practical steps.
- Check answers from the questionnaires and give notice of intended action to any contractors below standard.
- Before any action is taken which may disqualify the contractor from tendering, provide the grounds for dissatisfaction and an opportunity to make representation to an elected member of the housing association.
- Provide feedback to contractors recognising positive steps already taken as well as areas for improvement.

Suggested review questions

- Has your equal opportunities policy been agreed with trade unions and employee representatives?
- How is the equal opportunities policy communicated?
- What type of training is provided for managers and supervisors regarding Race Relations Act and Sex Discrimination Act?
- How often do you review recruitment, promotion, transfer and training procedures? Have you been involved in any positive steps to encourage ethnic minorities/women to apply for jobs?
- Do you offer special training for ethnic minorities or women?

Pre-tender documentation

The provision of pre-tender documentation is an excellent opportunity for housing associations to highlight the importance of equal opportunities to the contract. It is also a chance for housing associations to question contractors about practical steps to ensure equality.

The contractors' responses to the questions about equality have to be evaluated prior to the invitation to tender and used as a basis on which to work with the contractor, rather than to award the contract. Nevertheless, housing associations should consider using the answers provided by contractors and/or a contractor's performance during the contract to give general feedback and advice about equal opportunities during subsequent meetings.

Promoting good practice

Housing associations should use their financial muscle and experience to encourage building contractors to take practical steps to promote equality. Just as importantly, housing associations should help contractors to benefit from action to promote equality, especially in the shorter-term.

- Communicate commitment to equality using a clear statement drawing attention to the housing association's objectives regarding equality, the local population and harassment.
- Ask building contractors to produce a similar set of statement or policies, or to comply with the association's policy.
- Demonstrate housing association good practice for example by pro-active advertising, training and employment and including pictures of women and other under-represented groups in annual reports.
- Hold regular meetings for contractors designed to promote good practice regarding equal opportunities. These could be run by an individual housing associations and/or a group of housing associations, thus minimising costs and maximising impact.
- Provide guidance about practical steps for contractors and measurement tools, using the report of CIB Working Group 8 and this document.
- Invite contractors to write to the housing association with details of their achievements regarding equal opportunities, providing supporting evidence and appropriate measurement and monitoring tools.
- Review material provided by contractors to identify good practice as well as areas for improvement.
- Provide appropriate positive feedback.

Recognition and reward

Contractors who perform well in terms of equal opportunities should be recognised and/or rewarded by housing associations. It is important for housing associations to acknowledge the effectiveness of the business case argument for equal opportunities amongst building contractors, and not to view it as less valid than social pressure. There are various ways in which this can be done, without using financial mechanisms.

- Recognise that contractors personally committed to equal opportunities cannot risk jeopardising company productivity by devoting time to activities at the expense of efficient management.

- Provide a platform for contractors to give short presentations about equal opportunities at meetings and conferences.
- Provide opportunities to discuss association objectives/plans, partnerships and current thinking and practice with senior managers.
- Report building contractor good practice in the housing association's annual report.
- Contribute to joint articles for publication.
- Consider entry on to approved lists.
- Provide opportunities to tender, especially for construction projects requiring recruitment, training, equal opportunities and local labour.

Remedies

Housing associations cannot and would not want to terminate a contract solely for breach of equality in employment conditions. However, the housing association can refuse to place further work with the offending contractor until it introduces satisfactory policies and procedures.

Working in partnership

Housing associations should aim to work in partnership with building contractors to develop specific projects to promote equality for men and women. In addition to promoting equality, the partnerships could help to provide training, improve recruitment procedures, increase co-operation, reduce disputes, lower costs and improve quality.

The process of setting up and working in partnership involves meetings, gaining commitment, assessment of contractors, feedback, rewards and an ongoing dialogue.

Setting up partnerships

The mechanism for promoting equality will help housing associations identify building contractors committed to equal opportunities and prepared to take practical steps.

The housing associations may be able to choose a contractor with whom to work in partnership or may have to invite them to tender for the contract. This will depend upon the nature of the work and the contract.

- Invite contractors to a meeting to discuss working in partnership on a specific project, with the aim of adopting specific practical steps and measurement tools.
- Create an informal and participative atmosphere in meetings to help participants feel relaxed and gain some insight into individuals' attitudes and interests.

DIY and maintenance courses were run successfully in Rochdale in a predominantly Asian area and in France. Two other housing associations are planning to provide DIY training for tenants, one intends to use a woman from the directory of minority-led firms and the other a mainstream male contractor.

Working together

Work in partnership with the contractor to develop strategies and take action.

- Specify practical steps to promote equality in the contract, which are perceived as achievable by both parties.
- Include targets for the recruitment of under-represented groups for monitoring purposes, rather than contractual requirements.
- Adopt and develop measurement tools and monitor progress.
- Record project results using the measuring tools, verbal accounts, photographs and videos. The successes and learning point should then be publicised in annual reports, research documentation and trade articles.
- Arrange regular meetings to discuss the work, outline specific actions to promote equality.

Housing associations estimate the cost of a taster course in the manual trades at £2,000–£3,000. Funding could be funded by the building contractors and/or housing associations through 'Housing Plus' or 'New Deal'.

Types of partnerships

There are various arrangements and objectives possible in a partnering initiative:

- Education – business – working with contractors to organize and run visits to schools, work placements, site visits, shadowing and sponsorship.
- Training and recruitment – working with mainstream and minority-led contractors to recruit, train and employ under-represented groups.
- Employment conditions – helping contractors provide better working conditions by, for example, extending contract times for contractors who provide flexible hours.
- Directory of minority-led firms – encouraging and helping contractors to subcontract work to minority-led firms.
- Tenant training, employment and involvement – working with mainstream and/or minority-led contractors to provide, for example, DIY/maintenance training.

A training scheme offering apprenticeships was successfully set up by a housing association in Bolton, using two houses provided by the local authority. A local contractor provided the training.

Sources of information

Dr Andrew Gale, School of Mechanical, Aerospace and Civil Engineering, The University of Manchester, Pariser Building, Sackville Street, PO Box 88, Manchester M60 1QD, Tel: 0161 306 4236.

Commission for Racial Equality, Maybrook House (5th Floor), 40 Blackfriars Street, Manchester M3 2EG, Tel: 0161 835 5500.

Construction Industry Training Board, 10 Waterside Court, St Helens Technology Campus, Pocket Nook Street, St Helens, Merseyside, WA9 1HA, Tel: 01744 616004.

The Contractor Code, The National Housing Federation, Manchester Regional Centre, City Point, 701 Chester Road, Manchester, M32 0RW, Tel: 0161 848 8132.

Department of Environment, Transport and the Regions, Construction Sponsorship Directorate, Floor 3/J1 Eland House, Bressenden Place, London, SW1E 5DUT, Tel: 0207 890 3000.

Directory of Minority Led Firms, Ms Chris Root, Places for People Group, North British Housing Association, 4 The Pavilions, Portway, Preston, PR2 2UB, Tel: 01772 897 200.

Equal Opportunities Commission, Arndale House, Arndale Centre, Manchester, M4 3EQ, Tel: 0845 601 5901, Fax: 0161 838 1733, Email: info@coc.org.uk

Moss Side and Hulme Women's Action Forum (MOSHWAF), 97 Princess Road, Moss Side, Manchester, M14 4TH, Tel: 0161 232 0545.

Women and Manual Trades, 52–54 Featherstone Street, London, EC1Y 8RT, Tel: 0207 251 9192.

Building equality in construction – Part II

Good practice guidelines for race and gender equality – for building contractors and housing associations

Somerville, P., Davey, C., Sodhi, D., Steele, A.S., Gale, A.W., Davidson, M.J. and Rhys Jones, S.

These guidelines were part funded by the Department of Environment Transport and the Regions (DETR) and delivered in partnership with UMIST, RhysJones Consultants and the University of Salford. The Guidelines do not necessarily represent the views of the DETR.

These guidelines should be read in conjunction with:

Building Equality in Construction Good Practice Guidelines for Building Contractors and Housing Associations.
C. Davey, M.J. Davidson, A.W. Gale, A. Hopley, S. Rhys Jones, edited by S. Rhys Jones OBE, Working Paper No: 9901, Manchester Business School, The University of Manchester. ISBN: 1–86115–050–4.

Research project and team

These Guidelines represent some of the outcome from a research project funded by the DETR. The project forms part of the work of the Centre for Diversity, Inclusion and Equality in Construction Centre at the University of Manchester (formerly UMIST). The research project was delivered in partnership with UMIST, Rhys Jones Consultants and the University of Salford. Project team, in alphabetical order: Professor M.J. Davidson, Dr S. Fielden, Dr A.W. Gale, S. Gill, H. Lince, S. Rhys Jones OBE, D. Sodhi, Professor P. Somerville, Professor A. Steele, Dr D. Vick.

Introduction

This document provides guidance on the development of an equal opportunities policy and issues relating to the implementation of equal opportunities.

It also highlights examples of good practice to promote race equality in the construction industry. It compliments and builds on an earlier report produced to promote sex equality in the construction industry (1998), which should be read in conjunction with these Guidelines. The first phase of the research which led to the report on women was funded by the DETR under the 1996 Partners in Technology programme. It was carried out by UMIST in partnership with Building Positive Action and Rhys Jones Consultants. The second phase of the project was also funded by the DETR. It was carried out by UMIST, Rhys Jones Consultants and the University of Salford with other collaborating organizations including Teesside University and the Housing Corporation. It shows how the development of a partnership methodology between clients Registered Social Landlords (RSLs) and the building contractors they employ, can help promote equality in the construction industry.

The development of these guidelines has been an incremental process incorporating feedback from those working in the construction industry and the social housing sector at various stages throughout the project. This has been achieved primarily through dissemination and discussion in the Action-Learning Sets in Salford, London and Middlesborough and a Conference, held in March 2001.

The report is divided into five sections, Section 1 sets out the business case for promoting equality generally. Section 2 emphasizes the importance of developing an equal opportunities policy suggesting who should be involved, what the policy should say and provides an example of an equal opportunities policy. Section 3 details the process of implementation of equal opportunities. Section 4 sets out the case for reviewing an equal opportunities policy and Section 5 provides examples of good practice in promoting race equality in the construction industry and social housing sector. It also highlights the achievements of the Action-Learning Sets in promoting race equality during the lifetime of the project.

The first guidelines produced entitled: *Building Equality in Construction Good Practice Guidelines for Building Contractors and Housing Associations* (1998), do not contain guidance on the implementation of equal opportunities policy development including a stepwise approach involving base-line monitoring techniques, toolkits and so forth. However, this volume compliments and builds on the former document.

Building equality in the construction industry: the dimension of race

The construction industry has the worst record for the employment of women and black and minority ethnic workers. In 1998, only 1.9 per cent of chartered civil engineers and 0.9 per cent of chartered builders were women and only 3.4 per cent of construction professionals were female.

Evidence of racial and particularly institutional discrimination within the public sector was highlighted by the Macpherson Report in 1999 on the enquiry into the death of the Black teenager, Stephen Lawrence. The report was critical of the police as well as social housing providers. However, at the same time, it is widely recognized that the private sector is also not beyond criticism in respect of its approach to equality of opportunity and racial discrimination. A report commissioned by the Construction Industry Training Board (CITB) on the under-representation of ethnic minorities in the industry found that less than 2 per cent of the construction industry, Britain's largest industry, was Black or Asian. This compares with a national average of 6.4 per cent (of working age). The report further details the extent of racial abuse within this sector: 39 per cent of ethnic minority construction employees had experienced racial remarks and although the majority of firms (97 per cent) had an equal opportunities policy, only 50 per cent actively monitored the policy.

Recent research on the employment and career opportunities of ethnic minorities within the RSLs sector (Somerville *et al.*, 2000) also highlighted the discriminatory practices within these organizations, despite evidence of equal opportunities policies. The report found that only 1.6 per cent of the senior management teams were from the Black and Minority Ethnic community and that generally, they tended to be located in the lower level jobs within the non-technical disciplines. It has been argued that 'this suggests a lack of commitment to the principle of equality of opportunity and has been linked to the failure of RSLs to set in motion the cultural change required to make equal opportunities a core value (Sodhi and Steele, 2000). A starting point for organizations addressing the issue of racial discrimination is to develop comprehensive and effective equal opportunities policies. This guide is intended to assist with the practical development and implementation of such policies, step-by-step, and with the benefit of 'good practice' case studies.

References

CITB and Royal Holloway University of London (1999) 'The under representation of Black and Asian people in Construction', CITB.

Somerville, P., Steele, A. and Sodhi, D. (2000) 'A question of diversity: black and minority ethnic staff in the RSL sector', Housing Corporation Research Source Findings, No. 43, Housing Corporation.

Sodhi, D. and Steele, A. (2000) 'Contracts of Exclusion: a study of black and minority ethnic outputs from registered social landlords contracting power, London Equal Opportunities Federation.

How to use the guidelines

The report is divided into the following sections:

Section 1   The business case
Section 2   Developing an equal opportunities policy
Section 3   Implementing equal opportunities

Section 4   Reviewing your equal opportunities policy
Section 5   Good practice in promoting race equality

It is recognized that organizations will be at different stages in the development and implementation of an equal opportunities policy. For this reason, the following questions will help you assess which sections within this guide are of most relevance to you.

Questions

Section 1
Why are equal opportunities important?
Why does my organization need to develop an equal opportunities policy?

Section 2
Who should be involved in developing an equal opportunities policy?
What should an equal opportunities statement say?
What does an equal opportunities policy look like?
How can I find out more about developing/reviewing the organization's equal opportunities policy?

Section 3
How does an organization implement equal opportunities?

Section 4
Why is it important for an organization to review its equal opportunities policy?
How does an organization review its equal opportunities policy?
Who should be involved in reviewing the equal opportunities policy?

Section 5
What constitutes good practice in race equality?
What can construction companies do to promote race equality?
What can RSLs do to promote race equality?
How can construction companies and RSLs collaborate to promote race equality?

Section 1   The business case

Introduction

Ethically and legally there are good grounds for ensuring that your organization both promotes and implements equal opportunities in relation to all its activities, but equally there are also sound business reasons for doing so. In this section we outline the business case for equal opportunities.

QUESTION

Why are equal opportunities important?

ANSWER

- A good track record of providing equality of opportunity will enable your organization to widen the recruitment net and increase applications from quality applicants, especially black and minority ethnic (BME) groups. Given the current and increasing skills gaps in the industry can your organization afford not to take action to promote equality?
- A diverse workforce provides a range of skills and experience. This combination can improve creativity, innovation, problem solving and decision making and lead to increased productivity, quality and customer service.
- Equal opportunities represents good employment practice and can improve morale and job satisfaction leading to savings from lower staff turnover and absenteeism.
- A diverse organization has more opportunities to develop its role in the community, and to explore its full business potential.
- Equal opportunities is a significant component of construction projects offered by housing associations and local authorities. Action by contractors to promote equality has the potential to improve performance in construction projects stipulating local labour, apprenticeship schemes, tenant satisfaction and involvement and the regeneration of the local economy.
- RSLs are a significant client group – in the North West alone the sector spends approximately £1 billion over 5 years. They are concerned to improve their equal opportunities performance and widen the recruitment pool – despite a degree of cynicism about the effectiveness of monitoring equal opportunities compliance with requirements.
- A positive response to the issues of equality will help to position construction companies in the forefront of change – change necessary to enhance the long term competitiveness of the construction industry and promote equality for all.

Section 2   Developing an equal opportunities policy

Introduction

In this section we outline why your organization needs to develop an equal opportunities policy and set out the process for doing so as well as providing details of what the policy should say and an example of an equal opportunities policy.

QUESTION

Why does my organization need to develop an equal opportunities policy?

ANSWER

- If employers take steps to prevent unlawful discrimination within their organization, that is, through the development and implementation of an equal opportunities policy they may avoid liability for acts of discrimination in any legal proceedings brought against them.
- There are strong links between equal opportunity and good employment practice, which means that an effective policy will lead to improvements in recruitment, retention and performance overall.

The first priority then for any organization is to ensure compliance with the law and then to formulate an equal opportunities policy, which provides improvements to the working environment for all those working for and with the organization.

QUESTION

Who should be involved in developing the policy?

ANSWER

- To ensure cooperation for the policy across the whole organization it is important that staff at all levels, including ethnic minority staff, women and any relevant trade unions are consulted on the development of the policy in terms of the content and implementation.

Example

Bramall Construction joined the Kickstart Programme, run by Kirklees and Calderdale TEC and established a multi-disciplinary, cross-sectional steering group drawn from a cross-section of the company's staff: different job functions, work locations, ages, gender, race, etc. to lead the company's equality drive. On completing the programme, Bramall Construction became the first UK construction company to achieve the 'Kickstart Equality Award'.

Ideally community organizations should also be consulted on the policy, that is, on ethnic monitoring categories to be used, these should be compatible with those used by other organizations. Such organizations could include, the Race Equality Council or local RSLs and the Equal Opportunities Commission.

Consultation is inevitably a time consuming process and for smaller companies this may prove difficult, however, at a minimum, staff should be consulted.

QUESTION

What should an equal opportunities policy say?

ANSWER

- An equal opportunities policy is a basic statement of the aims and objectives for the organization on equal opportunities. It underpins specific measures aimed at ensuring equality of opportunity for present and potential employees and should embrace recruitment, promotion and training at a minimum, but could also include clients and suppliers.

The exact contents can be decided by those involved in developing the policy but the report by the CITB/CIB (1996) recommends that all companies as a bare minimum should have a statement which sets out their policy. The policy should outline the overall strategy for implementing equal opportunities. They suggest that this should include:

- a general statement of the company's commitment to equal opportunities and to removing discriminatory barriers;
- definition of the types of discrimination which the company is trying to avoid;
- explanation of the legal requirements;
- how the equal opportunities policy will be communicated and implemented;
- names of the people responsible for implementing the policy;
- obligations of individual employees;
- commitment to regular review of the policy.

The CITB/CIB (1995) also provide an example of an equal opportunities policy statement, from the construction group Lovell Partnerships.

Example

Equal opportunities policy statement

It is the policy of Lovell Partnerships Limited that there shall be no discrimination or less favourable treatment of employees or job applicants in respect of race, colour, ethnic or national origins, religion, sex, disability, political beliefs or marital status. It is the company's policy to engage, promote and train staff on the basis of their capabilities, qualifications and experience, without discrimination and all employees will receive equal opportunity to progress with the company.

In order to put this policy of equal opportunities into practice in the day-to-day operations of the business, we will:

- monitor decisions on recruitment, selection, training and promotion to ensure that they are based solely on objective and job-related criteria;

- provide training for managers to ensure that they understand the nature of discrimination and are fully aware of their responsibilities in implementing our equal opportunities policy;
- provide information and advice on the implications of the relevant legislation and on assistance to help in the employment of people with disabilities;
- ensure that any grievance involving discrimination or harassment is considered seriously, thoroughly and fairly;
- encourage our subcontractors to adopt policies and working practices which reflect our own views on equal opportunities;
- bring this policy to the attention of all our employees and our subcontractors.

The directors and senior managers of the company fully support this policy statement. All employees are responsible for playing their part in achieving its objectives. The Company Personnel and Training Manager is responsible for monitoring the implementation of this policy. The policy will be reviewed on a regular basis.

The policy statement could be more detailed than this and include other types of discrimination not covered by the legislation, for example, discrimination on the grounds of sexuality, HIV or AIDS or otherwise.

QUESTION
How can I find out more about developing an equal opportunities policy?
ANSWER
See for example:

- CRE (1995) *Racial Equality Means Business: A Standard for Racial Equality for Employers*: This can be used to develop racial equality strategies and measure their impact. It covers six areas where organizations can apply and use a race equality programme: policy and planning; selection; developing and retaining staff; communication and corporate image; corporate citizenship; and auditing for racial equality. It also sets five levels in implementing a programme, which can also be used to assess progress.
- Employment Department (1992) *Equal Opportunities: Ten Point Plan for Employers*: Provides guidance on setting up an equal opportunities policy and a basic toolkit, which gives practical advice on how you can offer equality of opportunity to people from ethnic minorities, women and people with disabilities.
- CITB and CIB (1996) *Constructive Guidelines in Equal Opportunities, a Simple Checklist for Construction Companies*, the Construction Industry Board Working Group 8: Provides guidelines on what construction companies can do to set up an equal opportunities policy

from scratch and to help those with equal opportunities policies in place to review them.

Section 3   Implementing equal opportunities

Introduction

An equal opportunities policy will only be effective if it is properly implemented. An action plan is an effective tool to ensure implementation. In this section we provide guidance on developing an action plan and describe a range of activities, which can be taken to implement equal opportunities.

QUESTION
How can the organization implement equal opportunities?

ANSWER
A documented procedure is required for successful implementation of an equal opportunities policy and an action plan, which relates to the policy statement.

An action plan should be developed which:

1   identifies those responsible for implementing the equal opportunities policy;
2   identifies resources needed for implementation;
3   sets specific targets for action to ensure effective implementation;
4   provides a timetable for action;
5   outlines how objectives will be measured and assessed;
6   provides a deadline for assessment.

1   Identifies those responsible for implementing the equal opportunities policy
     Who should have responsibility for the policy?
     To show that the policy is supported at the top the Chief Executive or Managing Director should take the lead, but the overall responsibility for establishing and implementing the policy should be allocated to a named person on the management team. This could be the human resource manager or a specialist post created to implement the policy, but other senior management should also be involved in implementing the policy and should be named. Equal opportunities should be part of management performance assessment criteria.

2   Identifies resources needed for implementation
     What type of resources will be needed?
     Staff time and effort will be the main resources, but clearly there will be cost implications. These are likely to be higher in large organizations

where a new post may be required to ensure effective administration. Costs will depend on the level of activity and for smaller companies these may be limited. However, what ever the size of the company resources should be set aside for implementation since they will undoubtedly be less than the potential costs of not doing so.

3   Sets specific targets for action to ensure effective implementation
These could relate to a host of activities, which will be necessary to ensure implementation of the equal opportunities policy. The action plan should detail when and how these activities will be conducted and the person(s) responsible. The key priorities should be:

A   *Communication of the policy*   All existing and potential employees should be given a copy of the equal opportunities statement and agency staff, self-employed workers and clients and suppliers. This sends out a positive message about the company's commitment to equal opportunities and informs staff and others with whom the organization works of their rights and obligations under the terms of the policy.

Community organizations should also be advised about the policy, in particular schools, colleges and employment agencies and voluntary organisations working in the BME communities.

B   *Training on equal opportunities*   Training on equality plays a major part in the effective implementation of an equal opportunities policy and needs to be part of an overall training strategy for staff.

This will need to include staff at all levels of the organization, but additional training will be needed for those who recruit, select and train employees, that is managers, human resources or personnel managers, equal opportunity staff, recruitment and selection personnel, trainers.

The overriding objective should be to ensure that the behaviour of staff is in line with the organizations legal duties under the legislation (i.e. the Race Relations Act 1976, the Sex Discrimination Act 1975, the Disability Discrimination Act 1995) and its own equal opportunities policy. Staff need to understand the law, how discrimination occurs, including how their own actions at work can disadvantage particular groups of people and what the organizations policies to combat unlawful discrimination are. If your organization is not large enough to run its own training programme, you should consider combining with another employer or buying training from an outside company/consultant.

(i) *Setting targets and performance indicators*
Targets can be set in relation to employment, training and promotion. They should take account of the availability of the targeted group in the area of operation, their representation in the various levels of the workforce, and estimated business growth and anticipated staffing needs in relation to recruitment and promotion.

It may be appropriate to set targets in relation to subcontracting, clients and suppliers in some instances. Clearly for some small organizations setting any kind of targets may not be appropriate or realistic, for example, where staff turnover is low and there is no expected growth which will lead to job creation.

Targets should be set in consultation with community organizations and staff within the organization. They should be realistic and achievable and will provide a yardstick by which progress can be analysed.

The Employment Department (1992) defines a target as a system of measurement and emphasises that this should not be confused with a pre-determined quota. A pre-determined quota is a fixed number or percentage imposed for a particular area and is unlawful under the Race Relations Act. A target is not an absolute minimum or maximum, it is a yardstick for measuring success. Targets should not involve unlawful positive discrimination. All appointments will need to be made on merit.

C   Develop or review policies, procedures which promote good practice
Other policies and procedures will be important in promoting and implementing equal opportunities and will need to be developed or reviewed. They can provide teeth to equal opportunities policies, particularly where there are sanctions for non-compliance, that is, disciplinary action. Examples of policies/procedures which organisations should consider reviewing include:

Racial harassment
Grievance
Recruitment and selection
Recruitment channels
Training
Promotion
Staff appraisals
Dismissal
Redundancy and redeployment
Religious and cultural needs
Health and safety

(See Section 5 for examples on good practice)

D   Promotion of the policy   Key themes of the policy statement can be used in marketing material and when advertising. It is also important to include positive images of people from different minority ethnic groups and women. Publicity and training should be a continuous process to ensure promotion of the policy.

E   Implementing positive action   The Race Relations Act 1976 allows for positive action to be taken in some circumstances by organizations.

Under Section 37 where members of particular racial groups have been under-represented over the previous 12 months in particular work, employers and specific training bodies are allowed under the Act to encourage them to take advantage of opportunities for doing that work and to provide training to enable them to attain the skills needed for it. In the case of employers, such training can be provided for persons currently in their employment (as defined by the Act) and in certain circumstances for others too, for example, if they have been designated as training bodies.

Larger organizations could take advantage of these provisions. There are also a number of positive action measures, which they and smaller organizations could consider:

- advertising job vacancies in the ethnic minority press, as well as other newspapers;
- use of employment agencies and careers offices in areas where these groups are concentrated;
- encouragement to employees from those groups to apply for promotion or transfer opportunities;
- training for promotion or skill training for employees of these groups who lack particular expertise but show potential.

F Triggering change more widely in the industry   Organizations could encourage those with whom they work to develop policies, or make them a requirement. There will be a resource cost in terms of monitoring compliance, which will make it unviable for smaller firms, but for larger firms this could be a consideration.

Larger organizations should also consider offering their experience in developing a racial equality programme as a benchmarking resource to other organizations.

Furthermore, they could also consider involvement in or setting up initiatives to promote equality in the industry generally in partnership with clients and others in the industry. As part of the programme for developing these guidelines three Action-Learning Sets were set up in London, Salford and Middlesborough. These focus on the promotion and dissemination of the Good Practice Guidelines through the involvement and collaboration of organizations such as trade bodies, industry groups, construction companies, RSLs and community organisations.

(See Section 5 for examples on good practice initiatives)

G Setting up systems to monitor equal opportunities   Monitoring the success of the equal opportunities policy is an essential part of implementation. The action plan should identify who will be responsible for monitoring, what data will be analysed and the information needed to enable this and how the information will be collected. For smaller organizations, the process may be less formalised than detailed here.

Employees and where appropriate their representatives will need to be consulted on the needs and requirements of the organization. Staff will need to be informed about the monitoring system.

(i)   *What type of information might the organisation need?*
In terms of race equality, the first step should be to collect base-line information on the ethnicity of the workforce, that is, the number employed and level at which they are employed.

Information should then be collected on the ethnic composition of the labour market, customers, suppliers and subcontractors.

Information will also need to be collected on recruitment and selection (ideally on job advertisements, job applications, shortlisting and selection), transfers, promotions, grievances, training opportunities and career development and staff turnover including dismissals and redundancies and the reasons for decisions on these processes.

(ii)   *How can I get the information?*
Personnel records should provide information on the ethnicity of staff and the position held etc., otherwise it may be necessary to conduct a survey of staff.

Data on the local labour market may be available from the local authority or Race Equality Council. Data on ethnic minority companies can be obtained from various directories which have been set up. For example, the London Equal Opportunities Federation in London, Building Positive Action in the North West and in Bristol, Elm Housing Association and partners have set up a directory.

Job applicants should be asked to complete an equal opportunities monitoring form, which includes information on ethnic origin and job grade, other data may also be collected at the same time, that is, on gender, sexuality, age and disability. This information should be treated as strictly confidential and separated from the application on receipt so that those shortlisting do not see this.

Those that are involved in recruitment and selection and other personnel procedures should complete assessment forms which explain their decisions that is, on shortlisting, selection, promotions, etc.

(iii)   *Will this information be useful for the organisation?*
Information on the local ethnic minority population and staff within the organization could be compared to assess the level of under-representation of ethnic minorities and targets could then be set to enable measurement of the success/failure of the equal opportunities policy and any initiatives undertaken in pursuit of equality.

The information obtained will help to identify successful techniques and where improvements are needed and to identify where there are barriers or problems for example, in career development.

(iv)   *What categories should be used for data collection on ethnicity?*
For the purposes of monitoring, the Commission for Racial Equality (CRE) recommends that employers should use the 1991 Census categories. These are white, Black Caribbean, Black African, Black Other (please describe), Indian, Pakistani, Bangladeshi, Chinese and any other ethnic group (please describe). The consultation process may mean that this is revised slightly, but it is important to ensure a uniform approach with others in the area.

The CRE's handbook on ethnic monitoring 'Accounting for Equality' (1991) outlines seven principles of ethnic monitoring:

Consultation – to gain public trust and confidence;
Confidentiality – it should not be possible to identify individuals;
Self-classification – individuals should decide for themselves which ethnic group they belong to;
Effectiveness – the system must be clearly defined, staff must be trained, performance measures must be set and response rates must be monitored;
Monitoring – ethnic records must be monitored regularly;
Action – problems revealed by monitoring reports should be followed up by action accountability and regular reports should be produced and made available publicly.

4   Provides a timetable for action
    The timetable for action could be short-term, that is, detailing actions to be taken over the next 6–12 months, or more long term, that is, over a period of two/three years. The programme will need to be continuous. It should detail when information will be analysed and reported and to whom. It should also detail when the policy or specific procedures will be reviewed.
5   Outlines how objectives will be measured and assessed
    Monitoring information is a pre-requisite for measuring and assessing whether the objectives of a policy have been met. The information collected should be carefully and regularly analysed and a number of key questions will need to be asked.

CRE Code of Practice on Employment (1996) recommends the following questions:
    Is there evidence that individual's from any particular racial group

•   do not apply for employment or promotion, or that fewer apply than might be expected?

- are not recruited or promoted at all, or are appointed in a significantly lower proportion than their rate of application?
- are under-represented in training or jobs carrying higher pay, status or authority?
- are concentrated in certain shifts, sections or departments?

The reasons for this will need to be investigated if the answer is yes.

6   Provides a deadline for assessment
It is important to set out when monitoring and evaluation of the policy will be carried out. How often, at what stage, who is to carry out monitoring and evaluation and who will receive these reports.

Analytical reports should be produced at a minimum of every 6 months to check whether equal opportunities are being achieved. Reports should be available to staff and local community organisations. Where problems are identified a programme of action will need to be developed to rectify this.

Section 4   Reviewing your equal opportunities policy

Introduction

Reviewing your equal opportunities policy should be a continuous process, but periodically you should carry out a full-scale review that is, every two/three years. The amount of work involved in this process will clearly depend on the size of the organization and the effectiveness of the policy already in place. This section outlines why your organization needs to review its equal opportunities policy, who should be involved and how.

QUESTION
Why is it important for an organization to review its equal opportunities policy?

ANSWER

- To ensure that the policy is being fully implemented and that the aims and objectives set out in the policy are being achieved.
- We work in a changing environment and it is important that your equal opportunities policy reflects this.
- A system of regular reviews needs to be established to ensure that unlawful practices are not introduced inadvertently.

Organizations should consider reviewing their policy to ensure that all the actions detailed in the section on implementing equal opportunities in this report are part of their policy where this is appropriate.

Monitoring will help you to identify the areas where your equal opportunities policy is successful and those in which there is still cause for concern.

QUESTION
Who should be involved in reviewing the policy?

ANSWER
To ensure cooperation for the policy across the whole organization it is important that staff at all levels, including ethnic minority staff and any relevant trade unions are consulted on the review.

Ideally community organizations should also be consulted, for example, on ethnic monitoring categories to be used, these should be compatible with those used by other organizations, that is, the Race Equality Council or local RSLs and the Equal Opportunities Commission.

Consultation is inevitably a time-consuming process and for smaller companies this may prove difficult, however, at a minimum staff should be consulted on the review.

QUESTION
How does an organization review its equal opportunities policy?

ANSWER
The steps for reviewing your organizations equal opportunities policy will be similar to those identified in Section 3 on implementing equal opportunities.

Section 5 Good practice in promoting race equality

Introduction

In this section we provide examples of good practice for construction companies and RSLs in promoting racial equality. We also illustrate how change within the construction industry can be achieved through collaboration between construction companies, RSLs, industry groups, Universities and TECs.

QUESTION
What constitutes good practice in equal opportunities?

ANSWER
Any activity, which promotes equality of opportunity for BME groups, such as those described in this section, can be viewed as good practice. The Change the Face of Construction Toolkits provides complimentary examples of good practice (see Appendix 1).

Good practice examples for construction companies

1   Attracting young people into the industry

The building industry needs to attract more good people. The current skills shortage in many areas has emphasised the difficulties of recruiting people with the basic skills, knowledge and motivation to build a competent workforce. In particular, there is concern that very few BME people are attracted to building as a career, at a time when the number of such people entering other non-traditional careers is growing.

There is therefore a need to raise awareness of the opportunities offered by the construction industry, to improve conditions and to present an image and a reputation that will attract and keep high-quality recruits – black and white.

Royal Holloway University of London (1998) found low levels of awareness about construction particularly amongst Pakistani and Indian potential entrants as the following comment illustrates.

> all you do is stick bricks on top of bricks. You only need cement and brick, right, and put the doors on and screw the things in.
>
> (Black school student, 13 years old)

Members of community groups commented on the need for information:

> Knowledge of the promotion prospects was virtually non-existent with some groups left with the opinion that the work is of low status and you do not need to have qualifications to get into the industry.
>
> (Local Asian Councillor, Manager of Indian Community Centre)

School visits

Visits to primary schools in the local area are an effective method of encouraging interest in construction and improving the construction industry's image and understanding of the industry, not only amongst children and young people but also amongst the people who influence them, their parents and teachers. It is important to remember that whilst there is a need to warn of the dangers inherent on a construction site, this should be balanced by creating a sense of excitement and interest in the process itself. Schools in areas with high concentrations of members of the BME communities should be targeted.

It is important to define the objectives of visits and to ensure that all children have an equal opportunity to participate.

- Highlight interesting and positive aspects of the construction industry – site visits by health and safety officers focus on the dangers of the industry.

- Encourage all children to consider opportunities within the industry by providing information about the wide range of jobs.
- Make the most of BME role models – ideally, BME people from the contracting company who are employed as engineers, quantity surveyors and trades people.
- BME staff working in building companies on the administrative or support side can also take part in school visits and careers events, as they can provide information about construction. Their feedback can help the company identify and overcome any problems in encouraging BME children to consider working within construction.
- Offer appropriate training to employees interested in attending careers events and visiting schools to help them to talk confidently and appropriately about a range of opportunities within construction. This training could be provided in-house or through the CITB. Training and school visits will help BME employees develop their own careers and may also encourage some to consider working in fields that they would not otherwise have contemplated.
- Arrange for a BME role model through London Equal Opportunities Federation (LEOF) and the Federation of Black Housing Organisations (FBHO) to take part in careers events, host visits and attend speech days.

Careers events

These are often organised by schools and colleges and are an opportunity for contractors to encourage new recruits. Such events are also opportunities to meet potential clients – schools and universities frequently commission building work.

- Encourage interest by ensuring information is presented in a form, which appeals to children of all ethnic groups. Design material using photographs of BME workers on site or articles from the media about BME people in construction.
- Make sure that those on exhibition or career stands react positively when BME students ask about opportunities in construction and follow up their interest in an encouraging way. Ideally those on career stands should be from the BME community.
- Provide activities of interest and relevance which children can take part in, with appropriate guidance.
- Present a positive image of the industry by highlighting impressive construction projects such as bridges, the Channel Tunnel, the Millennium Dome, company buildings or local amenities such as housing and school buildings.
- Talk about the challenges of building safely, for example, reinforcement against earthquakes, increasing personal security and building for the high-risk industries.

- Encourage children to discuss the design of a building in terms of effective use of space, image and maintenance. For example, ask the children to work in small groups to redesign the school or playground, with particular emphasis upon the needs of children, teachers or parents. Research shows that school students are attracted by the concept of constructing a 'socially useful' building.

Curriculum centres

The CITB Curriculum Centres enable schools, colleges, local enterprise councils and education business partnerships to work together to develop construction as a context for learning. The CITB is involved in organising site visits, in-service training for teachers, school–college partnerships, curriculum developments, visits to primary and secondary schools, the provision of mentors, the delivery of (National Vocational Qualifications) NVQs for 14–19-year olds and university liaison.

- Get involved with CITB curriculum centres to demonstrate commitment to the long-term success of the industry;
- Encourage more BME and white employees to get involved in the centres; don't just leave it to senior managers.

Example

The CITB in the North West organised a 2 day residential conference for sixth form girls in October 2000 which provided opportunities to visit sites, talk to people in the industry and obtain information on courses and sponsorship. Feedback from the girls that took part was very positive. Of the forty girls that attended, 100 per cent said that it had been a valuable experience and that if it was repeated they would recommend it to a friend. Over 50 per cent expressed an interest in joining the industry.

A similar event could be organised for BME sixth form students.

The CITB welcomes involvement in curriculum centres by contractors and, in return, offers participants courses in presentation skills – an excellent way to promote the industry.

University sponsorship, placement and liaison

- Contractors interested in forging links should contact their local universities and colleges.
- Provide sponsorships for BME students at university and ensure that they work with the company during their course and probably afterwards.
- Consider giving guest lectures, getting involved in research projects and advertising jobs at the firm to students and university staff.

Site visits

Site visits not only stimulate children's interest in construction, but also provide opportunities for parents and teachers to learn about construction and thus offer more informed guidance. Site visits can be made to new buildings, rehabilitation or specialist construction projects. Contractors can also arrange visits to manufacturers and suppliers.

- Provide lively information about a range of jobs – both professional and manual – give a short talk and encourage interaction with employees on site.
- If possible, provide opportunities for children to see BME employees working on site. This can be achieved by arranging visits to sites where BME bricklayers, quantity surveyors, safety officers and site managers are based.
- Consider arranging for the visit to be hosted by a BME office-based employee who can provide information about BME groups and construction.
- Organise a short, interactive visit to a local factory, which is of relevance to the pupils and teachers.

Shadowing and work placements

Placements are an opportunity for the contractor to learn more about potential new recruits as well as to encourage BME and white people to enter the industry.

The CITB run Employers Seminars to provide information on how Work Experience programmes are organised, to explain what Employers can do to maximise the experience of students on placements and to offer advice on key issues, for example, Health and Safety.

- Provide BME young people with the opportunity to shadow staff for two or three days. This is an effective way of highlighting the range of occupations on site, the challenging nature of the industry and its suitability for BME groups.
- Offer work experience for BME young people before they choose their A levels, preferably in the summer after their GCSEs.
- Consider paid holiday work for A level students.
- Provide short placements for teachers, careers advisors and youth leaders so that they can find out about the industry and advise children appropriately.

Research undertaken by the CITB (2000) suggests that there is limited availability of information on construction in schools.

I don't think the school has any literature about the construction industry, I have not come across any...it isn't a job I would be recommending to my students, we hope that most of our students will follow a planned career path and construction is a job not a career.

(Female Head of Year, Blackpool)

2   Widening the recruitment net

There are specific steps that employers can take to widen their recruitment base and encourage diversity in the workforce. Advertising and selection criteria are key to this.

Advertising

- Advertise all vacancies externally using sources which allow the widest possible range of applicants to see the advert.
- Consider using BME press and BME community organizations for advertising.
- Use the Web site set up for BME staff in the social housing movement – Career Opportunities for Ethnic Minorities (COFEM).
- Think about attracting a wide range of people when designing advertisements, through the use of pictures, profiles of BME and other minority groups.
- Include statements on the organizations commitment to equal opportunities, such as 'working towards equal opportunities' and 'we welcome applications from under-represented groups' in advertisements, brochures and so on.
- Highlight the need for experience in management, project management, consultancy-gained working for other industries, housing associations or local authorities rather than focusing solely on construction.
- Highlight possibilities for job share, flexible hours, part-time work and training.
- If mentoring and support networks are in place, talk about them.
- Consider flood lighting a site for evening visits, as happens in France.

Selection criteria

When recruiting, pay particular attention to criteria regarding age, construction experience, professional qualifications and technical skill.

- Use a structured procedure.
- Ensure that selection criteria do not discriminate unfairly, criteria should be clear, relevant and objective.
- Confine questions to the requirements of the job and ensure consistency.
- Accept equivalent experience and qualifications, where possible.
- Include BME staff on recruitment panels wherever possible.
- Ensure that more than one person carries out interviews, ideally there should be three people on the panel.

- Keep a record at each stage and the reasons for the decision.
- Use the formal recruitment methods described by the CIB's Working Group 8.

Encouraging minority groups into the manual trades

Innovative ideas are needed to reach under-represented groups and encourage them to consider employment and training in building, particularly in the manual trades. One effective method is to develop 'dynamic duos', black and white people working in pairs, who appreciate the need to attract under-represented groups, who can visit target areas and who are able to answer questions about the training and the company. These people should be in a position to offer practical and financial help for anyone interested in attending any meetings and/or arrange a home visit. They should also be present at subsequent meetings so that potential new recruits are reassured that a friendly face will be present.

- Design documentation to appeal to BME as well as white people in order to overcome reluctance to apply for training in occupations where they are under-represented.
- Find appropriate channels, languages and media to communicate to a wide range of groups.
- Extend the distribution of leaflets to housing estates, local amenities such as launderettes, the post office, mosques, temples and community centres used by the BME communities, careers events, sixth form colleges, housing association properties and Do it Yourself (DIY) courses.
- Send leaflets to people prepared to recruit under-represented groups such as the contractor's own employees, housing association representatives, people from the directory of minority led firms, participants in 'New Deal' and others involved in equal opportunities initiatives.
- Consider an incentive scheme to encourage people to either recommend potential recruits and/or hand out leaflets. The RAC and the Co-operative Bank offer vouchers, payments and free gifts.
- Set up a task force within the company specifically charged with the recruitment of under-represented people.
- Encourage BME apprentices to tell their friends about opportunities in construction.
- Give a realistic, but positive, account of training and future employment prospects during the recruitment phase.
- Publicise training schemes for BME apprentices in the press.

Directory of minority-led firms

As well as recruiting individuals to the industry, building contractors can help improve equal opportunities by giving minority-led firms the opportunity to obtain subcontract work. For example, there is evidence to suggest that many more BME people are setting up their own businesses, partly as a result of being unable to progress in traditional employment in the construction industry. To help in identifying minority-led firms a number of directories have been developed. For example, in the North West, Building Positive Action (this is currently held by North British Housing Association); in London, LEOF; and in Bristol, Elm Housing Association. Organizations should:

- Consider subcontracting work to firms from the directory, especially when providing housing or services for BME and other under-represented groups.
- Consider providing financial support for the directory of minority-led firms.

Supporting the directory of minority-led firms demonstrates commitment to equality, gains publicity for the company and potentially can put them in contact with housing associations.

Taster courses

Short or taster courses designed to provide basic skills can provide an effective method of attracting BME people on to courses and give them some experience of construction. A short course also gives trainees a good start in a trade of their choice and enables established BME tradespersons to become multi-skilled.

- Offer one-day a week courses in electrical installation, painting and decorating, plastering or carpentry, lasting for approximately ten weeks.
- Give help with finding training and/or employment in construction once the course has finished. This would be in the form of passing on information about job opportunities, help with job searches or advice about curricula vitae.
- Put on displays and exhibitions of the work done and the achievements of those taking part, to give others an opportunity to view the work of BME trainees and employees.
- Offer evening courses one evening per week, again in a ten-week block or taster courses lasting for six weeks, with trainees spending two weeks on each trade.
- Provide BME-only courses, as these are an effective method of attracting BME applicants and countering the white-dominated culture.
- Mixed courses are satisfactory as long as there are a reasonable number of BME trainees on the course and racist behaviour is not tolerated.

- Provide a BME instructor from your own organisation or from the directory of minority-led firms.
- Examine the costs and opportunities for funding taster courses with housing associations.

3   Innovative approaches to training

Providing attractive packages

The construction industry needs to increase the quantity and quality of training, especially for BME groups. This will help the industry provide qualified professional employees and tradespeople. Improving opportunities for local people, reduces vandalism on site, reduces property mistreatment, increases tenant involvement and improves building design. It will also enable contractors and housing associations to work together in partnership and gain a competitive edge for future contracts.

- Lift age restrictions and open up courses to a whole range of BME groups.
- Offer supplementary courses in maths, English and communications/ negotiation skills for work in the construction industry; English as a second language; and computing and multimedia studies.
- Highlight the provision of training opportunities and good links with local training providers in promotional material or discussions with BME applicants.
- Draw attention to possibilities for BME tradespersons to attend college, take courses offered by the CITB and/or supervise apprentices.
- Ensure that BME workers are not isolated by recruiting other people belonging to the same ethnic group, offering them a BME mentor and/or encouraging networking with other BME staff in the organization, that is, through a BME support group.

Bramall Construction have a policy of pairing BME trainees to avoid isolation on site.

- Provide opportunities for apprentices to gain qualifications, which will make them employable and competent. Ideally these should be City and Guilds or National Vocational Qualifications in the manual trades at level III – not just I and II.
- Monitor the training programme of the contractor, client and/or CITB by regular, personal contact in addition to forms and questionnaires.
- Monitor the cost and quality of training by setting aside houses specifically for training purposes and nominating a training supervisor.

- Display the work produced by trainees to senior managers in the company, the client, partners and visitors such as potential new recruits, local residents and school children.
- Consider flexible hours and taster/short courses.
- Consider culturally specific activities, for example, prayer time.

DIY and maintenance for tenants

DIY courses run with housing associations enable contractors to work in partnership with housing associations, develop a pool of BME people potentially interested in working in construction and gain contact with people listed in the directory of minority-led firms. Funding could initially be provided through 'Housing Plus' or 'New Deal' or Innovation and Good Practice Grants, but the savings in maintenance costs arising from fewer unnecessary call outs and less property mistreatment would make the courses self-financing in the longer term.

- Consider working with housing associations to provide DIY and maintenance courses for BME people.
- Investigate the costs and benefits of DIY. Housing associations estimate the cost to be between £2,000 and £3,000.
- Highlight the potential for courses to increase the quality of the service provided for tenants by increasing levels of tenant involvement, training and potentially employment.

4   Keeping employees

Building contractors should aim not only to build a good workforce by recruiting BME staff, but also encourage them to stay by helping them to develop their careers – good people are a valuable resource.

Preparing for apprentices

An apprentice may find it daunting to arrive on site for the first time and have little idea about what to expect. When BME workers arrive on a site, supervisors, foremen and co-workers may react with varying degrees of curiosity, friendliness, resentment, surprise, indifference or hostility.

- Involve and inform the workforce by outlining the organization's commitment to hiring BME staff, the practical steps which have been taken to ensure equality and the company's belief in the ability of BME and other minority groups to do the job.
- Provide opportunities for new apprentices to meet other apprentices and to learn about the company, the type of work and the working conditions before starting on site.

- Produce a trainee's handbook with introductory information, an outline of training and health and safety structures.
- Ask the supervisor to ensure follow-up of any complaints of harassment.
- Start training on site during the spring or summer, rather than the winter.

Example

A contractor in the Manchester area worked with the CITB to provide training for young local bricklayers and joiners which encouraged team spirit, reduced vandalism (e.g. fewer windows broken) and increased the respect shown by local people. Other contractors worked closely with housing associations and local authorities to provide apprenticeship schemes.

Keeping apprentices

Building contractors often experience a high drop-out rate amongst apprentices, both BME and white, for various reasons. Whilst some apprentices recognize that the manual trades provide good employment opportunities and wages in the longer term, building contractors have to attempt to satisfy the short-term expectations, needs and aspirations of apprentices. To reduce wasted time and effort in recruitment, it is important to find out why apprentices leave and try to resolve problems where appropriate.

- Carry out exit monitoring to find out why apprentices leave and whether they have been lost to the industry – recognizing that it might be necessary to keep some reasons confidential.
- Provide apprentices with a reasonable wage/allowance and if possible contribute to travel costs, equipment, overalls and luncheon vouchers.
- Where appropriate, help with the cost of childcare.
- Offer apprentices opportunities to supplement their income by doing paid work for the company.
- Identify benefits of training already provided and costs of leaving.
- Remedy any problems revealed by the exit monitoring.

An apprentice leaving does not necessarily represent a failure, especially where a contractor introduced professional recruitment methods and/or the individual has benefited from the training.

Example

One contractor based near Stockport suspects that apprentices have been poached by construction companies prepared to pay higher wages.

He intends to introduce exit monitoring to find out if this is the case or if there are problems with training and/or recruitment.

Improving working conditions

The world of work is changing dramatically. More people are rejecting poor conditions and long hours, and technology is offering different ways of carrying out traditional tasks. Learn from the manufacturing process, using system building or the off-site production of components where appropriate, so that work is conducted indoors and flexible hours can be provided more easily. Also compare what other industries and professions are providing in terms of working hours and support mechanisms.

- Provide flexible and/or shorter hours. This will be a real benefit to BME female employees who care for children and/or elderly relatives. It will also help those who have to observe regular prayer times.
- Consider flexible starting and finishing times, part-time employment, shift system, temporary work during the school holidays, job share and home working.
- Adapt shifts and piece-work to suit family commitments and religious requirements that is, fasting during Ramadan.
- Maintain a dialogue with housing associations about lengthening project deadlines in order to provide flexible or shorter hours for employees.
- Provide and maintain good toilet and washing facilities.
- Provide facilities for prayer.
- Investigate the possibility of providing childcare facilities.
- Examine the possibility of helping with the costs of emergency childcare cover for work required outside of normal office hours and/or at short notice.
- Support BME and other minority groups by encouraging them to participate in networks or mentoring schemes. These may be specifically for BME groups, for those who wish to meet and support others of the same group, or mixed organisations such as the unions and professional bodies.

5   Starting the process

Running a pilot project

Achieving results and bringing about change demand that actions are measured against targets. Identify activities proposed in these guidelines that are feasible and appropriate to undertake.

- Develop a pilot project, likely to attract media attention, and a series of supporting actions.

- Meet with other managers to discuss possible pilot projects in more detail and identify further action.
- Extend meetings to potential clients and/or partners.
- Encourage open discussion, welcoming ideas and minimising criticism, to encourage innovation and change.
- Sustain momentum by arranging follow-up actions and meetings.
- Once a pilot project and supporting practical steps have been agreed, define objectives, specific activities, timetable for completion, outcomes and benefits.

Developing and using measurement tools

Develop appropriate measurement tools through discussions with participants. This involves defining parameters, identifying practical steps, developing specific measures, testing and incorporating the results into overall performance measures.

- Establish a base line, against which progress can be measured.
- Identify ways of measuring outcomes, for example, numbers of applications from BME groups; number of apprenticeships; and number of careers events etc.
- Keep a qualitative record of innovations, successes and problems. The records are useful for illustrating accounts during discussions and presentations to clients, the media and work colleagues.
- Use the measurement tools to gain feedback about the successes and costs of action to promote equality.
- Identify and build upon successes.
- Establish learning points and take remedial action.
- Encourage people to admit to problems and to learn from difficulties.
- Increase motivation by recognising progress, giving recognition, giving praise where due and avoiding criticism.
- Encourage a sense of shared responsibility, especially when problems are encountered. Avoid allocating blame.
- Communicate successes to employees, clients and potential new recruits.
- Keep material to publicise interest and action such as photographs, articles in the building press and letters of recommendation.

Promoting the results

Getting involved in equal opportunities initiatives and activities enables contractors to gain insight into current thinking and behaviour within housing associations, contributing to informed decisions about future business and marketing activities. Taking practical steps to promote equality enables building contractors to demonstrate commitment and initiative to housing associations, other clients and the public.

- Contact housing associations, the Housing Corporation, the National Housing Federation and those involved in Housing Plus, New Deal and Local Labour schemes to ask about working in partnership.
- Inform potential partners about action taken to promote equality.
- Show measurement tools and supporting evidence to demonstrate commitment.
- Improve the industry's image by highlighting opportunities for BME people.
- Try to obtain media coverage to promote the company and the industry, especially when involved in specific projects.

Good practice examples for registered social landlords to promote race equality

The Race Relations Amendment Act 2000 means that from April 2001 RSLs will have an enforceable positive duty to eliminate unlawful discrimination and promote equality of opportunity and good race relations in respect of all functions. This new duty will be enforceable and they will be audited and inspected by the Housing Corporation to ensure compliance.

It is important, therefore, that RSLs develop and fully implement equal opportunities policies in relation to all their activities. In most cases, policies have been developed in terms of recruitment and selection of staff and allocating and letting properties, although there is evidence that much still needs to be done. However, in other areas such as their commissioning activities in relation to construction and maintenance work, very few have developed policies or monitor these. '*Contracts of exclusion: a study of black and minority ethnic outputs from registered social landlords contracting power*' (Sodhi and Steele, 2000) sets out what the policy should include, the monitoring stages and the role of RSLs in promoting equality through its commissioning activities. Good practice examples are outlined here.

1   Contractual mechanisms

Promoting change through contractual mechanisms is one way in which RSLs can demonstrate that they are serious about promoting equal opportunities and encourages compliance by contractors.

Contract clauses

Of course, RSLs and building contractors are bound to conform to statutory obligations, but the importance of doing so can be highlighted in contract documentation. More detailed information about promoting equality through contracts is available from the CRE.

- Remind building contractors in the contract to abide by their statutory obligations under the Race Relations Act, for example, in order to demonstrate reasonable steps to ensure the widest response from all sections of the community to employment opportunities within the undertaking and that all potential applicants receive equal encouragement to apply.
- Consider including a clause on discrimination in employment, for example: 'The housing association has adopted equal opportunities policies in its working practice and is opposed to any discrimination on the basis of gender, colour, race, ethnic origin or disability by contractors or subcontractors'.
- Consider including a clause on neighbourly relations. For example:

> The housing association wishes to promote neighbourly relations with the areas in which it is working. Contractors must ensure that their employees and subcontractors do not cause abuse or harassment while engaged by the association. Contractors are required to take prompt and firm action against any employee causing abuse or harassment.

- Remind building contractors of their obligations under the Disability Discrimination Act (1995), which requires employers operating in the United Kingdom with twenty or more employees to employ a quota of registered disabled people, currently 3 per cent. Employers who are 'below quota' are required to recruit only suitable registered disabled people when vacancies arise, or to obtain a permit allowing it to recruit other than disabled people.

Approved lists

RSLs should evaluate building contractors' ability to meet the criteria for entry on to approved lists using questions such as, 'Do you comply with the Race Relations Act?' Open questions will elicit more useful responses than simple yes or no questions.

- Emphasise the importance of equal opportunities by providing guidance to contractors about the approved questions. Ideally, a face-to-face meeting should be held with contractors but the guidance can be in written form.
- Suggest contractors refer to guides and articles such as: *Racial Equality and Council Contractors* by the Commission for Racial Equality (1995) and *Equal Opportunities in Development* by George Barlow (1989).

- Ask for documentary evidence to support claims of meeting the criteria, such as recruitment manuals, recruitment advertisements, promotional brochures, flyers about training opportunities, application forms, relevant extracts from disciplinary and grievance procedures and equal opportunities policies and statements.
- Seek confirmation that a senior person from the firm is responsible for the equal opportunities policy and its operation.
- Check that the text and illustration of documentation shows that people are welcomed from different backgrounds and ethnic groups.
- Identify good practice and areas for improvement and feed the information back to the contractor, either individually or collectively.
- Follow up the forms and the documentary evidence submitted for entry to approved lists by checking a number of contractors.

RSLs should also ensure that they provide opportunities for BME contractors/consultants to join their approved lists.

> Ethnic minority contractors should be encouraged to join the associations approved list for tenders.
>
> (The Housing Corporation)

They should consider:

- Setting targets for the use of BME contractors/consultants by value and number of contracts.
- Encouraging larger contractors to subcontract to BME companies.
- Provision of training and assistance to BME companies to enable them to compete successfully for contracts.
- Flexibility in the treatment of BME/SME contractors/consultants in terms of payment and continuity of work.

See 'Contracts of exclusion: a study of black and minority ethnic outputs from registered social landlords contracting power' (Sodhi and Steele, 2000, for further details on the role of RSLs).

> [Our association] is a grassroots BME organisation; it has always endeavoured to use BME contractors/consultants. It is essential that we use our purchasing power to combat discrimination.
>
> (Sodhi and Steele, 2000: 21)

Using Constructionline

Constructionline is the United Kingdom's largest register of qualified construction services, having over 9,000 contractors and consultants registered with them (7,000 contractors and 2,000 consultants) operating across the

United Kingdom. They cover the full spectrum of construction activities from architecture to demolition and range in size from small specialist contractors to the largest main ones. The service is free to the public sector and can be accessed online. The companies conform to strict financial and technical criteria set and audited by the DETR allowing clients to bypass the pre-qualification stage and generate tender lists from the database.

The use of Constructionline has become a mandatory requirement for RSLs under the Housing Corporations Scheme Development Standards. Using Constructionline means that RSLs can bypass the pre-qualification stage before selecting contractors/consultants. They will however, continue to have responsibility for equal opportunities; this will not be transferred to Constructionline. The procedures described here are therefore no less relevant.

Checking compliance

It is easy for building contractors to show apparent compliance with equal opportunities criteria by simply ticking a box saying 'yes' and producing appropriate documentation. RSLs should aim to ensure that building contractors are checked regularly and effectively and are given help to meet the standards where appropriate.

- Check to see whether contractors have recently been involved in any industrial tribunals by contacting the Commission for Racial Equality.
- Talk to people responsible for equal opportunities, recruitment, training and front-line management to evaluate successful implementation of the policy.
- Find out about meetings and training sessions to discuss equality and specific steps to ensure the fair representation of BME employees.
- Examine figures broken down by ethnic group for recruitment, training and employment.

Annual review of approved lists

An annual review gives RSLs the chance to demonstrate their equal opportunities objectives and goals. It also provides an opportunity to ask contractors for a commitment to continue to meet the housing association criteria as a condition of entry or retention on the approved list.

A meeting for all the contractors on the association's approved list is an effective method of motivating and guiding action, rather than issuing a written questionnaire, and gives contractors an opportunity to ask questions. An RSL might consider a joint meeting organized by several RSLs in order to minimise costs and maximise impact. Whether by interview or by mail, a standardised questionnaire ensures consistency.

- Review building contractors on the approved lists using a questionnaire designed by the RSLs or a standard questionnaire published by the Joint Consultative Committee for Building.
- Ask 'how' rather than 'if' the contractor complies with requirements such as the Race Relations Act to encourage contractors to engage with the issues and provide more specific details about practical steps.
- Check answers from the questionnaires and give notice of intended action to any contractors below standard.
- Before any action is taken which may disqualify the contractor from tendering, provide the grounds for dissatisfaction and an opportunity to make representation to a member of the RSLs' governing body.
- Provide feedback to contractors recognising positive steps already taken as well as areas for improvement.

Suggested review questions

- Has your equal opportunities policy been agreed with trade unions and employee representatives?
- How is the equal opportunities policy communicated?
- What type of training is provided for managers and supervisors regarding the Race Relations Act?
- How often do you review recruitment, promotion, transfer and training procedures?
- Have you been involved in any positive steps to encourage ethnic minorities to apply for jobs?
- Do you offer special training for ethnic minorities or women?

Pre-tender documentation

The provision of pre-tender documentation is an excellent opportunity for RSLs to highlight the importance of equal opportunities to the contract. It is also a chance for them to question contractors about practical steps to ensure equality.

The contractors' responses to the questions about equality have to be evaluated prior to the invitation to tender and used as a basis on which to work with the contractor, rather than to award the contract. Nevertheless, RSLs should consider using the answers provided by contractors and/or a contractor's performance during the contract to give general feedback and advice about equal opportunities during subsequent meetings.

2 Promoting good practice

RSLs should use their financial muscle and experience to encourage building contractors to take practical steps to promote equality. Just as importantly, housing associations should help contractors to benefit from action to promote equality, especially in the shorter term.

- Communicate commitment to equality using a clear statement drawing attention to the RSLs' objectives regarding equality, the local population and harassment.
- Ask building contractors to produce a similar set of statement or policies, or to comply with the association's policy.
- Demonstrate housing association good practice, for example, by proactive advertising, training and employment and including pictures of BME and other under-represented groups in annual reports.
- Hold regular meetings for contractors designed to promote good practice regarding equal opportunities. These could be run by individual RSLs and/or a group of RSLs, thus minimising costs and maximising impact.
- Provide guidance about practical steps for contractors and measurement tools, using the report of CIB Working Group 8 and this document.
- Invite contractors to write to the housing association with details of their achievements regarding equal opportunities, providing supporting evidence and appropriate measurement and monitoring tools.
- Review material provided by contractors to identify good practice as well as areas for improvement.
- Provide appropriate positive feedback.

Recognition and reward

Contractors who perform well in terms of equal opportunities should be recognized and/or rewarded by RSLs. It is important that they acknowledge the effectiveness of the business case argument for equal opportunities amongst building contractors, and not to view it as less valid than social pressure. There are various ways in which this can be done, without using financial mechanisms.

- Provide a platform for contractors to give short presentations about equal opportunities at meetings and conferences.
- Provide opportunities to discuss association objectives/plans, partnerships and current thinking and practice with senior managers.
- Report building contractor good practice in the RSLs annual report.
- Contribute to joint articles for publication.
- Consider entry on to approved lists.
- Provide opportunities to tender, especially for construction projects requiring recruitment, training, equal opportunities and local labour.

Remedies

RSLs cannot and would not want to terminate a contract solely for breach of equality in employment conditions. However, they can refuse to place further work with the offending contractor until it introduces satisfactory policies and procedures.

3   Working in partnership

RSLs should aim to work in partnership with building contractors to develop specific projects to promote racial equality. In addition to promoting equality, the partnerships could help to provide training, improve recruitment procedures, increase cooperation, reduce disputes, lower costs and improve quality.

The process of setting up and working in partnership involves meetings, gaining commitment, assessment of contractors, feedback, rewards and on ongoing dialogue.

Setting up partnerships

This mechanism for promoting equality will help RSLs to identify building contractors committed to equal opportunities and prepared to take practical steps. The RSLs may be able to choose a contractor with whom to work in partnership or may have to invite them to tender for the contract. This will depend upon the nature of the work and the contract.

- Invite contractors to a meeting to discuss working in partnership on a specific project, with the aim of adopting specific practical steps and measurement tools.
- Create an informal and participative atmosphere in meetings to help participants feel relaxed and gain some insight into individuals' attitudes and interests.

Working together

- Work in partnership with the contractor to develop strategies and take action.
- Specify practical steps to promote equality in the contract, which are perceived as achievable by both parties.
- Include targets for the recruitment of under-represented groups for monitoring purposes, rather than contractual requirements.
- Adopt and develop measurement tools and monitor progress.
- Record project results using the measuring tools, verbal accounts, photographs and videos. The successes and learning points should then be publicised in annual reports, research documentation and trade articles.

- Arrange regular meetings to discuss the work, and outline specific actions to promote equality.

RSLs estimate the cost of a taster course in the manual trades at £2,000–£3,000. Funding could be provided by building contractors/RSLs through 'Housing Plus' or 'New Deal'.

Types of partnerships

There are various arrangements and objectives possible in a partnering initiative:

- Education–business – working with contractors to organize and run visits to schools, work placements, site visits, shadowing and sponsorship.
- Training and recruitment – working with mainstream and minority-led contractors to recruit, train and employ under-represented groups.
- Employment conditions – helping contractors provide better working conditions by, for example, extending contract times for contractors who provide flexible hours.
- Directory of minority-led firms – encouraging and helping contractors to subcontract work to minority-led firms.
- Tenant training, employment and involvement – working with mainstream and/or minority-led contractors to provide, for example, DIY/maintenance training.

Example

A partnership between Riverside Housing Association, Bramall Construction and Shokoya Eleshin Construction (a BME contractor) on a stock transfer scheme (Carter Thackeray, Liverpool) led to the development of a local labour initiative targeted at unemployed youth, BME groups and women. BME groups were the major beneficiaries and the initiative led to the formation of 11 businesses (8 of which were BME). Following completion of the scheme four businesses were up and running.

Collaboration on promoting race equality in the construction industry – Action-Learning Sets

This project has enabled the development of active industry networks through the development of Action-Learning Sets in Salford, Middlesborough and London. These sets are actively seeking to promote race equality within the construction industry and provide exemplars of good practice in promoting equality. They involve collaboration between construction companies and their clients, RSLs, industry groups such as the CITB and Universities and TECs.

How have the sets promoted race equality
within the industry?

THE SALFORD ACTION-LEARNING SET

The Salford set started off by looking at the draft guidelines for race equality in the construction industry, this led to discussions on a range of problems/issues around the promotion of equality within the industry and for construction clients. The set felt that one of the key problems in terms of promoting equality was that some companies just did not know how to establish their own policies and procedures on equal opportunities. It was felt that this was an area where the set could make a positive contribution and the group decided to hold a breakfast seminar to gauge interest in this and to consult on the requirements of those in the industry for support. The group proposed to follow up the seminar with a number of workshops on establishing procedures. The breakfast seminar was also seen as an opportunity to develop an awareness of the issues and to disseminate the good practice guidelines on sex equality in the construction industry. The workshops would also enable the guidelines on race equality to be disseminated and developed. The breakfast seminar was held in November 2000 and senior people in construction companies and RSLs were invited to attend. The event was supported by the Housing Corporation and was well attended, there was a good deal of interest in the idea of workshops to provide guidance on establishing equal opportunities and those who attended were asked to indicate topic areas from a list provided. Following on from this a series of half-day workshops were set up covering the areas described here:

- Widening the net, improving recruitment and selection.
- Equal opportunities: case studies and good practice.
- Training and retaining: creating a skilled and flexible workforce.

So far two of these workshops have been held and a great deal of interest has been generated in the work of the Salford Action-Learning Set, but also in terms of developing initiatives to promote race equality. They have provided those working in the industry and the social housing sector with an opportunity to network, share experiences and build new working relationships.

The set was successful in securing funding for these events from the Revans Centre at the University of Salford and the CITB which means that the workshops have been free.

THE LONDON ACTION-LEARNING SET

The London Action-Learning Set is in the early stages of development. Their approach is slightly different but no less valid. Members have focused on the issues/problems associated with promoting equality often relating this to problems that they are experiencing in their own organizations. This sharing of experiences has helped some members to overcome difficulties, benefiting from advice and support that this opportunity affords. Information is also shared about initiatives being undertaken to promote equality nationally and within the organizations of set members. Encouraged by the discussions some members have been spurred on to develop initiatives within their own organizations.

For example, one member from Walter Llewellyn's and Sons following a meeting took a step back to consider why he himself had joined the construction industry – and realized that it was accidental rather than planned. This prompted him to encourage his organization to undertake a survey to discover the reasons people had joined the industry and what had prompted their interest. From a recruitment point of view he felt that it would be valuable to discover how people became interested in the industry since this would be an indicator of how to engage new people. Initial findings suggest that most people had first been exposed to the industry through school 'taster days' rather than adverts. Their entry into the industry seems to be mainly based on experience of the industry in one way or another. The findings will be used to get youth and women into the industry.

Another member from Lovell's applied for funding to set up a mentoring scheme from the DETR funding under Partners in Innovation following a suggestion at set meeting. Each regional office of the company will sign up to a company mentoring scheme, focusing on local schools. The aim is to gain an ethnic and gender mix in recruitment. Both academic and craft scholarships are being offered, from GCSE level up to degree level. It is similar to a youth enterprise scheme since it offers scholarships and prizes for students to work towards. It is based on a partnering system with schools, whereby the students selected will be able to have first-hand access at how an office and site environment works – the teachers are encouraged to experience this as well. The students will be offered work shadowing placements and they can compete for student awards. The programme will be attractive to schools because it will fit into the national curriculum and timetables.

THE MIDDLESBOROUGH ACTION-LEARNING SET

Early discussions about the draft guidelines on race equality in the construction industry spurred the Middlesborough set to seek to increase equal opportunities for women and BME groups through the development of a promotional video to encourage young people to enter the construction industry. The video will be aimed at young people in schools, primarily years nine and ten (i.e. those aged 14 and 15).

It will provide guidance on the range of jobs available in the industry, career paths and entry requirements and will be accompanied by a pack which will replicate this information. The CITB have provided funding for the project and the University of Northumbria are carrying out the filming.

It is intended to be amusing and will show architects, surveyors, site managers, engineers and bricklayers at work. It will also feature famous and interesting buildings such as the Millennium Dome and the Sydney Opera House. It will also show the different stages of construction from laying foundations to completion. Women and BME people will be included in the video wherever this is possible to provide role models.

Appendix I

Change the face of construction – Toolkit 1

Attract the right people through making the most of diversity

Why?

Can you find enough of the right people for your business? Are you suffering from a skills shortage? Is staff turnover high? Start thinking differently.

Did you know?

- The construction industry needs 350,000 people in the next 4 years – where are you going to find the ones you need?
- Women make up 48 per cent of the working population – but only 10 per cent of construction workers are female?
- More and more building and engineering students are choosing other careers – how are you going to make your firm attractive to them?

How?

In brief

1 Take a fresh look at the recruitment pool.
2 Assess the effectiveness of your current recruitment policies and practice.
3 Find out what your target recruits want.
4 Deliver what they want within the capacity of the business.
5 Make sure your target recruits know about you and your offer.

1 Take a fresh look at the recruitment pool: Acknowledging that jobs can be done in different ways by different people, for example, jobshare – flexible working, women returners, disabled people, black and Asian people; identifying who will be responsible for gathering information

about equal opportunities, its impact on the business and putting plans into action; ensuring that everyone responsible for recruitment has received formal training in equal opportunities and diversity issues.

2 Assess the effectiveness of your current recruitment policies and practice: Surveying existing employees, past employees (if practicable) and people who turned down offers of employment. Find out:

- how they knew about the company,
- their first impressions,
- comments about the application procedure,
- reactions to the interview experience,
- opinions about your competitors.

Reviewing links with schools and colleges to generate applicants and potential trainees.

3 Find out what your target recruits want: Analysing the feedback from the surveys to identify key messages and issues; gathering information from trade associations, professional institutions and business organizations on market trends; investigating how the content of your offer measures up in the marketplace.

4 Deliver what they want within the capacity of the business: Identifying gaps and opportunities to bring about practical improvement; establishing plan, schedule and frequency of review, for example, annual, biennial, every five years. Add value to your offer – without adding significantly to payroll costs – to attract a wider and more diverse workforce. For example, does the job lend itself to home working, job share or flexible hours, thus appealing to women returners or older workers. Assessing resources required and appointing someone to implement action.

5 Make sure your target recruits know about you and your offer: Outlining objectives to employees; reviewing advertising, public relations and recruitment practice to reach different groups; measuring and communicating results.

Who?

Whatever the size of your business, senior management should be involved in recruitment policy and practice, as well as the human resources and personnel departments. Attracting the right people is also an important part of marketing and business development, helping companies to become employers of choice and to build brand.

This is one of a series of change the face of construction toolkits being developed to help organizations put equal opportunities policy into practice. For more information and guidance on diversity and equal opportunities, go to the website at *www.change-construction.org* or contact by fax on 020 8853 3281.

Change the face of construction – Toolkit 2

Keeping the right people by building the right culture

Why?

Do you spend too much time getting people up to speed, only to have them leave? Is there a sense of loyalty and commitment in your company? Are you losing people to your competitors?
Start thinking differently.

Did you know?

- Replacing a qualified person costs £20,000 on average?
- More women managers leave their job to progress their career than to have children?
- Valuable skills and competencies are being lost through racial discrimination?

How?

In brief

1 Find out who is leaving and why.
2 Assess your current and future resource needs.
3 Find out if the people you need feel welcomed and motivated.
4 Acknowledge the need for change and develop a plan.
5 Build a distinctive, appealing and fair workplace culture.

1 Find out who is leaving and why by:

- Reviewing employee turnover figures, for example, by gender, race, age and length of service.
- Using exit monitoring records where available.
- Talking to employees – if appropriate using a third party for confidential feedback.

2 Assess your current and future resource needs by:

- Reviewing workload and resources, by volume and skills.
- Looking at market trends to predict the skills you will need.
- Thinking beyond traditional skills and profile (see Toolkit 1 in this series).

3   Find out if the people you need feel welcomed and motivated by:
Reviewing management systems relating to employees, with questions
such as:

- How are staff recognized and valued, for example, regular perfor-
  mance feedback, rewards, training?
- Are the company's values defined and clearly expressed?
- Are there safe and comfortable working conditions for all?
- What training and support schemes are in place, for example, men-
  toring, support systems, networks?
- Is there a confidential system for reporting/resolving complaints –
  and do staff know about it?

Reviewing workplace policies, for example, employment of the disabled, maternity leave, flexible working, social structure and activities.
    Identifying what benefits your company provides, such as bonus sche-
mes, crèche provision, vouchers, teleworking, compassionate leave, etc.

4   Acknowledge the need for change and develop a plan by:

- Consulting with employees on the reasons and proposals for
  change.
- Defining reasonable length of service and establish how to reward it.
- Reviewing benefits and incentives, identifying those which employees value and those which are in the capacity of the business to offer.
- Devising a plan to develop and support a more diverse workforce,
  for example, balancing gender, race, age, varied competencies,
  multi-skills.

5   Build a distinctive, appealing and fair workplace culture by:

- Delivering the changes and benefits defined in the plan.
- Involving employees in innovative approaches to problem solving
  and improving efficiency, informing them about the company's per-
  formance.
- Managing out the wrong people by applying correct procedures in
  a fair way.
- Training all staff in diversity.
- Devising succession plans based on competence.
- Regularly finding out if expectations and objectives are being met.

Who?

Senior management should take the lead in establishing the company's
values and communicate them throughout the company, using external
specialists if necessary. All managers should contribute to building the

culture by ensuring that the new plan and mechanisms are recognized and implemented throughout the workforce on a day-to-day basis.

This is one of a series of change the face of construction toolkits being developed to help organisations put equal opportunities policy into practice. For more information and guidance on diversity and equal opportunities, go to our Web site at *www.change-construction.org* or contact by fax on 020 8853 3281.

Change the face of construction – Toolkit 3

Satisfying customers by becoming more like them

Why?

Successful business depends on identifying good customers, understanding their needs and satisfying them at a profit. A more diverse workforce will help the construction industry to be more aligned with the customers it serves – and provide a bigger pool of the skills and competencies needed in a changing marketplace. Start thinking differently.

Did you know?

- The purchasing power of ethnic minorities was £14.9 billion in 1997 and is rising?
- In 10 years 41 per cent of managers and administrators and 42 per cent of professionals will be women?
- Brand leaders in other industries see diversity as a key tool in improving performance?

How?

In brief

1 Identify your customers.
2 Review current performance and satisfaction levels.
3 Respect the customer.
4 Ensure employees understand and support business objectives.
5 Deliver quality work through a more diverse approach.

1 Identify your customers by:

- Drawing up a list of key customers – and the ones you would like.
- Asking yourself honestly how well you understand their business and what they are trying to achieve – and improve your understanding by asking them.

- Looking at the entire supply chain as potential customers, for example, developers, contractors, consultants, house purchasers, funders, materials suppliers, etc.

2 Review current performance and satisfaction levels by:

- Analysing how much work is repeat business, how many jobs are one-offs.
- Tracking recurring themes or patterns, for example, individuals or teams who consistently win work, manage projects effectively, resolve problems constructively.
- Looking at profitability, for example, who pays promptly, who briefs efficiently.
- Reviewing how complaints and disputes are managed and tracked.
- Identifying whether work is being lost through failing to prequalify on diversity criteria.

3 Respect the customer by:

- Helping your client in achieving the project to their standards.
- Training all employees in customer care, ensuring that all people are treated with respect.
- Not relying on industry norms to excuse bad behaviour.
- Respecting differences and responding appropriately.

4 Ensure employees understand and support business objectives by:

- Developing a welcoming and responsive culture, without preconceptions.
- Resisting stereotypes, for example, the woman at the meeting may be the MD, not the secretary.
- Spotting business opportunities and winning work – traditional markets are changing, clients are not always white males.
- Recognizing the importance of internal customers, others in the supply chain, the end user.
- Holding team reviews on project completion – as important as the next job.

5 Deliver quality work through a more diverse approach by:

- Developing a customer focused strategy.
- Looking at your main competitors and the image/brand they project.
- Keeping in touch with the changing world of work, demographic shifts, purchasing/user trends.

- Learning from other sectors where companies are actively seeking improvement through diversity.
- Recognizing the contributions that different people can make, for example, women often have excellent communication skills; black and ethnic minorities can improve a company's cultural awareness.
- Defining your brand through core values, staff attitudes, quality of delivery, feedback. Finding and developing the mix of people you need to understand, deliver and build the brand.

Who?

Customer satisfaction should be the aim of everyone in an organization. Leadership from the top is needed to create a culture of pride in the job, fair dealing and respect for customers and colleagues. Building a diverse workforce can result in new sources of work and improved workplace culture.

This is one of a series of change the face of construction toolkits being developed to help organizations put equal opportunities policy into practice. For more information and guidance on diversity and equal opportunities, go to our Web site at *www.change-construction.org* or contact by fax on 020 8853 3281.

Sources of information

Commission for Racial Equality, Maybrook House (5th Floor), 40 Blackfriars Street, Manchester, M3 2EG, Tel: 0161 831 7782.

Construction Industry Training Board, 10 Waterside Court, St Helens Technology Campus, Pocket Nook Street, Merseyside, WA9 1HA, Tel: 01744 616004.

The Contractor Code, The National Housing Federation, North West Office, City Point, 701 Chester Road, Manchester, M32 0RW, Tel: 0161 848 8132.

DETR, Construction Sponsorship Directorate, Floor 3/J1, Eland House, Bressenden Place, London, SW1E 5DU, Tel: 0171 890 5718.

Directory of Minority led Firms, Ms Chris Root, North British Housing Association, 4 The Pavilions, Portway, Preston PR2 2YB, Tel: 01772 897 200.

Equal Opportunities Commission, Overseas House, Quay Street, Manchester M3 3HN, Tel: 0161 833 9244.

London Women and Manual Trades, 52-54 Featherstone Street, London, EC1Y 8RT, Tel: 0171 251 9192.

Moss Side and Hulme Women's Action Forum (MOSHWAF), 97 Princess Road, Moss Side, Manchester, M14 4TH, Tel: 0161 232 0545.

Women as Role Models, Fiona Crehan, c/o EEP Ltd, The View, 6th Floor, 32-36 Hanover Street, Liverpool, L1 4LN, Tel: 0151 708 7103.

Women's Education in Building (WEB), 12–14 Malton Road, London, W10 5UP, Tel: 0181 968 9139.

Index